Lecture Notes in Mathematics 1881

Editors:
J.-M. Morel, Cachan
F. Takens, Groningen
B. Teissier, Paris

T0281351

S. Attal · A. Joye · C.-A. Pillet (Eds.)

Open Quantum Systems II

The Markovian Approach

 Springer

Editors

Stéphane Attal
Institut Camille Jordan
Université Claude Bernard Lyon 1
21 av. Claude Bernard
69622 Villeurbanne Cedex
France
e-mail: attal@math.univ-lyon1.fr

Alain Joye
Institut Fourier
Université de Grenoble 1
BP 74
38402 Saint-Martin d'Hères Cedex
France
e-mail: alain.joye@ujf-grenoble.fr

Claude-Alain Pillet
CPT-CNRS, UMR 6207
Université du Sud Toulon-Var
BP 20132
83957 La Garde Cedex
France
e-mail: pillet@univ-tln.fr

Library of Congress Control Number: 2006923432

Mathematics Subject Classification (2000): 37A60, 37A30, 47A05, 47D06, 47L30, 47L90, 60H10, 60J25, 81Q10, 81S25, 82C10, 82C70

ISSN print edition: 0075-8434
ISSN electronic edition: 1617-9692
ISBN-10 3-540-30992-6 Springer Berlin Heidelberg New York
ISBN-13 978-3-540-30992-5 Springer Berlin Heidelberg New York

DOI 10.1007/b128451

Springer is a part of Springer Science+Business Media
springer.com
© Springer-Verlag Berlin Heidelberg 2006
Printed in The Netherlands

Typesetting: by the authors and SPI Publisher Services using a Springer LATEX package
Cover design: *design & production* GmbH, Heidelberg

Printed on acid-free paper   SPIN: 11602620   VA41/3100/SPI   5 4 3 2 1 0

Preface

This volume is the second in a series of three volumes dedicated to the lecture notes of the summer school "Open Quantum Systems" which took place in the Institut Fourier in Grenoble, from June 16th to July 4th 2003. The contributions presented in these volumes are revised and expanded versions of the notes provided to the students during the school. After the first volume, developing the Hamiltonian approach of open quantum systems, this second volume is dedicated to the Markovian approach. The third volume presents both approaches, but at the recent research level.

Open quantum systems

A quantum open system is a quantum system which is interacting with another one. This is a general definition, but in general, it is understood that one of the systems is rather "small" or "simple" compared to the other one which is supposed to be huge, to be the environment, a gas of particles, a beam of photons, a heat bath ...

The aim of quantum open system theory is to study the behaviour of this coupled system and in particular the dissipation of the small system in favour of the large one. One expects behaviours of the small system such as convergence to an equilibrium state, thermalization ... The main questions one tries to answer are: Is there a unique invariant state for the small system (or for the coupled system)? Does one always converge towards this state (whatever the initial state is)? What speed of convergence can we expect ? What are the physical properties of this equilibrium state ?

One can distinguish two schools in the way of studying such a situation. This is true in physics as well as in mathematics. They represent in general, different groups of researchers with, up to now, rather few contacts and collaborations. We call these two approaches the *Hamiltonian approach* and the *Markovian approach*.

In the Hamiltonian approach, one tries to give a full description of the coupled system. That is, both quantum systems are described, with their state spaces, with their own Hamiltonians and their interaction is described through an explicit interaction Hamiltonian. On the tensor product of Hilbert spaces we end up with a total Hamiltonian, and the goal is then to study the behaviour of the system under this dynamics. This approach is presented in details in the volume I of this series.

In the Markovian approach, one gives up trying to describe the large system. The idea is that it may be too complicated, or more realistically we do not know it completely. The study then concentrates on the effective dynamics which is induced on the small system. This dynamics is not a usual reversible Hamiltonian dynamics, but is described by a particular semigroup acting on the states of the small system.

Before entering into the heart of the Markovian approach and all its development, in the next courses, let us have here an informal discussion on what this approach exactly is.

The Markovian approach

We consider a simple quantum system $\mathcal{H}$ which evolves as if it were in contact with an exterior quantum system. We do not try to describe this exterior system. It is maybe too complicated, or more realistically we do not quite know it. We observe on the evolution of the system $\mathcal{H}$ that it is evolving like being in contact with something else, like an open system (by opposition with the usual notion of closed Hamiltonian system in quantum mechanics). But we do not quite know what is effectively acting on $\mathcal{H}$. We have to deal with the efffective dynamics which is observed on $\mathcal{H}$.

By such a dynamics, we mean that we look at the evolution of the states of the system $\mathcal{H}$. That is, for an initial density matrix ρ_0 at time 0 on $\mathcal{H}$, we consider the state ρ_t at time t on $\mathcal{H}$. The main assumption here is that this evolution

$$\rho_t = P_t(\rho_0)$$

is given by a semigroup. This is to say that the state ρ_t at time t determines the future states ρ_{t+h}, without needing to know the whole past $(\rho_s)_{s \leq t}$.

Each of the mapping P_t is a general state transform $\rho_0 \mapsto \rho_t$. Such a map should be in particular trace-preserving and positivity-preserving. Actually these assumptions are not quite enough and the positivity-preserving property should be slightly extended to a natural notion of *completely positive map* (see R. Rebolledo's course). We end up with a semigroup $(P_t)_{t \geq 0}$ of completely positive maps. Under some continuity conditions, the famous Lindblad theorem (see R. Rebolledo's course), shows that the infinitesimal generator of such a semigroup is of the form

$$\mathcal{L}(\rho) = i[H, \rho] + \sum_i \left(L_i \rho L_i^* - \frac{1}{2} L_i^* L_i \rho - \frac{1}{2} \rho L_i^* L_i \right)$$

for some self-adjoint bounded operator H on $\mathcal{H}$ and some bounded operators L_i on $\mathcal{H}$. The evolution equation for the states of the system can be summarized into

$$\frac{d}{dt} \rho_t = \mathcal{L}(\rho_t).$$

This is the so-called *quantum master equation* in physics. It is actually the starting point in many physical articles on open quantum systems: a specific system to be studied is described by its master equation with a given explicit Linblad generator $\mathcal{L}$.

The specific form of the generator $\mathcal{L}$ has to understood as follows. It is similar to the decomposition of a Feller process generator (see L. Rey-Bellet's first course) into a first order differential part plus a second order differential part. Indeed, the first term

$$i[H, \cdot]$$

is typical of a derivation on an operator algebra. If $\mathcal{L}$ were reduced to that term only, then $P_t = e^{t\mathcal{L}}$ is easily seen to act as follows:

$$P_t(X) = e^{itH} X e^{-itH}.$$

That is, this semigroup extends into a group of automorphisms and describes a usual Hamiltonian evolution. In particular it describes a closed quantum system, there is no exterior system interacting with it.

The second type of terms have to be understood as follows. If $L = L^*$ then

$$LXL^* - \frac{1}{2}L^*LX - \frac{1}{2}XL^*L = [L, [L, X]].$$

It is a double commutator, it is a typical second order differential operator on the operator algebra. It carries the diffusive part of the dissipation of the small system in favor of the exterior, like a Laplacian term in a Feller process generator.

When L does not satisfy $L = L^*$ we are left with a more complicated term which is more difficult to interpret in classical terms. It has to be compared with the jumping measure term in a general Feller process generator.

Now, that the semigroup and the generator are given, the quantum noises (see S. Attal's course) enter into the game in order to provide a *dilation* of the semigroup (F. Fagnola's course). That is, one can construct an appropriate Hilbert space $\mathcal{F}$ on which quantum noises $da_j^i(t)$ live, and one can solve a differential equation on the space $\mathcal{H} \otimes \mathcal{F}$ which is of the form of a Schrödinger equation perturbed by quantum noises terms:

$$dU_t = LU_t\, dt + \sum_{i,j} K_j^i U_t\, da_j^i(t). \tag{1}$$

This equation is an evolution equation, whose solutions are unitary operators on $\mathcal{H} \otimes \mathcal{F}$, so it describes a closed system (in interaction picture actually). Furthermore it dilates the semigroup $(P_t)_{t \geq 0}$ in the sense that, there exists a (pure) state Ω on $\mathcal{F}$ such that if ρ is any state on $\mathcal{H}$ then

$$< \Omega, U_t(\rho \otimes I)U_t^* \Omega > = P_t(\rho).$$

This is to say that the effective dynamics $(P_t)_{t \geq 0}$ we started with on $\mathcal{H}$, which we did not know what exact exterior system was the cause of, is obtained as follows: the small system $\mathcal{H}$ is actually coupled to another system $\mathcal{F}$ and they interact according to the evolution equation (1). That is, $\mathcal{F}$ acts like a source of (quantum) noises on $\mathcal{H}$. The effective dynamics on $\mathcal{H}$ is then obtained when averaging over the noises through a certain state Ω.

This is exactly the same situation as the one of Markov processes with respect to stochastic differential equations (L. Rey-Bellet's first course). A Markov semigroup is given on some function algebra. This is a completely deterministic dynamics which describes an irreversible evolution. The typical generator, in the diffusive case say, contains two types of terms.

First order differential terms which carry the ordinary part of the dynamics. If the generator contains only such terms the dynamics is carried by an ordinary differential equation and extends to a reversible dynamics.

Second order differential operator terms which carry the dissipative part of the dynamics. These terms represent the negative part of the generator, the loss of energy in favor of some exterior.

But in such a description of a dissipative system, the environment is not described. The semigroup only focuses on the effective dynamics induced on some system by an environment. With the help of stochastic differential equations one can give a model of the action of the environment. It is possible to solve an adequat stochastic differential equation, involving Brownian motions, such that the resulting stochastic process be a Markov process with same semigroup as the one given at the begining. Such a construction is nowadays natural and one often use it without thinking what this really means. To the state space where the function algebra acts, we have to add a probability space which carries the noises (the Brownian motion). We have enlarged the initial space, the noise does not come naturally with the function algebra. The resolution of the stochastic differential equation gives rise to a solution living in this extended space (it is a stochastic process, a function of the Brownian motions). It is only when avering over the noise (taking the expectation) that one recovers the action of the semigroup on the function algebra.

We have described exactly the same situation as for quantum systems, as above.

Organization of the volume

The aim of this volume is to present this quantum theory in details, together with its classical counterpart.

The volume actually starts with a first course by L. Rey-Bellet which presents the classical theory of Markov processes, stochastic differential equations and ergodic theory of Markov processes.

The second course by L. Rey-Bellet applies these techniques to a family of classical open systems. The associated stochastic differential equation is derived from an Hamiltonian description of the model.

The course by S. Attal presents an introduction to the quantum theory of noises and their connections with classical ones. It constructs the quantum stochastic integrals and proves the quantum Ito formula, which are the cornerstones of quantum Langevin equations.

R. Rebolledo's course presents the theory of completely positive maps, their representation theorems and the semigroup theory attached to them. This ends up with the celebrated Lindblad's theorem and the notion of quantum master equations.

Finally, F. Fagnola's course develops the theory of quantum Langevin equations (existence, unitarity) and shows how quantum master equations can be dilated by such equations.

Lyon, Grenoble, Toulon
September 2005

Stéphane Attal
Alain Joye
Claude-Alain Pillet

Contents

Ergodic Properties of Markov Processes
Luc Rey-Bellet 1
1   Introduction 1
2   Stochastic Processes 2
3   Markov Processes and Ergodic Theory 4
    3.1   Transition probabilities and generators 4
    3.2   Stationary Markov processes and Ergodic Theory 7
4   Brownian Motion 12
5   Stochastic Differential Equations 14
6   Control Theory and Irreducibility 24
7   Hypoellipticity and Strong-Feller Property 26
8   Liapunov Functions and Ergodic Properties 28
References 39

Open Classical Systems
Luc Rey-Bellet 41
1   Introduction 41
2   Derivation of the model 44
    2.1   How to make a heat reservoir 44
    2.2   Markovian Gaussian stochastic processes 48
    2.3   How to make a Markovian reservoir 50
3   Ergodic properties: the chain 52
    3.1   Irreducibility 56
    3.2   Strong Feller Property 57
    3.3   Liapunov Function 58
4   Heat Flow and Entropy Production 66
    4.1   Positivity of entropy production 69
    4.2   Fluctuation theorem 71
    4.3   Kubo Formula and Central Limit Theorem 75
References 77

Quantum Noises

Stéphane Attal 79
1   Introduction 80
2   Discrete time 81
   2.1   Repeated quantum interactions 81
   2.2   The Toy Fock space 83
   2.3   Higher multiplicities 89
3   Itô calculus on Fock space 93
   3.1   The continuous version of the spin chain: heuristics 93
   3.2   The Guichardet space 94
   3.3   Abstract Itô calculus on Fock space 97
   3.4   Probabilistic interpretations of Fock space 105
4   Quantum stochastic calculus 110
   4.1   An heuristic approach to quantum noise 110
   4.2   Quantum stochastic integrals 113
   4.3   Back to probabilistic interpretations 122
5   The algebra of regular quantum semimartingales 123
   5.1   Everywhere defined quantum stochastic integrals 124
   5.2   The algebra of regular quantum semimartingales 127
6   Approximation by the toy Fock space 130
   6.1   Embedding the toy Fock space into the Fock space 130
   6.2   Projections on the toy Fock space 132
   6.3   Approximations 136
   6.4   Probabilistic interpretations 138
   6.5   The Itô tables 139
7   Back to repeated interactions 139
   7.1   Unitary dilations of completely positive semigroups 140
   7.2   Convergence to Quantum Stochastic Differential Equations 142
8   Bibliographical comments 145
References 145

Complete Positivity and the Markov structure of Open Quantum Systems

Rolando Rebolledo 149
1   Introduction: a preview of open systems in Classical Mechanics 149
   1.1   Introducing probabilities 152
   1.2   An algebraic view on Probability 154
2   Completely positive maps 157
3   Completely bounded maps 162
4   Dilations of CP and CB maps 163
5   Quantum Dynamical Semigroups and Markov Flows 168
6   Dilations of quantum Markov semigroups 173
   6.1   A view on classical dilations of QMS 174
   6.2   Towards quantum dilations of QMS 180
References 181

Quantum Stochastic Differential Equations and Dilation of Completely Positive Semigroups

Franco Fagnola 183
1   Introduction 183
2   Fock space notation and preliminaries 184
3   Existence and uniqueness 188
4   Unitary solutions 191
5   Emergence of H-P equations in physical applications 193
6   Cocycle property 196
7   Regularity 199
8   The left equation: unbounded G_β^α 203
9   Dilation of quantum Markov semigroups 208
10 The left equation with unbounded G_β^α: isometry 213
11 The right equation with unbounded $\hat{F}_\beta^\alpha$ 216
References 218

Index of Volume II 221

Information about the other two volumes
Contents of Volume I 224
Index of Volume I 228
Contents of Volume III 232
Index of Volume III 236

List of Contributors

Stéphane Attal
Institut Camille Jordan
Université Claude Bernard Lyon 1
21 av Claude Bernard
69622 Villeurbanne Cedex, France
email: attal@math.univ-lyon1.fr

Franco Fagnola
Politecnico di Milano
Dipartimento di Matematica "F.
Brioschi"
Piazza Leonardo da Vinci 32
20133 Milano Italy
email: franco.fagnola@polimi.it

Rolando Rebolledo
Facultad de Matemáticas
Universidad Católica de Chile
Casilla 306 Santiago 22, Chile
email: rrebolle@puc.cl

Luc Rey-Bellet
Department of Mathematics and
Statistics
University of Massachusetts
Amherst, MA 01003, USA
email: lr7q@math.umass.edu

Ergodic Properties of Markov Processes

Luc Rey-Bellet

Department of Mathematics and Statistics, University of Massachusetts,
Amherst, MA 01003, USA
e-mail: lr7q@math.umass.edu

1   Introduction   1
2   Stochastic Processes   2
3   Markov Processes and Ergodic Theory   4
    3.1   Transition probabilities and generators   4
    3.2   Stationary Markov processes and Ergodic Theory   7
4   Brownian Motion 12
5   Stochastic Differential Equations 14
6   Control Theory and Irreducibility 24
7   Hypoellipticity and Strong-Feller Property 26
8   Liapunov Functions and Ergodic Properties 28
References 39

1 Introduction

In these notes we discuss Markov processes, in particular stochastic differential equations (SDE) and develop some tools to analyze their long-time behavior. There are several ways to analyze such properties, and our point of view will be to use systematically Liapunov functions which allow a nice characterization of the ergodic properties. In this we follow, at least in spirit, the excellent book of Meyn and Tweedie [7]. In general a Liapunov function W is a positive function which grows at infinity and satisfies an inequality involving the generator of the Markov process L: roughly speaking we have the implications (α and β are positive constants)

1. $LW \leq \alpha + \beta W$ implies existence of solutions for all times.
2. $LW \leq -\alpha$ implies the existence of an invariant measure.
3. $LW \leq \alpha - \beta W$ implies exponential convergence to the invariant. measure

For (2) and (3), one should assume in addition, for example smoothness of the transition probabilities (i.e the semigroup e^{tL} is smoothing) and irreducibility of the process (ergodicity of the motion). The smoothing property for generator of SDE's is naturally linked with hypoellipticity of L and the irreducibility is naturally expressed in terms of control theory.

In sufficiently simple situations one might just guess a Liapunov function. For interesting problems, however, proving the existence of a Liapunov functions requires both a good guess and a quite substantial understanding of the dynamics. In these notes we will discuss simple examples only and in the companion lecture [11] we will apply these techniques to a model of heat conduction in anharmonic lattices. A simple set of equations that the reader should keep in mind here are the Langevin equations

$$dq = pdt ,$$
$$dp = (-\nabla V(q) - \lambda p)dt + \sqrt{2\lambda T}dB_t ,$$

where, $p, q \in \mathbf{R}^n$, $V(q)$ is a smooth potential growing at infinity, and B_t is Brownian motion. This equation is a model a particle with Hamiltonian $p^2/2 + V(q)$ in contact with a thermal reservoir at temperature T. In our lectures on open classical systems [11] we will show how to derive similar and more general equations from Hamiltonian dynamics. This simple model already has the feature that the noise is degenerate by which we mean that the noise is acting only on the p variable. Degeneracy (usually even worse than in these equations) is the rule and not the exception in mechanical systems interacting with reservoirs.

The notes served as a crash course in stochastic differential equations for an audience consisting mostly of mathematical physicists. Our goal was to provide the reader with a short guide to the theory of stochastic differential equations with an emphasis long-time (ergodic) properties. Some proofs are given here, which will, we hope, give a flavor of the subject, but many important results are simply mentioned without proof.

Our list of references is brief and does not do justice to the very large body of literature on the subject, but simply reflects some ideas we have tried to conveyed in these lectures. For Brownian motion, stochastic calculus and Markov processes we recommend the book of Oksendal [10], Kunita [15], Karatzas and Shreve [3] and the lecture notes of Varadhan [13, 14]. For Liapunov function we recommend the books of Has'minskii [2] and Meyn and Tweedie [7]. For hypoellipticity and control theory we recommend the articles of Kliemann [4], Kunita [6], Norris [8], and Stroock and Varadhan [12] and the book of Hörmander [1].

2 Stochastic Processes

A *stochastic process* is a parametrized collection of random variables

$$\{x_t(\omega)\}_{t \in T} \tag{1}$$

defined on a probability space $(\tilde{\Omega}, \mathcal{B}, \mathbf{P})$. In these notes we will take $T = \mathbf{R}^+$ or $T = \mathbf{R}$. To fix the ideas we will assume that x_t takes value in $X = \mathbf{R}^n$ equipped with the Borel σ-algebra, but much of what we will say has a straightforward generalization to more general state space. For a fixed $\omega \in \tilde{\Omega}$ the map

$$t \mapsto x_t(\omega) \tag{2}$$

is a *path* or a *realization* of the stochastic process, i.e. a random function from T into $\mathbf{R}^n$. For fixed $t \in T$

$$\omega \mapsto x_t(\omega) \tag{3}$$

is a random variable ("the state of the system at time t"). We can also think of $x_t(\omega)$ as a function of two variables (t, ω) and it is natural to assume that $x_t(\omega)$ is jointly measurable in (t, ω). We may identify each ω with the corresponding path $t \mapsto x_t(\omega)$ and so we can always think of $\tilde{\Omega}$ as a subset of the set $\Omega = (\mathbf{R}^n)^T$ of all functions from T into $\mathbf{R}^n$. The σ-algebra $\mathcal{B}$ will then contain the σ-algebra $\mathcal{F}$ generated by sets of the form

$$\{\omega ; x_{t_1}(\omega) \in F_1, \cdots, x_{t_n}(\omega) \in F_n\}, \tag{4}$$

where F_i are Borel sets of $\mathbf{R}^n$. The σ-algebra $\mathcal{F}$ is simply the Borel σ-algebra on Ω equipped with the product topology. From now on we take the point of view that a stochastic process is a probability measure on the measurable (function) space $(\Omega, \mathcal{F})$.

One can seldom describe explicitly the full probability measure describing a stochastic process. Usually one gives the *finite-dimensional* distributions of the process x_t which are probability measures $\mu_{t_1, \cdots, t_k}$ on $\mathbf{R}^{nk}$ defined by

$$\mu_{t_1, \cdots, t_k}(F_1 \times \cdots \times F_k) = \mathbf{P}\{x_{t_1} \in F_1, \cdots, x_{t_k} \in F_k\}, \tag{5}$$

where $t_1, \cdots, t_k \in T$ and the F_i are Borel sets of $\mathbf{R}^n$.

A useful fact, known as Kolmogorov Consistency Theorem, allows us to construct a stochastic process given a family of compatible finite-dimensional distributions.

Theorem 2.1. (Kolmogorov Consistency Theorem) *For $t_1, \cdots, t_k \in T$ and $k \in \mathbf{N}$ let $\mu_{t_1, \cdots, t_k}$ be probability measures on $\mathbf{R}^{nk}$ such that*

1. For all permutations σ of $\{1, \cdots, k\}$

$$\mu_{t_{\sigma(1)}, \cdots, t_{\sigma(k)}}(F_1 \times \cdots \times F_k) = \mu_{t_1, \cdots, t_k}(F_{\sigma^{-1}(1)} \times \cdots \times F_{\sigma^{-1}(k)}). \tag{6}$$

2. For all $m \in \mathbf{N}$

$$\mu_{t_1, \cdots, t_k}(F_1 \times \cdots \times F_k) = \mu_{t_1, \cdots, t_{k+m}}(F_1 \times \cdots \times F_k \times \mathbf{R}^n \times \cdots \times \mathbf{R}^n). \tag{7}$$

Then there exists a probability space $(\Omega, \mathcal{F}, \mathbf{P})$ and a stochastic process x_t on Ω such that

$$\mu_{t_1, \cdots, t_k}(F_1 \times \cdots \times F_k) = \mathbf{P}\{x_{t_1} \in F_1, \cdots, x_{t_k} \in F_k\}, \tag{8}$$

for all $t_i \in T$ and all Borel sets $F_i \subset \mathbf{R}^n$.

3 Markov Processes and Ergodic Theory

3.1 Transition probabilities and generators

A Markov process is a stochastic process which satisfies the condition that the future depends only on the present and not on the past, i.e., for any $s_1 \leq \cdots \leq s_k \leq t$ and any measurable sets $F_1, \cdots, F_k$, and F

$$\mathbf{P}\{x_t(\omega) \in F | x_{s_1}(\omega) \in F_1, \cdots, x_{s_k}(\omega) \in F_k\} = \mathbf{P}\{x_t(\omega) \in F | x_{s_k}(\omega) \in F_k\}. \tag{9}$$

More formally let $\mathcal{F}_t^s$ be the subalgebra of $\mathcal{F}$ generated by all events of the form $\{x_u(\omega) \in F\}$ where F is a Borel set and $s \leq u \leq t$. A stochastic process x_t is a *Markov process* if for all Borel sets F, and all $0 \leq s \leq t$ we have almost surely

$$\mathbf{P}\{x_t(\omega) \in F | \mathcal{F}_s^0\} = \mathbf{P}\{x_t(\omega) \in F | \mathcal{F}_s^s\} = \mathbf{P}\{x_t(\omega) \in F | x(s, \omega)\}. \tag{10}$$

We will use later an equivalent way of describing the Markov property. Let us consider 3 subsequent times $t_1 < t_2 < t_3$. The Markov property means that for any g bounded measurable

$$E[g(x_{t_3}) | \mathcal{F}_{t_2}^{t_2} \times \mathcal{F}_{t_1}^{t_1}] = E[g(x_{t_3}) | \mathcal{F}_{t_2}^{t_2}]. \tag{11}$$

The time reversed Markov property that for any bounded measurable function f

$$E[f(x_{t_1}) | \mathcal{F}_{t_3}^{t_3} \times \mathcal{F}_{t_2}^{t_2}] = E[f(x_{t_1}) | \mathcal{F}_{t_2}^{t_2}], \tag{12}$$

which says that the past depends only on the present and not on the future. These two properties are in fact equivalent, since we will show that they are both equivalent to the symmetric condition

$$E[g(x_{t_3}) f(x_{t_1}) | \mathcal{F}_{t_2}^{t_2}] = E[g(x_{t_3}) | \mathcal{F}_{t_2}^{t_2}] E[f(x_{t_1}) \mathcal{F}_{t_2}^{t_2}], \tag{13}$$

which asserts that given the present, past and future are conditionally independent. By symmetry it is enough to prove

Lemma 3.1. *The relations (11) and (13) are equivalent.*

Proof. Let us fix f and g and let us set $x_{t_i} = x_i$ and $\mathcal{F}_{t_i}^{t_i} \equiv \mathcal{F}_i$, for $i = 1, 2, 3$. Let us assume that Eq. (11) holds and denote by $\hat{g}(x_2)$ the common value of (11). Then we have

$$\begin{aligned}
\mathbf{E}[g(x_3) f(x_1) | \mathcal{F}_2] &= \mathbf{E}\left[\mathbf{E}[g(x_3) f(x_1) | \mathcal{F}_2 \times \mathcal{F}_1] | \mathcal{F}_2\right] \\
&= \mathbf{E}\left[f(x_1) \mathbf{E}[g(x_3) | \mathcal{F}_2 \times \mathcal{F}_1] | \mathcal{F}_2\right] = \mathbf{E}\left[f(x_1) \hat{g}(x_2) | \mathcal{F}_2\right] \\
&= \mathbf{E}\left[f(x_1) | \mathcal{F}_2\right] \hat{g}(x_2) = \mathbf{E}\left[f(x_1) | \mathcal{F}_2\right] \mathbf{E}[g(x_3) | \mathcal{F}_2],
\end{aligned} \tag{14}$$

which is Eq. (13). Conversely let us assume that Eq. (13) holds and let us denote by $\overline{g}(x_1, x_2)$ and by $\hat{g}(x_2)$ the left side and the right side of (11). Let $h(x_2)$ be any bounded measurable function. We have

$$\mathbf{E}\left[f(x_1)h(x_2)\bar{g}(x_1,x_2)\right] = \mathbf{E}\left[f(x_1)h(x_2)\mathbf{E}[g(x_3)|\mathcal{F}_2 \times \mathcal{F}_1]\right]$$

$$= \mathbf{E}\left[f(x_1)h(x_2)g(x_3)\right] = \mathbf{E}\left[h(x_2)\mathbf{E}[f(x_1)g(x_3)\,|\,\mathcal{F}_2]\right]$$

$$= \mathbf{E}\left[h(x_2)\left(\mathbf{E}[g(x_3)\,|\,\mathcal{F}_2]\right)\left(\mathbf{E}[f(x_1)\,|\,\mathcal{F}_2]\right)\right]$$

$$= \mathbf{E}\left[h(x_2)\hat{g}(x_2)\mathbf{E}[f(x_1)\,|\,\mathcal{F}_2]\right] = \mathbf{E}\left[f(x_1)h(x_2)\hat{g}(x_2)\right]. \qquad (15)$$

Since f and h are arbitrary this implies that $\bar{g}(x_1,x_2) = \hat{g}(x_2)$ a.s. □

A natural way to construct a Markov process is via a *transition probability function*

$$P_t(x, F), \qquad t \in T, \quad x \in \mathbf{R}^n, \quad F \text{ a Borel set}, \qquad (16)$$

where $(t, x) \mapsto P_t(x, F)$ is a measurable function for any Borel set F and $F \mapsto P_t(x, F)$ is a probability measure on $\mathbf{R}^n$ for all (t, x). One defines

$$\mathbf{P}\{x_t(\omega) \in F \,|\, \mathcal{F}_s^0\} = \mathbf{P}\{x_t(\omega) \in F \,|\, x_s(\omega)\} = P_{t-s}(x_s(\omega), F). \qquad (17)$$

The finite dimensional distribution for a *Markov process starting at x at time 0* are then given by

$$\mathbf{P}\{x_{t_1} \in F\} = P_{t_1}(x, F_1),$$

$$\mathbf{P}\{x_{t_1} \in F_1, x_{t_2} \in F_2\} = \int_{F_1} P_{t_1}(x, dx_1) P_{t_2-t_1}(x_1, F_2), \qquad (18)$$

$$\vdots$$

$$\mathbf{P}\{x_{t_1} \in F_1, \cdots, x_{t_k} \in F_k\} = \int_{F_1} \cdots \int_{F_{k-1}} P_{t_1}(x, dx_1) \cdots P_{t_k-t_{k-1}}(x_{k-1}, F_k).$$

By the Kolmogorov Consistency Theorem this defines a stochastic process x_t for which $\mathbf{P}\{x_0 = x\} = 1$. We denote $\mathbf{P}_x$ and $\mathbf{E}_x$ the corresponding probability distribution and expectation.

One can also give an *initial distribution* π, where π is a probability measure on $\mathbf{R}^n$ which describe the initial state of the system at $t = 0$. In this case the finite dimensional probability distributions have the form

$$\int_{\mathbf{R}^n} \int_{F_1} \cdots \int_{F_{k-1}} \pi(dx) P_t(x, dx_1) P_{t_2-t_1}(x_1, dx_2) \cdots P_{t_k-t_{k-1}}(x_{k-1}, F_k), \qquad (19)$$

and we denote $\mathbf{P}_\pi$ and $\mathbf{E}_\pi$ the corresponding probability distribution expectation.

Remark 3.2. We have considered here only time homogeneous process, i.e., processes for which $\mathbf{P}_x\{x_t(\omega) \in F \,|\, x_s(\omega)\}$ depends only on $t - s$. This can generalized this by considering transition functions $P(t, s, x, A)$.

The following property is a immediate consequence of the fact that the future depends only on the present and not on the past.

Lemma 3.3. (Chapman-Kolmogorov equation) *For $0 \leq s \leq t$ we have*

$$P_t(x, A) = \int_{\mathbf{R}^n} P_s(x, dy) P_{t-s}(y, A). \tag{20}$$

Proof. : We have

$$P_t(x, A) = P\{x_0 = x, x_t \in A\} = P\{x_0 = x, x_s \in \mathbf{R}^n, x_t \in A\}$$

$$= \int_{\mathbf{R}^n} P_s(x, dy) P_{t-s}(y, A). \tag{21}$$

□

For a measurable function $f(x)$, $x \in \mathbf{R}^n$, we have

$$\mathbf{E}_x[f(x_t)] = \int_{\mathbf{R}^n} P_t(x, dy) f(y). \tag{22}$$

and we can associate to a transition probability a linear operator acting on measurable function by

$$T_t f(x) = \int_{\mathbf{R}^n} P_t(x, dy) f(y) = \mathbf{E}_x[f(x_t)]. \tag{23}$$

From the Chapman-Kolmogorov equation it follows immediately that T_t is a semigroup: for all $s, t \geq 0$ we have

$$T_{t+s} = T_t T_s. \tag{24}$$

We have also a dual semigroup acting on σ-finite measures on $\mathbf{R}^n$:

$$S_t \mu(A) = \int_{\mathbf{R}^n} \mu(dx) P_t(x, A). \tag{25}$$

The semigroup T_t has the following properties which are easy to verify.

1. T_t preserves the constant, if $1(x)$ denotes the constant function then

$$T_t 1(x) = 1(x). \tag{26}$$

2. T_t is positive in the sense that

$$T_t f(x) \geq 0 \quad \text{if} \quad f(x) \geq 0. \tag{27}$$

3. T_t is a contraction semigroup on $L^\infty(dx)$, the set of bounded measurable functions equipped with the sup-norm $\| \cdot \|_\infty$.

$$\|T_t f\|_\infty = \sup_x \left| \int_{\mathbf{R}^n} P_t(x, dy) f(y) \right|$$

$$\leq \sup_y |f(y)| \sup_x \int)\mathbf{R}^n P_t(x, dy) = \|f\|_\infty. \tag{28}$$

The spectral properties of the semigroup T_t are important to analyze the long-time (ergodic) properties of the Markov process x_t. In order to use method from functional analysis one needs to define these semigroups on function spaces which are more amenable to analysis than the space of measurable functions.

We say that the semigroup T_t is *weak-Feller* if it maps the set of bounded continuous function $C^b(\mathbf{R}^n)$ into itself. If the transition probabilities $P_t(x, A)$ are stochastically continuous, i.e., if $\lim_{t\to 0} P_t(x, B_\epsilon(x)) = 1$ for any $\epsilon > 0$ ($B_\epsilon(x)$ is the ϵ-neighborhood of x) then it is not difficult to show that $\lim_{t\to 0} T_t F(x) = f(x)$ for any $f(x) \in C^b(\mathbf{R}^n)$ (details are left to th reader) and then T_t is a contraction semigroup on $C^b(\mathbf{R}^n)$.

We say that the semigroup T_t is *strong-Feller* if it maps bounded measurable function into continuous function. This reflects the fact that T^t has a "smoothing effect". A way to show the strong-Feller property is to establish that the transition probabilities $P_t(x, A)$ have a density

$$P_t(x, dy) = p_t(x, y)dy, \tag{29}$$

where $p_t(x, y)$ is a sufficiently regular (e.g. continuous or differentiable) function of x, y and maybe also of t. We will discuss some tools to prove such properties in Section 7.

If T_t is weak-feller we define the *generator* L of T_t by

$$Lf(x) = \lim_{t\to 0} \frac{T_t f(x) - f(x)}{t}. \tag{30}$$

The domain of definition of L is set of all f for which the limit (30) exists for all x.

3.2 Stationary Markov processes and Ergodic Theory

We say that a stochastic process is *stationary* if the finite dimensional distributions

$$\mathbf{P}\{x_{t_1+h} \in F_1, \cdots, x_{t_k+h} \in F_k\} \tag{31}$$

are independent of h, for all $t_1 < \cdots < t_k$ and all measurable F_i. If the process is Markovian with initial distribution $\pi(dx)$ then (take $k = 1$)

$$\int_{\mathbf{R}^n} \pi(dx) P_t(x, F) = S_t \pi(F) \tag{32}$$

must be independent of t for any measurable F, i.e., we must have

$$S_t \pi = \pi, \tag{33}$$

for all $t \geq 0$. The condition (33) alone implies stationarity since it implies that

$$\mathbf{P}_\pi\{x_{t_1+h} \in F_1, \cdots, x_{t_k+h} \in F_k\}$$
$$= \int_{\mathbf{R}^n} \int_{F_1} \cdots \int_{F_{k-1}} \pi(dx) P_{t_1+h}(x, dx_1) \cdots P_{t_k-t_{k-1}}(x_{k-1}, F_k),$$
$$= \int_{F_1} \cdots \int_{F_{k-1}} \pi(dx) P_{t_1}(x, dx_1) \cdots P_{t_k-t_{k-1}}(x_{k-1}, F_k), \tag{34}$$

which is independent of h.

Intuitively stationary distribution describe the long-time behavior of x_t. Indeed let us suppose that the distribution of x_t with initial distribution μ converges in some sense to a distribution $\gamma = \gamma_\mu$ (a priori γ may depend on the initial distribution μ), i.e.,

$$\lim_{t \to \infty} \mathbf{P}_\mu \{x_t \in F\} = \gamma_\mu(F), \tag{35}$$

for all measurable F. Then we have, formally,

$$
\begin{aligned}
\gamma_\mu(F) &= \lim_{t \to \infty} \int_{\mathbf{R}^n} \mu(dx) P_t(x, F) \\
&= \lim_{t \to \infty} \int_{\mathbf{R}^n} \mu(dx) \int_{\mathbf{R}^n} P_{t-s}(x, dy) P_s(y, F) \\
&= \int \gamma_\mu(dy) \int P_s(y, F) = S_s \gamma_\mu(F),
\end{aligned}
\tag{36}
$$

i.e., γ_μ is a stationary distribution.

In order to make this more precise we recall some concepts and results from ergodic theory. Let $(X, \mathcal{F}, \mu)$ be a probability space and $\phi_t, t \in \mathbf{R}$ a group of measurable transformations of X. We say that ϕ_t is *measure preserving* if $\mu(\phi_{-t}(A)) = \mu(A)$ for all $t \in \mathbf{R}$ and all $A \in \mathcal{F}$. We also say that μ is an invariant measure for ϕ_t. A basic result in ergodic theory is the pointwise Birkhoff ergodic theorem.

Theorem 3.4. (Birkhoff Ergodic Theorem) *Let ϕ_t be a group of measure preserving transformations of $(X, \mathcal{F}, \mu)$. Then for any $f \in L^1(\mu)$ the limit*

$$\lim_{t \to \infty} \frac{1}{t} \int_0^t f(\phi_s(x)) \, ds = f^*(x) \tag{37}$$

exists μ-a.s. The limit $f^*(x)$ is ϕ_t invariant, $f(\phi^t(x)) = f(x)$ for all $t \in \mathbf{R}$, and $\int_X f \, d\mu = \int_X f^* \, d\mu$.

The group of transformation ϕ_t is said to be *ergodic* if $f^*(x)$ is constant μ-a.s. and in that case $f^*(x) = \int f \, d\mu$, μ-a.s. Ergodicity can be also expressed in terms of the σ-field of invariant subsets. Let $\mathcal{G} \subset \mathcal{F}$ be the σ-field given by $\mathcal{G} = \{A \in \mathcal{F} : \phi^{-t}(A) = A \text{ for all } t\}$. Then in Theorem 3.4 $f^*(x)$ is given by the conditional expectation

$$f^*(x) = E[f \,|\, \mathcal{G}]. \tag{38}$$

The ergodicity of ϕ_t is equivalent to the statement that $\mathcal{G}$ is the trivial σ-field, i.e., if $A \in \mathcal{G}$ then $\mu(A) = 0$ or 1.

Given a measurable group of transformation ϕ_t of a measurable space, let us denote by $\mathcal{M}$ the set of invariant measure. It is easy to see that $\mathcal{M}$ is a convex set and we have

Proposition 3.5. *The probability measure μ is an extreme point of $\mathcal{M}$ if and only if μ is ergodic.*

Proof. Let us suppose that μ is not extremal. Then there exists μ_1, $\mu_2 \in \mathcal{M}$ with $\mu_1 \neq \mu_2$ and $0 < a < 1$ such that $\mu = a\mu_1 + (1-a)\mu_2$. We claim that μ is not ergodic. It μ were ergodic then $\mu(A) = 0$ or 1 for all $A \in \mathcal{G}$. If $\mu(A) = 0$ or 1, then $\mu_1(A) = \mu_2(A) = 0$ or $\mu_1(A) = \mu_2(A) = 1$. Therefore μ_1 and μ_2 agree on the σ-field $\mathcal{G}$. Let now f be a bounded measurable function and let us consider the function

$$f^*(x) = \lim_{t \to \infty} \frac{1}{t} \int_0^t f(\phi_s x)\,ds, \tag{39}$$

which is defined on the set E where the limit exists. By the ergodic theorem $\mu_1(E) = \mu_2(E) = 1$ and f^* is measurable with respect to $\mathcal{G}$. We have

$$\int_E f\,d\mu_i = \int_E f^*\,d\mu_i, \qquad i = 1, 2. \tag{40}$$

Since $\mu_1 = \mu_2$ on $\mathcal{G}$, f^* is $\mathcal{G}$-measurable, and $\mu_i(E) = 1$ for $i = 1, 2$, we see that

$$\int_X f\,d\mu_1 = \int_X f\,d\mu_2. \tag{41}$$

Since f is arbitrary this implies that $\mu_1 = \mu_2$ and this is a contradiction.

Conversely if μ is not ergodic, then there exists $A \in \mathcal{G}$ with $0 < \mu(A) < 1$. Let us define

$$\mu_1(B) = \frac{\mu(A \cap B)}{\mu(A)}, \qquad \mu_2(B) = \frac{\mu(A^c \cap B)}{\mu(A^c)}. \tag{42}$$

Since $A \in \mathcal{G}$, it follows that μ_i are invariant and that $\mu = \mu(A)\mu_1 + \mu(A^c)\mu_2$. Thus μ is not an extreme point. $\square$

A stronger property than ergodicity is the property of *mixing* . In order to formulate it we first note that we have

Lemma 3.6. *μ is ergodic if and only if*

$$\lim_{t \to \infty} \frac{1}{t} \int_0^t \mu(\phi_{-s}(A) \cap B) = \mu(A)\mu(B), \tag{43}$$

for all A, $B \in \mathcal{F}$

Proof. If μ is ergodic, let $f = \chi_A$ be the characteristic function of A in the ergodic theorem, multiply by the characteristic function of B and use the bounded convergence theorem to show that Eq. (43) holds. Conversely let $E \in \mathcal{G}$ and set $A = B = E$ in Eq. (43). This shows that $\mu(E) = \mu(E)^2$ and therefore $\mu(E) = 0$ or 1. $\square$

We say that an invariant measure μ is *mixing* if we have

$$\lim_{t \to \infty} \mu(\phi_{-t}(A) \cap B) = \mu(A)\mu(B) \tag{44}$$

for all $A, B \in \mathcal{F}$, i.e., we have convergence in Eq. (44) instead of convergence in the sense of Cesaro in Eq. (43).

Mixing can also be expressed in terms of the triviality of a suitable σ-algebra. We define the remote future σ-field, denoted $\mathcal{F}_\infty$, by

$$\mathcal{F}_\infty = \bigcup_{t \geq 0} \phi_{-t}(\mathcal{F}). \tag{45}$$

Notice that a set $A \in \mathcal{F}_\infty$ if and only if for every t there exists a set $A_t \in \mathcal{F}$ such that $A = \phi_{-t} A_t$. Therefore the σ-field of invariant subsets $\mathcal{G}$ is a sub-σ-field of $\mathcal{F}^\infty$. We have

Lemma 3.7. μ is mixing if and only if the σ-field $\mathcal{F}_\infty$ is trivial.

Proof. Let us assume first that $\mathcal{F}_\infty$ is not trivial. There exists a set $A \in \mathcal{F}_\infty$ with $0 < \mu(A) < 1$ or $\mu(A)^2 \neq \mu(A)$ and for any t there exists a set A_t such that $A = \phi_{-t}(A_t)$. If μ were mixing we would have $\lim_{t \to \infty} \mu(\phi_{-t}(A) \cap A) = \mu(A)^2$. On the other hand

$$\mu(\phi_{-t}(A) \cap A) = \mu(\phi_{-t}(A) \cap \phi_{-t}(A_t)) = \mu(A \cap A_t) \tag{46}$$

and this converge to $\mu(A)$ as $t \to \infty$. This is a contradiction.

Let us assume that $\mathcal{F}_\infty$ is trivial. We have

$$\mu(\phi_{-t}(A) \cap B) - \mu(A)\mu(B) = \mu(B \mid \phi_{-t}(A))\mu(\phi_{-t}(A)) - \mu(A)\mu(B)$$

$$= (\mu(B \mid \phi_{-t}(A)) - \mu(B))\,\mu(A) \tag{47}$$

The triviality of $\mathcal{F}_\infty$ implies that $\lim_{t \to \infty} \mu(B \mid \phi_{-t}(A)) = \mu(B)$. $\square$

Given a stationary Markov process with a stationary distribution π one constructs a stationary Markov process with probability measure $\mathbf{P}_\pi$. We can extend this process in a natural way on $-\infty < t < \infty$. The marginal of $\mathbf{P}_\pi$ at any time t is π. Let Θ_s denote the shift transformation on Ω given by $\Theta_s(x_t(\omega)) = x_{t+s}(\omega)$. The stationarity of the Markov process means that Θ_s is a measure preserving transformation of $(\Omega, \mathcal{F}, \mathbf{P}_\pi)$.

In general given transition probabilities $P_t(x, dy)$ we can have several stationary distributions π and several corresponding stationary Markov processes. Let $\tilde{\mathcal{M}}$ denote the set of stationary distributions for $P_t(x, dy)$, i.e.,

$$\tilde{\mathcal{M}} = \{\pi : S_t \pi = \pi\}. \tag{48}$$

Clearly $\tilde{\mathcal{M}}$ is a convex set of probability measures. We have

Theorem 3.8. *A stationary distribution π for the Markov process with transition probabilities $P_t(x, dy)$ is an extremal point of $\tilde{\mathcal{M}}$ if and only if $\mathbf{P}_\pi$ is ergodic, i.e., an extremal point in the set of all invariant measures for the shift Θ_t.*

Proof. If P_π is ergodic then, by the linearity of the map $\pi \mapsto P_\pi$, π must be an extreme point of $\mathcal{M}$.

To prove the converse let E be a nontrivial set in the σ-field of invariant subsets. Let $\mathcal{F}_\infty$ denote the far remote future σ-field and $\mathcal{F}^{-\infty}$ the far remote past σ-field which is defined similarly. Let also $\mathcal{F}_0^0$ be the σ-field generated by x_0 (this is the present). An invariant set is both in the remote future $\mathcal{F}_\infty$ as well as in the remote past $\mathcal{F}_\infty$. By Lemma 3.1 the past and the future are conditionally independent given the present. Therefore

$$\mathbf{P}_\pi[E \mid \mathcal{F}_0^0] = \mathbf{P}_\pi[E \cap E \mid \mathcal{F}_0^0] = \mathbf{P}_\pi[E \mid \mathcal{F}_0^0]\mathbf{P}_\pi[E \mid \mathcal{F}_0^0]. \tag{49}$$

and therefore it must be equal either to 0 or 1. This implies that for any invariant set E there exists a measurable set $A \subset \mathbf{R}^n$ such that $E = \{\omega : x_t(\omega) \in A \text{ for all } t \in \mathbf{R}\}$ up to a set of $\mathbf{P}_\pi$ measure 0. If the Markov process start in A or A^c it does not ever leaves it. This means that $0 < \pi(A) < 1$ and $P_t(x, A^c) = 0$ for π a.e. $x \in A$ and $P_t(x, A) = 0$ for π a.e. $x \in A^c$. This implies that π is not extremal.

Remark 3.9. Theorem 3.8 describes completely the structure of the σ-field of invariant subsets for a stationary Markov process with transition probabilities $P_t(x, dy)$ and stationary distribution π. Suppose that the state space can be partitioned non trivially, i.e., there exists a set A with $0 < \pi(A) < 1$ such that $P_t(x, A) = 1$ for π almost every $x \in A$ and for any $t > 0$ and $P_t(x, A^c) = 1$ for π almost every $x \in A^c$ and for any $t > 0$. Then the event

$$E = \{\omega ; x_t(\omega) \in A \text{ for all } t \in \mathbf{R}\} \tag{50}$$

is a nontrivial set in the invariant σ-field. What we have proved is just the converse the statement.

We can therefore look at the extremal points of the sets of all stationary distribution, $S_t\pi = \pi$. Since they correspond to ergodic stationary processes, it is natural to call them ergodic stationary distributions. If π is ergodic then, by the ergodic theorem we have

$$\lim_{t \to \infty} \frac{1}{t} \int_0^t F(\theta_s(x.(\omega))) \, ds = E_\pi \left[F(x.(\omega)) \right]. \tag{51}$$

for $\mathbf{P}_\pi$ almost all ω. If $F(x.) = f(x_0)$ depends only on the state at time 0 and is bounded and measurable then we have

$$\lim_{t \to \infty} \frac{1}{t} \int_0^t f(x_s(\omega)) \, ds = \int f(x) d\pi(x). \tag{52}$$

for π almost all x and almost all ω. Integrating over ω gives that

$$\lim_{t \to \infty} \frac{1}{t} \int_0^t T_s f(x) \, ds = \int f(x) d\pi(x). \tag{53}$$

for π almost all x.

The property of mixing is implied by the convergence of the probability measure $P_t(x, dy)$ to $\mu(dy)$. In which sense we have convergence depends on the problem under consideration, and various topologies can be used. We consider here the total variation norm (and variants of it later): let μ be a signed measure on $\mathbf{R}^n$, the *total variation norm* $\|\mu\|$ is defined as

$$\|\mu\| = \sup_{|f| \leq 1} |\mu(f)| = \sup_A \mu(A) - \inf_A \mu(A). \tag{54}$$

Clearly convergence in total variation norm implies weak convergence.

Let us assume that there exists a stationary distribution π for the Markov process with transition probabilities $P_t(x, dy)$ and that

$$\lim_{t \to \infty} \|P_t(x, \cdot) - \pi\| = 0, \tag{55}$$

for all x. The condition (55) implies mixing. By a simple density argument it is enough to show mixing for $E \in \mathcal{F}_s^{-\infty}$ and $F \in \mathcal{F}_\infty^t$. Since $\Theta_{-t}(\mathcal{F}_s^{-\infty}) = \mathcal{F}_{s-t}^{-\infty}$ we simply have to show that as $k = t - s$ goes to ∞, $\mu(E \cap F)$ converges to $\mu(E)\mu(F)$. We have

$$\mu(E)\mu(F) = \int_E \left(\int_{\mathbf{R}^n} \mathbf{P}_x(\Theta_{-t_1} F) d\pi(x) \right) d\mathbf{P}_\pi(\omega),$$

$$\mu(E \cap F) = \int_E \left(\int_{\mathbf{R}^n} \mathbf{P}_x(\Theta_{-t_1} F) P_k(x_{s_2}(\omega), dx) \right) d\mathbf{P}_\pi, \tag{56}$$

and therefore

$$\mu(E \cap F) - \mu(E)\mu(F)$$
$$= \int_E \left(\int_{\mathbf{R}^n} \mathbf{P}_x(\Theta_{-t_1} F) \left(P_k(x_{s_2}(\omega), dx) - \pi(dx) \right) \right) d\mathbf{P}_\pi, \tag{57}$$

from which we conclude mixing.

4 Brownian Motion

An important example of a Markov process is the Brownian motion. We will take as a initial distribution the delta mass at x, i.e., the process starts at x. The transition probability function of the process has the density $p_t(x, y)$ given by

$$p_t(x, y) = \frac{1}{(2\pi t)^{n/2}} \exp\left(-\frac{(x - y)^2}{2t} \right). \tag{58}$$

Then for $0 \leq t_1 < t_2 < \cdots < t_k$ and for Borel sets F_i we define the finite dimensional distributions by

$$\nu_{t_1, \ldots, t_x}(F_1 \times \cdots \times F_x)$$
$$= \int p_{t_1}(x, x_1) p_{t_2 - t_1}(x_1, x_2) \cdots p_{t_x - t_{x-1}}(x_{x-1}, x_x) dx_1 \cdots dx_x, \tag{59}$$

with the convention

$$p_0(x, x_1) = \delta_x(x_1).\qquad(60)$$

By Kolmogorov Consistency Theorem this defines a stochastic process which we denote by B_t with probability distribution $\mathbf{P}_x$ and expectation $\mathbf{E}_x$. This process is the *Brownian motion starting at* x.

We list now some properties of the Brownian motion. Most proofs are left as exercises (use your knowledge of Gaussian random variables).

(a) The Brownian motion is a *Gaussian process*, i.e., for any $k \geq 1$, the random variable $Z \equiv (B_{t_1}, \cdots, B_{t_k})$ is a $\mathbf{R}^{nk}$-valued normal random variable. This is clear since the density of the finite dimensional distribution (59) is a product of Gaussian (the initial distribution is a degenerate Gaussian). To compute the mean and variance consider the characteristic function which is given for $\alpha \in \mathbf{R}^{nk}$ by

$$\mathbf{E}_x\left[\exp(i\alpha^T Z)\right] = \exp\left(-\frac{1}{2}\alpha^T C\alpha + i\alpha^T M\right),\qquad(61)$$

where

$$M = \mathbf{E}_x[Z] = (x, \cdots, x),\qquad(62)$$

is the mean of Z and the covariance matrix $C_{ij} = \mathbf{E}_x[Z_i Z_j]$ is given by

$$C = \begin{pmatrix} t_1\mathbf{I}_n & t_1\mathbf{I}_n & \cdots & t_1\mathbf{I}_n \\ t_1\mathbf{I}_n & t_2\mathbf{I}_n & \cdots & t_2\mathbf{I}_n \\ \vdots & \vdots & \cdots & \vdots \\ t_1\mathbf{I}_n & t_2\mathbf{I}_n & \cdots & t_k\mathbf{I}_n \end{pmatrix},\qquad(63)$$

where $\mathbf{I}_n$ is n by n identity matrix. We thus find

$$\mathbf{E}_x[B_t] = x,\qquad(64)$$
$$\mathbf{E}_x[(B_t - x)(B_s - x)] = n\min(t, s),\qquad(65)$$
$$\mathbf{E}_x[(B_t - B_s)^2] = n|t - s|,\qquad(66)$$

(b) If $B_t = (B_t^{(1)}, \cdots, B_t^{(n)})$ is a m-dimensional Brownian motion, $B_t^{(j)}$ are independent one-dimensional Brownian motions.

(c) The Brownian motion B_t has *independent increments*, i.e., for $0 \leq t_1 < t_2 < \cdots < t_k$ the random variables $B_{t_1}, B_{t_2} - B_{t_1}, \cdots B_{t_k} - B_{t_{k-1}}$ are independent. This easy to verify since for Gaussian random variables it is enough to show that the correlation $\mathbf{E}_x[(B_{t_i} - B_{t_{i-1}})(B_{t_j} - B_{t_{j-1}})]$ vanishes.

(d) The Brownian motion has *stationary increments*, i.e., $B_{t+h} - B_t$ has a distribution which is independent of t. Since it is Gaussian it suffices to check $\mathbf{E}_x[B_{t+h} - B_t] = 0$ and $\mathbf{E}_x[(B_{t+h} - B_t)^2]$ is independent of t.

(d) A stochastic process $\tilde{x}_t$ is called *a modification* of x_t if $\mathbf{P}\{x_t = \tilde{x}_t\}$ holds for all t. Usually one does not distinguish between a stochastic process and its modification.

However the properties of the paths can depend on the choice of the modification, and for us it is appropriate to choose a modification with particular properties, i.e., the paths are continuous functions of t. A criterion which allows us to do this is given by (another) famous theorem from Kolmogorov

Theorem 4.1. (Kolmogorov Continuity Theorem) *Suppose that there exists positive constants α, β, and C such that*

$$\mathbf{E}[|x_t - x_s|^\alpha] \leq C|t - s|^{1+\beta} . \tag{67}$$

Then there exists a modification of x_t such that $t \mapsto x_t$ is continuous a.s.

In the case of Brownian motion it is not hard to verify (use the characteristic function) that we have

$$\mathbf{E}[|B_t - B_s|^4] = 3|t - s|^2 , \tag{68}$$

so that the Brownian motion has a continuous version, i.e. we may (and will) assume that $x_t(\omega) \in \mathcal{C}([0, \infty); \mathbf{R}^n)$ and will consider the measure $\mathbf{P}_x$ as a measure on the function space $\mathcal{C}([0, \infty); \mathbf{R}^n)$ (this is a complete topological space when equipped with uniform convergence on compact sets). This version of Brownian motion is called the *canonical* Brownian motion.

5 Stochastic Differential Equations

We start with a few purely formal remarks. From the properties of Brownian motion it follows, formally, that its time derivative $\xi_t = \dot{B}_t$ satisfies $\mathbf{E}[\xi_t] = 0$, $\mathbf{E}[(\xi_t)^2] = \infty$, and $\mathbf{E}[\xi_t\xi_s] = 0$ if $t \neq s$, so that we have formally, $\mathbf{E}[\xi_t\xi_s] = \delta(t - s)$. So, intuitively, $\xi(t)$ models an time-uncorrelated random noise. It is a fact however that the paths of B_t are a.s. nowhere differentiable so that ξ_t cannot be defined as a random process on $(\mathbf{R}^n)^T$ (it can be defined if we allow the paths to be distributions instead of functions, but we will not discuss this here). But let us consider anyway an equation of the form

$$\dot{x}_t = b(x_t) + \sigma(x_t)\dot{B}_t , \tag{69}$$

where, $x \in \mathbf{R}^n$, $b(x)$ is a vector field, $\sigma(x)$ a $n \times m$ matrix, and B_t a m-dimensional Brownian motion. We rewrite it as integral equation we have

$$x_t(\omega) = x_0(\omega) + \int_0^t b(x_s(\omega))ds + \int_0^t \sigma(x_s(\omega))\dot{B}_s ds . \tag{70}$$

Since $\dot{B}_u$ is uncorrelated $x_t(\omega)$ will depend on the present, $x_0(\omega)$, but not on the past and the solution of such equation should be a Markov process. The goal of this chapter is to make sense of such differential equation and derive its properties. We rewrite (69) with the help of differentials as

$$dx_t = b(x_t)dt + \sigma(x_t)dB_t , \tag{71}$$

by which one really means a solution to the integral equation

$$x_t - x_0 = \int_0^t b(x_s)ds + \int_0^t \sigma(x_s)dB_s .$$ (72)

The first step to make sense of this integral equation is to define *Ito integrals* or *stochastic integrals*, i.e., integrals of the form

$$\int_0^t f(s,\omega)dB_s(\omega),$$ (73)

for a suitable class of functions. Since, as mentioned before B_t is nowhere differentiable, it is not of bounded variation and thus Eq. (73) cannot be defined as an ordinary Riemann-Stieljes integral.

We will consider the class of functions $f(t,\omega)$ which satisfy the following three conditions

1. The map $(s,\omega) \mapsto f(s,\omega)$ is measurable for $0 \le s \le t$.
2. For $0 \le s \le t$, the function $f(s,\omega)$ depends only upon the history of B_s up to time s, i.e., $f(s,\omega)$ is measurable with respect to the σ-algebra $\mathcal{N}_s^0$ generated by sets of the form $\{B_{t_1}(\omega) \in F_1, \cdots, B_{t_k}(\omega) \in F_k\}$ with $0 \le t_1 < \cdots < t_k \le s$.
3. $\mathbf{E}\left[\int_0^t f(s,\omega)^2 ds\right] < \infty$.

The set of functions $f(s,\omega)$ which satisfy these three conditions is denoted by $\mathcal{V}[0,t]$.

It is natural, in a theory of integration, to start with *elementary functions* of the form

$$f(t,\omega) = \sum_j f(t_j^*,\omega)\mathbf{1}_{[t_j,t_{j+1})}(t),$$ (74)

where $t_j^* \in [t_j, t_{j+1}]$. In order to satisfy Condition 2. one chooses the right-end point $t_j^* = t_j$ and we then write

$$f(t,\omega) = \sum_j e_j(\omega)\mathbf{1}_{[t_j,t_{j+1})}(t),$$ (75)

and $e_j(\omega)$ is $\mathcal{N}_{t_j}$ measurable. We define the stochastic integral to be

$$\int_0^t f(s,\omega)dB_s(\omega) = \sum_j e_j(\omega)(B_{t_{j+1}} - B_{t_j}).$$ (76)

This is the *Ito integral*. To extend this integral from elementary functions to general functions, one uses Condition 3. together with the so called Ito isometry

Lemma 5.1. (Ito isometry) *If $\phi(s,\omega)$ is bounded and elementary*

$$\mathbf{E}\left[\left(\int_0^t \phi(s,\omega)dB_s(\omega)\right)^2\right] = \mathbf{E}\left[\int_0^t f(s,\omega)^2 ds\right].$$ (77)

Proof. Set $\Delta B_j = B_{t_{j+1}} - B_{t_j}$. Then we have

$$\mathbf{E}\left[e_i e_j \Delta B_i \Delta B_j\right] = \begin{cases} 0 & i \neq j \\ \mathbf{E}[e_j^2](t_{j+1} - t_j) & i = j \end{cases}, \tag{78}$$

using that $e_j e_i \Delta B_i$ is independent of ΔB_j for $j > i$ and that e_j is independent of B_j by Condition 2. We have then

$$\mathbf{E}\left[\left(\int_0^t \phi(s,w)dB_s(w)\right)^2\right] = \sum_{i,j} \mathbf{E}\left[e_i e_j \Delta B_i \Delta B_j\right]$$

$$= \sum_j \mathbf{E}\left[e_j^2\right](t_{j+1} - t_j)$$

$$= \mathbf{E}\left[\int_0^t f(s,w)^2 dt\right]. \tag{79}$$

$\square$

Using the Ito isometry one extends the Ito integral to functions which satisfy conditions (a)-(c). One first shows that one can approximate such a function by elementary bounded functions, i.e., there exists a sequence $\{\phi_n\}$ of elementary bounded such that

$$\mathbf{E}\left[\int_0^t \left(f(s,w) - \phi_n(s,w)\right)^2 ds\right] \to 0. \tag{80}$$

This is a standard argument, approximate first f by a bounded, and then by a bounded continuous function. The details are left to the reader. Then one defines the stochastic integral by

$$\int_0^t f(s,w)dB_s(w) = \lim_{n\to\infty} \int_0^t \phi_n(s,w)dB_s(w), \tag{81}$$

where the limit is the $L^2(P)$-sense. The Ito isometry shows that the integral does not depend on the sequence of approximating elementary functions. It easy to verify that the Ito integral satisfy the usual properties of integrals and that

$$\mathbf{E}\left[\int_0^t f dB_s\right] = 0. \tag{82}$$

Next we discuss Ito formula which is a generalization of the chain rule. Let $v(t,w) \in \mathcal{V}[0,t]$ for all $t > 0$ and let $u(t,w)$ be a measurable function with respect to $\mathcal{N}_t^0$ for all $t > 0$ and such that $\int_0^t |u(s,w)| ds$ is a.s. finite. Then the *Ito process* x_t is the stochastic integral with differential

$$dx_t(w) = u(t,w)dt + v(t,w)dB_t(w). \tag{83}$$

Theorem 5.2. (Ito Formula) *Let x_t be an one-dimensional Ito process of the form (83). Let $g(x) \in \mathcal{C}^2(\mathbf{R})$ be bounded with bounded first and second derivatives. Then $y_t = g(x_t)$ is again an Ito process with differential*

$$dy_t(\omega) = \left(\frac{dg}{dx}(x_t)u(t,\omega)dt + \frac{1}{2}\frac{d^2g}{dx^2}(x_t)v^2(t,\omega) \right) dt + \frac{dg}{dx}(x_t)v(t,\omega)dB_t(\omega).$$

Proof. We can assume that u and v are elementary functions. We use the notations $\Delta t_j = t_{j+1} - t_j$, $\Delta x_j = x_{j+1} - x_j$, and $\Delta g(x_j) = g(x_{j+1}) - g(x_j)$. Since g is C^2 we use a Taylor expansion

$$g(x_t) = g(x_0) + \sum_j \Delta g(x_j)$$

$$= g(x_0) + \sum_j \frac{dg}{dx}(x_{t_j})\Delta x_j + \frac{1}{2}\sum_j \frac{d^2g}{d^2x}(x_{t_j})(\Delta x_j)^2 + R_j, \quad (84)$$

where $R_j = o((\Delta x_j)^2)$. For the second term on the r.h.s. of Eq. (84) we have

$$\lim_{\Delta t_j \to 0} \sum_j \frac{dg}{dx}(x_{t_j})\Delta x_j = \int \frac{dg}{dx}(x_s)dx_s$$

$$= \int \frac{dg}{dx}(x_s)u(s,\omega)ds + \int \frac{dg}{dx}(x_s)v(s,\omega)dB_s. \quad (85)$$

We can rewrite the third term on the r.h.s. of Eq. (84) as

$$\sum_j \frac{d^2g}{d^2x}(\Delta x_j)^2 = \sum_j \frac{d^2g}{d^2x}\left(u_j^2(\Delta t_j)^2 + 2u_jv_j\Delta t_j\Delta B_j + v_j^2(\Delta B)_j^2 \right). \quad (86)$$

The first two terms on the r.h.s. of Eq. (86) go to zero as $\Delta t_j \to 0$. For the first it is obvious while for the second one uses

$$\mathbf{E}\left[\left(\frac{d^2g}{d^2x}(x_{t_j})u_jv_j\Delta t_j\Delta B_j \right)^2 \right] = \mathbf{E}\left[\left(\frac{d^2g}{d^2x}(x_{t_j})u_jv_j \right)^2 \right](\Delta t_j)^3 \to 0, \quad (87)$$

as $\Delta t_j \to 0$. We claim that the third term on the r.h.s. of Eq. (86) converges to

$$\int_0^t \frac{d^2g}{d^2x}(x_s)v^2 ds, \quad (88)$$

in $L^2(P)$ as $\Delta t_j \to 0$. To prove this let us set $a(t) = \frac{d^2g}{d^2x}(x_t)v^2(t,\omega)$ and $a_i = a(t_i)$. We have

$$\mathbf{E}\left[\left(\sum_j a_j((\Delta B_j)^2 - \Delta t_j) \right)^2 \right] = \sum_{i,j} \mathbf{E}\left[a_ia_j((\Delta B_i)^2 - \Delta t_i)((\Delta B_j)^2 - \Delta t_j) \right] \quad (89)$$

If $i < j$, $a_ia_j((\Delta B_i)^2 - \Delta t_i)$ is independent of $((\Delta B_j)^2 - \Delta t_j)$. So we are left with

$$\sum_j \mathbf{E}\left[a_j^2((\Delta B_j)^2 - \Delta t_j)^2\right]$$

$$= \sum_j \mathbf{E}[a_j^2]\,\mathbf{E}[(\Delta B_j)^4 - 2(\Delta B_j)^2\Delta t_j + (\Delta t_j)^2]$$

$$= \sum_j \mathbf{E}[a_j^2](3(\Delta t_j)^2 - 2(\Delta t_j)^2 + (\Delta t_j)^2) = 2\sum_j \mathbf{E}[a_j^2]\Delta t_j^2, \qquad (90)$$

and this goes to zero as Δt_j goes to zero. □

Remark 5.3. Using an approximation argument, one can prove that it is enough to assume that $g \in \mathcal{C}^2$ without boundedness assumptions.

In dimension $n > 1$ one proceeds similarly. Let B_t be a m-dimensional Brownian motion, $u(t, w) \in \mathbf{R}^n$, and $v(t, w)$ an $n \times m$ matrix and let us consider the Ito differential

$$dx_t(w) = u(t, w)dt + v(t, w)dB_t(w) \qquad (91)$$

then $y_t = g(x_t)$ is a one dimensional Ito process with differential

$$dy_t(w) = \sum_j \left(\frac{\partial g}{\partial x_j}(x_t)u_j(t, w) + \frac{1}{2}\sum_{i,j}\frac{\partial^2 g}{\partial x_i \partial x_j}(x_t)(vv^T)_{ij}(t, w)\right) dt$$

$$+ \sum_{ij}\frac{\partial g}{\partial x_j}(x_t)v_{ij}(t, w)dB_t^{(i)}. \qquad (92)$$

We can apply this to a stochastic differential equation

$$dx_t(w) = b(x_t(w))dt + \sigma(x_t(w))dB_t(w), \qquad (93)$$

with

$$u(t, w) = b(x_t(w)), \quad v(t, w) = \sigma(x_t(w)), \qquad (94)$$

provided we can show that existence and uniqueness of the integral equation

$$x_t(w) = x_0 + \int_0^t b(x_s(w))ds + \int_0^t \sigma(x_t(w))dB_s(w). \qquad (95)$$

As for ordinary ODE's, if b and σ are locally Lipschitz one obtains uniqueness and existence of local solutions. If if one requires, in addition, that b and σ are linearly bounded

$$|b(x)| + |\sigma(x)| \leq C(1 + |x|), \qquad (96)$$

one obtains global in time solutions. This is proved using Picard iteration, and one obtains a solution x_t with continuous paths, each component of which belongs to $\mathcal{V}[0, T]$, in particular x_t is measurable with respect to $\mathcal{N}_t^0$.

Let us now introduce the probability distribution $\mathbf{Q}_x$ of the solution $x_t = x_t^x$ of (93) with initial condition $x_0 = x$. Let $\mathcal{F}$ be the σ-algebra generated by the random variables $x_t(\omega)$. We define $\mathbf{Q}_x$ by

$$\mathbf{Q}_x\left[x_{t_1} \in F_1, \cdots, x_{t_n} \in F_n\right] = \mathbf{P}\left[\omega\, ;\, x_{t_1} \in F_1, \cdots, x_{t_n} \in F_n\right] \qquad (97)$$

where $\mathbf{P}$ is the probability law of the Brownian motion (where the Brownian motion starts is irrelevant since only increments matter for x_t). Recall that $\mathcal{N}_t^0$ is the σ-algebra generated by $\{B_s, 0 \le s \le t\}$. Similarly we let $\mathcal{F}_t^0$ the σ-algebra generated by $\{x_s, 0 \le s \le t\}$. The existence and uniqueness theorem for SDE's proves in fact that x_t is measurable with respect to $\mathcal{N}_t$ so that we have $\mathcal{F}_t \subset \mathcal{N}_t$.

We show that the solution of a stochastic differential equation is a Markov process.

Proposition 5.4. (Markov property) *Let f be a bounded measurable function from $\mathbf{R}^n$ to $\mathbf{R}$. Then, for $t, h \ge 0$*

$$\mathbf{E}_x\left[f(x_{t+h}) \,|\, \mathcal{N}_t\right] = \mathbf{E}_{x_t(\omega)}\left[f(x_h)\right] . \qquad (98)$$

Here $\mathbf{E}_x$ denote the expectation w.r.t to $\mathbf{Q}_x$, that is $\mathbf{E}_y\left[f(x_h)\right]$ means $\mathbf{E}\left[f(x_h^y)\right]$ where $\mathbf{E}$ denotes the expectation w.r.t to the Brownian motion measure $\mathbf{P}$.

Proof. Let us write $x_t^{s,x}$ the solution a stochastic differential equation with initial condition $x_s = x$. Because of the uniqueness of solutions we have

$$x_{t+h}^{0,x} = x_{t+h}^{t,x_t} . \qquad (99)$$

Since x_{t+h}^{t,x_t} depends only x_t, it is measurable with respect to $\mathcal{F}_t^0$. The increments of the Brownian paths over the time interval $[t, t+h]$ are independent of $\mathcal{F}_t^0$, and the b and σ do not depend on t. Therefore

$$\mathbf{P}\left[x_{t+h} \in F \,|\, \mathcal{F}_t^0\right] = \mathbf{P}\left[x_{t+h}^{t,x_t} \in F \,|\, \mathcal{F}_t^0\right]$$

$$= \mathbf{P}\left[x_{t+h}^{t,y} \in F\right]\big|_{y=x_t(\omega)}$$

$$= \mathbf{P}\left[x_h^{0,y} \in F\right]\big|_{y=x_t(\omega)} . \qquad (100)$$

and this proves the claim. $\square$

Since $\mathcal{N}_t^0 \subset \mathcal{F}_t^0$ we have

Corollary 5.5. *Let f be a bounded measurable function from $\mathbf{R}^n$ to $\mathbf{R}$. Then, for $t, h \ge 0$*

$$\mathbf{E}_x\left[f(x_{t+h}) \,|\, \mathcal{F}_t^0\right] = \mathbf{E}_{x_t(\omega)}\left[f(x_h)\right]\big| , \qquad (101)$$

i.e. x_t is a Markov process.

Proof. Since $\mathcal{F}_t^0 \subset \mathcal{N}_t^0$ we have

$$
\begin{aligned}
\mathbf{E}_x\left[f(x_{t+h}) \,|\, \mathcal{F}_t^0\right] &= \mathbf{E}_x\left[\mathbf{E}_x\left[f(x_{t+h}) \,|\, \mathcal{N}_t^0\right] \,|\, \mathcal{F}_t^0\right] \\
&= \mathbf{E}_x\left[\mathbf{E}_{x_t}\left[f(x_h)\right] \,|\, \mathcal{F}_t^0\right] \\
&= \mathbf{E}_{x_t}\left[f(x_h)\right].
\end{aligned} \tag{102}
$$

□

Let $f \in \mathcal{C}_0^2$ (i.e. twice differentiable with compact support) and let L be the second order differential operator given by

$$
Lf = \sum_j b_j(x)\frac{\partial f}{\partial x_j}(x_t) + \frac{1}{2}\sum_{i,j} a_{ij}(x)\frac{\partial^2 f}{\partial x_i \partial x_j}(x), \tag{103}
$$

with $a_{ij}(x) = (\sigma(x)\sigma(x)^T)_{ij}$. Applying Ito formula to the solution of an SDE with $x_0 = x$, i.e. with $u(t,\omega) = b(x_t(\omega))$ and $v(t,\omega) = \sigma(x_t,\omega)$, we find

$$
\begin{aligned}
\mathbf{E}_x[f(x_t)] - f(x) &= \mathbf{E}_x\left[\int_0^t Lf(x_s)ds + \sum_{ij}\frac{\partial f}{\partial x_j}(x_s)\sigma_{ji}(x_s)dB_s^{(i)}\right] \\
&= \int_0^t \mathbf{E}_x\left[Lf(x_s)ds\right].
\end{aligned} \tag{104}
$$

Therefore

$$
Lf(x) = \lim_{t \to 0} \frac{\mathbf{E}_x\left[f(x_t)\right] - f(x)}{t}, \tag{105}
$$

i.e., L is the generator of the diffusion x_t. By the semigroup property we also have

$$
\frac{d}{dt}T_t f(x) = LT_f(x), \tag{106}
$$

so that L is the generator of the semigroup T_t and its domain contains $\mathcal{C}_0^2$.

Example 5.6. Let $p, q \in \mathbf{R}^n$ and let $V(q) : \mathbf{R}^n \to \mathbf{R}$ be a C^2 function and let B_t be a n-dimensional Brownian motion. The SDE

$$
\begin{aligned}
dq &= p\,dt, \\
dp &= \left(-\nabla V(q) - \lambda^2 p\right) dt + \lambda\sqrt{2T}dB_t,
\end{aligned} \tag{107}
$$

has unique local solutions, and has global solutions if $\|\nabla V(q)\| \le C(1 + \|q\|)$. The generator is given by the partial differential operator

$$
L = \lambda(T\nabla_p \cdot \nabla_p - p \cdot \nabla_p) + p \cdot \nabla_q - (\nabla_q V(q)) \cdot \nabla_p. \tag{108}
$$

We now introduce a strengthening of the Markov property, the *strong Markov property*. It says that the Markov property still holds provided we replace the time t

by a random time $\tau(\omega)$ in a class called stopping times. Given an increasing family of σ-algebra $\mathcal{M}_t$, a function $\tau : \Omega \rightarrow [0, \infty]$ is called a *stopping time* w.r.t to $\mathcal{M}_t$ if

$$\{\omega : \tau(\omega) \leq t\} \in \mathcal{M}_t, \text{ for all } t \geq 0. \tag{109}$$

This means that one should be able to decide whether or not $\tau \leq t$ has occurred based on the knowledge of $\mathcal{M}_t$.

A typical example is the *first exit time* of a set U for the solution of an SDE: Let U be an open set and

$$\sigma_U = \inf\{t > 0 ; x_t \notin U\} \tag{110}$$

Then σ^U is a stopping time w.r.t to either $\mathcal{N}_t^0$ or $\mathcal{F}_t^0$.

The Markov property and Ito's formula can be generalized to stopping times. We state here the results without proof.

Proposition 5.7. (Strong Markov property) *Let f be a bounded measurable function from $\mathbf{R}^n$ to $\mathbf{R}$ and let τ be a stopping time with respect to $\mathcal{F}_t^0$, $\tau < \infty$ a.s. Then*

$$\mathbf{E}_x \left[f(x_{\tau+h}) \, | \, \mathcal{F}_\tau^0 \right] = \mathbf{E}_{x_\tau} \left[f(x_h) \right] , \tag{111}$$

for all $h \geq 0$.

The Ito's formula with stopping time is called *Dynkin's formula*.

Theorem 5.8. (Dynkin's formula) *Let f be C^2 with compact support. Let τ be a stopping time with $\mathbf{E}_x [\tau] < \infty$. Then we have*

$$\mathbf{E}_x [f(x_\tau)] = f(x) + \mathbf{E}_x \left[\int_0^\tau LF(x_s) \, ds \right] . \tag{112}$$

As a first application of stopping time we show a method to extend local solutions to global solutions for problems where the coefficients of the equation are locally Lipschitz, but not linearly bounded. We call a function $W(x)$ a *Liapunov function* if $W(x) \geq 1$ and

$$\lim_{|x| \rightarrow \infty} W(x) = \infty \tag{113}$$

i.e., W has compact level sets.

Theorem 5.9. *Let us consider a SDE*

$$dx_t = b(x_t)dt + \sigma(x_t)dB_t , \quad x_0 = x , \tag{114}$$

with locally Lipschitz coefficients. Let us assume that there exists a Liapunov function W which satisfies

$$LW \leq cW , \tag{115}$$

for some constant c. Then the solution of Eq. (114) is defined for all time and satisfies

$$\mathbf{E} [W(x_t)] \leq W(x)e^{ct} . \tag{116}$$

Proof. Since b and σ are locally Lipschitz we have a local solution $x_t(\omega)$ which is defined at least for small time. We define

$$\tau_n(\omega) = \inf\{t > 0, W(x_t) \geq n\}, \tag{117}$$

i.e. τ_n is the first time exits the compact set $\{W \leq n\}$. It is easy to see that τ_n is a stopping time. We define

$$\tau_n(t) = \inf\{\tau_n, t\}. \tag{118}$$

We now consider a new process

$$\tilde{x}_t = x_{\tau_n(t)}, \tag{119}$$

We have $\tilde{x}_t = x_{\tau_n}$ for all $t > \tau_n$, i.e., $\tilde{x}_t$ is stopped when it reaches the boundary of $\{W \leq n\}$. Since τ_n is a stopping time, by Proposition 5.7 and Theorem 5.8, $\tilde{x}_t$ is a Markov process which is defined for all $t > 0$. Its Ito differential is given by

$$d\tilde{x}_t = 1_{\{\tau_n > t\}} b(\tilde{x}_t) dt + 1_{\{\tau_n > t\}} \sigma(\tilde{x}_t) dB_t. \tag{120}$$

From Eq. (115) we have

$$(\frac{\partial}{\partial t} + L) W e^{-ct} \leq 0, \tag{121}$$

and thus

$$\mathbf{E}\left[W(x_{\tau_n(t)}) e^{-c\tau_n(t)}\right] - W(x) = \mathbf{E}\left[\int_0^{\tau_n(t)} (\frac{\partial}{\partial s} + L) W(x_s) e^{-cs} ds\right] \leq 0. \tag{122}$$

Since $\tau_n(t) \leq t$, we obtain

$$\mathbf{E}\left[W(x_{\tau_n(t)})\right] \leq W(x) e^{ct}. \tag{123}$$

On the other hand we have

$$\mathbf{E}\left[W(x_{\tau_n(t)})\right] \geq \mathbf{E}\left[W(x_{\tau_n(t)}) 1_{\tau_n < t}\right] = n \mathbf{P}_x\{\tau_n < t\} \tag{124}$$

so that we obtain

$$\mathbf{P}_x\{\tau_n < t\} \leq \frac{e^{ct} W(x)}{n} \to 0, \tag{125}$$

as $n \to \infty$. This implies that the paths of the process almost surely do not reach infinity in a finite time, if $\tau = \lim_{n \to \infty} \tau_n$ then

$$\mathbf{P}_x\{\tau = \infty\} = 1. \tag{126}$$

Taking the limit $n \to \infty$ in Eq. (123) and using Fatou's lemma gives Eq. (116). □

Example 5.10. Consider the SDE of Example (5.6). If $V(q)$ is of class C^2 and $\lim_{\|q\| \to \infty} V(q) = \infty$, then the Hamiltonian $H(p, q) = p^2/2 + V(q)$ satisfy

$$LH(p, q) = \lambda(n - p^2) \leq \lambda n. \tag{127}$$

Since H is bounded below we can take $H + c$ as a Liapunov function, and by Theorem 115 the solutions exists for all time.

Finally we mention two important results of Ito calculus (without proof). The first result is a simple consequence of Ito's formula and give a probabilistic description of the semigroup $L - q$ where q is the multiplication operator by a function $q(x)$ and L is the generator of a Markov process x_t. The proof is not very hard and is an application of Ito's formula.

Theorem 5.11. (Feynman-Kac formula) *Let x_t is a solution of a SDE with generator L. If f is C^2 with bounded derivatives and if g is continuous and bounded. Put*

$$v(t, x) = \mathbf{E}_x \left[e^{-\int_0^t q(x_s)ds} f(x_t) \right]. \tag{128}$$

Then, for $t > 0$,

$$\frac{\partial v}{\partial t} = Lv - qv, \quad v(0, x) = f(x). \tag{129}$$

The second result describe the change of the probability distribution when the drift in an SDE is modified. The proof is more involved.

Theorem 5.12. (Girsanov formula) *Let x_t be the solution of the SDE*

$$dx_t = b(x_t)dt + \sigma(x_t)dB_t, \quad x_0 = x, \tag{130}$$

and let y_t be the solution of the SDE

$$dy_t = a(y_t)dt + \sigma(x_t)dB_t, \quad y_0 = x. \tag{131}$$

Suppose that there exist a function u such that

$$\sigma(x)u(x) = b(x) - a(x), \tag{132}$$

and u satisfy Novikov Condition

$$\mathbf{E} \left[\exp \left(\frac{1}{2} \int_0^t u^2(y_t(\omega)) \, ds \right) \right] < \infty. \tag{133}$$

Then on the interval $[0, t]$ the probability distribution $\mathbf{Q}_x^{[0,t]}$ of y_t is absolutely continuous with respect to the probability distribution $\mathbf{P}_x^{[0,t]}$ of x_t with a Radon-Nikodym derivative given by

$$d\mathbf{Q}_x^{[0,t]}(\omega) = e^{-\int_0^t u(y_s)dB_s - \frac{1}{2}\int_0^t u^2(y_s)ds} d\mathbf{P}_x^{[0,t]}(\omega). \tag{134}$$

6 Control Theory and Irreducibility

To study the ergodic properties of Markov process one needs to establish which sets can be reached from x in time t, i.e. to determine when $P_t(x, A) > 0$.

For solutions of stochastic differential equations there are useful tools which from control theory. For the SDE

$$dx_t = b(x_t)dt + \sigma(x_t)dB_t, \quad x_0 = x \tag{135}$$

let us replace the Brownian motion B_t by a piecewise polygonal approximation

$$B_t^{(N)} = B_{k/N} + N(t - \frac{k}{N})(B_{(k+1)/N} - B_{k/N}), \quad \frac{k}{N} \leq t \leq \frac{k+1}{N}. \tag{136}$$

Then its time derivative $\dot{B}_t^{(N)}$ is piecewise constant. One can show that the solutions of

$$dx_t^{(N)} = b(x_t^{(N)})dt + \sigma(x_t^{(N)})dB_t^{(N)}, \quad x_0 = x \tag{137}$$

converge almost surely to x_t uniformly on any compact interval $[t_1, t_2]$ to the solution of

$$dx_t = b(x_t) + \sigma\sigma'(x_t)dt + \sigma(x_t)dB_t, \tag{138}$$

The supplementary term in (138) is absent if $\sigma(x) = \sigma$ is independent of x and is related to Stratonovich integrals. Eq. (137) has the form

$$\dot{x} = b(x_t) + \sigma u_t, \quad x_0 = x, \tag{139}$$

where $t \mapsto u(t) = (u_1(t), \cdots, u_m(t))$ is a piecewise constant function. This is an ordinary (non-autonomous) differential equation. The function u is called a *control* and Eq. (139) a control system. The support theorem of Stroock and Varadhan shows that several properties of the SDE Eq. (135) (or (138)) can be studied and expressed in terms of the control system Eq. (139). The control system has the advantage of being a system of ordinary diffential equations.

Let us denote by $\mathcal{S}_x^{[0,t]}$ the support of the diffusion x_t, i.e., $\mathcal{S}_x$ is the smallest closed (in the uniform topology) subset of $\{f \in \mathcal{C}([0,t], \mathbf{R}^n), f(0) = x\}$ such that

$$\mathbf{P}\left\{x_s(\omega) \in \mathcal{S}_x^{[0,t]}\right\} = 1. \tag{140}$$

Note that Girsanov formula, Theorem 5.12 implies that the supports of (135) and (138) are identical.

A typical question of control theory is to determine for example the set of all possible points which can be reached in time t by choosing an appropriate control in a given class. For our purpose we will denote by $\mathcal{U}$ the set of all locally constant functions u. We will say a point y is accessible from x in time t if there exists a control $u \in \mathcal{U}$ such that the solution $x_t^{(u)}$ of the equation Eq. (139) satisfies $x^{(u)}(0) = x$ and $x^{(u)}(t) = y$. We denote by $A_t(x)$ the set of accessible points from x in time t. Further we define $C_x^{[0,t]}(\mathcal{U})$ to be the subset of all solutions of Eq. (139) as u varies in $\mathcal{U}$. This is a subset of $\{f \in \mathcal{C}([0,t], \mathbf{R}^n), f(0) = x\}$.

Theorem 6.1. (Stroock-Varadhan Support Theorem)

$$\mathcal{S}_x^{[0,t]} = \overline{C^{[0,t]}(\mathcal{U})} \tag{141}$$

where the bar indicates the closure in the uniform topology.

As an immediate consequence if we denote $\operatorname{supp}\mu$ the support of a measure μ on $\mathbf{R}^n$ we obtain

Corollary 6.2.

$$\operatorname{supp} P_t(x, \cdot) = \overline{A_t(x)}. \tag{142}$$

For example if can show that all (or a dense subset) of the points in a set F are accessible in time t, then we have

$$P_t(x, F) > 0, \tag{143}$$

that is the probability to reach F from x in the time t is positive.

Example 6.3. Let us consider the SDE

$$dx_t = b(x_t)\, dt + \sigma dB_t, \tag{144}$$

where b is such that there is a unique solution for all times. Assume further that $\sigma : \mathbf{R}^n \to \mathbf{R}^n$ is invertible. For any $t > 0$ and any $x \in \mathbf{R}^n$, the support of the diffusion $\mathcal{S}_x^{[0,t]} = \{f \in C([0,t], \mathbf{R}^n)\,,\, f(0) = x\}$ and, for all open set F, we have $P_t(x, F) > 0$. To see this, let ϕ_t be a C^1 path in $\mathbf{R}^n$ such that $\phi_0 = x$ and define define the (smooth) control $u_t = \sigma^{-1}\left(\dot{\phi}_t - b(\phi_t)\right)$. Clearly ϕ_t is a solution for the control system $\dot{x}_t = b(x_t) + \sigma u_t$. A simple approximation argument shows that any continuous paths can be approximated by a smooth one and then any smooth path can be approximated by replacing the smooth control by a piecewise constant one.

Example 6.4. Consider the SDE

$$\begin{aligned} dq &= p\, dt, \\ dp &= \left(-\nabla V(q) - \lambda^2 p\right) dt + \lambda\sqrt{2T}dB_t, \end{aligned} \tag{145}$$

under the same assumptions as in Example (6.4). Given $t > 0$ and two pair of points (q_0, p_0) and (q_t, p_t), let $\phi(s)$ be any C^2 path in $\mathbf{R}^n$ which satisfy $\phi(0) = q_0$, $\phi(t) = q_t$, $\phi'(0) = p_0$ and $\phi'(t) = p_t$. Consider the control u given by

$$u_t = \frac{1}{\lambda\sqrt{2T}}\left(\ddot{\phi}_t + \nabla V(\phi_t) + \lambda^2\dot{\phi}_t\right). \tag{146}$$

By definition $(\phi_t, \dot{\phi}_t)$ is a solution of the control system with control u_t, so that u_t drives the system from (q_0, p_0) to (q_t, p_t). This implies that $A_t(x, F) = \mathbf{R}^n$, for all $t > 0$ and all $x \in \mathbf{R}^n$. From the support theorem we conclude that $P_t(x, F) > 0$ for all $t > 0$, all $x \in \mathbf{R}^n$, and all open set F.

7 Hypoellipticity and Strong-Feller Property

Let x_t denote the solution of the SDE

$$dx_t = b(x_t)dt + \sigma(x_t)dB_t \,, \tag{147}$$

and we assume here that $b(x)$ and $\sigma(x)$ are $\mathcal{C}^\infty$ and such that the equation has global solutions.

The generator of the semigroup T_t is given, on sufficiently smooth function, by the second-order differential operator

$$L = \sum_i b_i(x)\frac{\partial}{\partial x_i} + \frac{1}{2}\sum_{ij} a_{ij}(x)\frac{\partial^2}{\partial x_i \partial x_j} \,, \tag{148}$$

where

$$a_{ij}(x) = (\sigma(x)\sigma^T(x))_{ij} \,. \tag{149}$$

The matrix $A(x) = (a_{ij}(x))$ is non-negative definite for all x, $A(x) \geq 0$. The adjoint (in $L^2(dx)$) operator L^* is given by

$$L^* = \sum_i \frac{\partial}{\partial x_i} b_i(x) + \frac{1}{2}\sum_{ij} \frac{\partial^2}{\partial x_i \partial x_j} a_{ij}(x) \,. \tag{150}$$

It is called the Fokker-Planck operator. We have

$$T_t f(x) = \mathbf{E}\left[f(x_t)\right] = \int_{\mathbf{R}^n} P_t(x, dy) f(y) \,, \tag{151}$$

and we write

$$P_t(x, dy) = p_t(x, y)\, dy \,. \tag{152}$$

Although, in general, the probability measure $P_t(x, dy)$ does not necessarily have a density with respect to the Lebesgue measure, we can always interpret Eq. (152) in the sense of distributions. Since L is the generator of the semigroup L we have, in the sense of distributions,

$$\frac{\partial}{\partial t} p_t(x, \cdot) = L p_t(x, \cdot) \,. \tag{153}$$

The dual S_t of T_t acts on probability measure and if we write, in the sense of distributions, $d\pi(x) = \rho(x)\, dx$ we have

$$d(S_t \pi)(x) = T_t^* \rho(x)\, dx \,, \tag{154}$$

so that

$$\frac{\partial}{\partial t} p_t(\cdot, y) = L^* p_t(\cdot, y) \,. \tag{155}$$

In particular if π is an invariant measure $S_t \pi = \pi$ and we obtain the equation

$$L^* \rho(x) = 0. \tag{156}$$

If $A(x)$ is positive definite, $A(x) \geq c(x)\mathbf{1}$, $c(x) > 0$, we say that L is *elliptic*. There is an well-known elliptic regularity result: Let H_s^{loc} denote the local Sobolev space of index s. If A is elliptic then we have

$$Lf = g \text{ and } g \in H_s^{\text{loc}} \implies f \in H_{s+2}^{\text{loc}}. \tag{157}$$

If L is elliptic then L^* is also elliptic. It follows, in particular that all eigenvectors of L and L^* are $\mathcal{C}^\infty$.

Let $X_i = \sum_j X_i^j(x) \frac{\partial}{\partial x_j}$, $i = 0, \cdots, M$ be $\mathcal{C}^\infty$ vectorfields. We denote by X_i^* its formal adjoint (on L^2). Let $f(x)$ be a $\mathcal{C}^\infty$ function. Let us consider operators K of the form

$$K = \sum_{j=1}^M X_j^*(x)X_j(x) + X_0(x) + f(x). \tag{158}$$

Note that L, L^*, $\frac{\partial}{\partial t} - L$, and $\frac{\partial}{\partial t} - L^*$ have this form.

In many interesting physical applications, the generator fails to be elliptic. There is a theorem due to Hörmander which gives a very useful criterion to obtain the regularity of $p_t(x, y)$. We say that the family of vector fields $\{X_j\}$ satisfy *Hörmander condition* if the Lie algebra generated by the family

$$\{X_i\}_{i=0}^M, \{[X_i, X_j]\}_{i,j=0}^M, \{[[X_i, X_j], X_k]\}_{i,j,k=0}^M, \cdots, \tag{159}$$

has maximal rank at every point x.

Theorem 7.1. (Hörmander theorem) *If the family of vector fields $\{X_j\}$ satisfy Hörmander condition then there exists $\epsilon > 0$ such that*

$$Kf = g \text{ and } g \in H_s^{\text{loc}} \implies f \in H_{s+\epsilon}^{\text{loc}}. \tag{160}$$

We call an operator which satisfies (160) an hypoelliptic operator. An analytic proof of Theorem 7.1 is given in [1], there are also probabilistic proofs which use Malliavin calculus, see [8] for a simple exposition.

As a consequence we have

Corollary 7.2. *Let $L = \sum_j Y_j(x)^* Y_j(x) + Y_0(x)$ be the generator of the diffusion x_t and let us assume that assume that (note that Y_0 is omitted!)*

$$\{Y_i\}_{i=1}^M, \{[Y_i, Y_j]\}_{i,j=0}^M, \{[[Y_i, Y_j], Y_k]\}_{i,j,k=0}^M, \cdots, \tag{161}$$

has rank n at every point x. Then L, L^*, $\frac{\partial}{\partial t} - L$, and $\frac{\partial}{\partial t} - L^*$ are hypoelliptic. The transition probabilities $P_t(x, y)$ have densities $p_t(x, y)$ which are $\mathcal{C}^\infty$ functions of (t, x, y) and the semigroup T_t is strong-Feller. The invariant measures, if they exist, have a $\mathcal{C}^\infty$ density $\rho(x)$.

Example 7.3. Consider the SDE

$$dq = p \, dt \, ,$$
$$dp = \left(-\nabla V(q) - \lambda^2 p \right) \, dt + \lambda \sqrt{2T} dB_t \, , \tag{162}$$

with generator

$$L = \lambda (T \nabla_p \cdot \nabla_p - p \cdot \nabla_p) + p \cdot \nabla_q - (\nabla_q V(q)) \cdot \nabla_p \, . \tag{163}$$

In order to put in the form (158) we set

$$X_j(p,q) = \lambda \sqrt{T} \frac{\partial}{\partial p_j} \, , \quad j = 1, 2, \cdots, n \, ,$$
$$X_0(p,q) = -\lambda p \cdot \nabla_p + p \cdot \nabla_q - (\nabla_q V(q)) \cdot \nabla_p \, , \tag{164}$$

so that $L = -\sum_{j=1}^{n} X_j^* X_j + X_0$. The operator L is not elliptic since the matrix a_{ij} has only rank n. But L satisfies condition (161) since we have

$$[X_j, X_0] = -\lambda^2 \sqrt{T} \frac{\partial}{\partial p_j} + \lambda \sqrt{T} \frac{\partial}{\partial q_j} \, , \tag{165}$$

and so the set $\{X_j, [X_j, X_0]\}_{j=1,n}$ has rank $2n$ at every point (p,q). This implies that L and L^* are hypoelliptic. The operator $\frac{\partial}{\partial t} - L$, and $\frac{\partial}{\partial t} - L^*$ are also hypoelliptic by considering the same set of vector fields together with X_0.

Therefore the transition probabilities $P_t(x, dy)$ have smooth densities $p_t(x, y)$. For that particular example it is easy to check that

$$\rho(x) = Z^{-1} e^{-\frac{1}{T} \left(\frac{p^2}{2} + V(q) \right)} \qquad Z = \int_{\mathbf{R}^{2n}} e^{-\frac{1}{T} \left(\frac{p^2}{2} + V(q) \right)} \, dp dq \, . \tag{166}$$

is the smooth density of an invariant measure, since it satisfies $L^* \rho = 0$. In general the explicit form of an invariant measure is not known and Theorem 7.1 implies that an invariant measure must have a smooth density, provided it exists.

8 Liapunov Functions and Ergodic Properties

In this section we will make the following standing assumptions

- **(H1)** The Markov process is irreducible aperiodic, i.e., there exists $t_0 > 0$ such that

$$P_{t_0}(x, A) > 0 \, , \tag{167}$$

  for all $x \in \mathbf{R}^n$ and all open sets A.
- **(H2)** The transition probability function $P_t(x, dy)$ has a density $p_t(x, y)$ which is a smooth function of (x, y). In particular T_t is strong-Feller, it maps bounded measurable functions into bounded continuous functions.

We are not optimal here and both condition **H1** can certainly be weakened. Note also that **H1** together with Chapman-Kolmogorov equations imply that (167) holds in fact for all $t > t_0$. We have discussed in Sections 6 and 7 some useful tools to establish **H1** and **H2**.

Proposition 8.1. *If conditions **H1** and **H2** are satisfied then the Markov process x_t has at most one stationary distribution. The stationary distribution, if it exists, has a smooth everywhere positive density.*

Proof. By **H2** the dual semigroup S_t acting on measures maps measures into measures with a smooth density with respect to Lebesgue measure: $S_t \pi(dx) = \rho_t(x)dx$ for some smooth function $\rho_t(x)$. If we assume there is a stationary distribution

$$S_t \pi(dx) = \pi(dx), \tag{168}$$

then clearly $\pi(dx) = \rho(x)dx$ must have a smooth density. Suppose that the invariant measure is not unique, then we might assume that π is not ergodic and thus, by Theorem 3.8 and Remark 3.9, there exists a nontrivial set A such that if the Markov process starts in A or A^c it never leaves it. Since π has a smooth density, we can assume that A is an open set and this contradicts (**H1**).

If π is the stationary distribution then by invariance

$$\pi(A) = \int \pi(dx)P_t(x, A), \tag{169}$$

and so $\pi(A) > 0$ for all open sets A. Therefore the density of π, denoted by ρ, is almost everywhere positive. Let us assume that for some y, $\rho(y) = 0$ then

$$\rho(y) = \int \rho(x)p_t(x, y)\, dx. \tag{170}$$

This implies that $p_t(x, y) = 0$ for almost all x and thus since it is smooth the function $p_t(\cdot, y)$ is identically 0 for all $t > 0$. On the other hand $p_t(x, y) \to \delta(x - y)$ as $t \to 0$ and this is a contradiction. So we have shown that $\rho(x) > 0$ for all x. $\square$

Condition **H1** and **H2** do not imply in general the existence of a stationary distribution. For example Brownian motion obviously satisfies both conditions, but has no finite invariant distribution.

Remark 8.2. If the Markov process x_t has a compact phase space X instead of $\mathbf{R}^n$, then x_t always has a stationary distribution. To see this choose an arbitrary x_0 and set

$$\pi_t(dy) = \frac{1}{t} \int_0^t P_s(x_0, dy)\, ds. \tag{171}$$

The sequence of measures π_t has accumulation points in the weak topology. (Use Riesz representation Theorem to identify Borel measure as linear functional on $\mathcal{C}(X)$ and use the fact that the set of positive normalized linear functional on $\mathcal{C}(x)$ is weak-$*$ compact if X is compact). Furthermore the accumulation points are invariants. The details are left to the reader.

If the phase is not compact as in many interesting physical problems we need to find conditions which ensure the existence of stationary distributions. The basic idea is to show the existence of a compact set on which the Markov process spends "most of his time" except for rare excursions outside. On can express these properties systematically in terms of hitting times and Liapunov functions.

Recall that a Liapunov function is, by definition a positive function $W \geq 1$ with compact level sets ($\lim_{|x| \to \infty} W(x) = +\infty$).

Our first results gives a condition which ensure the existence of an invariant measure. If K is a subset of $\mathbf{R}^n$ we denote by $\tau_K = \tau_K(\omega) = \inf\{t, x_t(\omega) \in K\}$ the first time the process x_t enters the set K, τ_K is a stopping time.

Theorem 8.3. *Let x_t be a Markov process with generator L which satisfies condition* **H1** *and* **H2**. *Let us assume that there exists positive constant b and c, a compact set K, and a Liapunov function W such that*

$$LW \leq -c + b\mathbf{1}_K x. \tag{172}$$

Then we have

$$\mathbf{E}_x[\tau_K] < \infty. \tag{173}$$

for all $x \in \mathbf{R}^n$ and the exists a unique stationary distribution π.

Remark 8.4. One can show the converse statement that the finiteness of the expected hitting time do imply the existence of a Liapunov function which satisfies Eq. (172).

Remark 8.5. It turns out that sometimes it is more convenient to show the existence of a Liapunov function expressed in terms of the semigroup T_{t_0} for a fixed time t_0 rather than in terms of the generator L. If we assume that there exists constants positive b, c, a compact set K, and a Liapunov function W which satisfies

$$T_{t_0} W - W \leq -c + b\mathbf{1}_K x. \tag{174}$$

then the conclusion of the theorem still hold true.

Proof. We first prove the assertion on the hitting time. If $x \in K$ then clearly $\mathbf{E}_x[\tau_K] = 0$. So let us assume that $x \notin K$. Let us choose n so large that $W(x) < n$. Now let us set

$$\tau_{K,n} = \inf\{t, x_t \in K \cup \{W(x) \geq n\}\}, \quad \tau_{K,n}(t) = \inf\{\tau_{K,n}, t\}. \tag{175}$$

Obviously $\tau_{K,n}$ and $\tau_{K,n}(t)$ are stopping time and using Ito formula with stopping time we have

$$
\begin{aligned}
\mathbf{E}_x\left[W(x_{\tau_{K,n}(t)})\right] - W(x) &= \mathbf{E}_x\left[\int_0^{\tau_{K,n}(t)} LW(x_s)\,ds\right] \\
&\leq \mathbf{E}_x\left[\int_0^{\tau_{K,n}(t)} -c + b\mathbf{1}_K(x)\,ds\right] \\
&\leq \mathbf{E}_x\left[\int_0^{\tau_{K,n}(t)} -c\right] \\
&\leq -c\mathbf{E}_x\left[\tau_{K,n}(t)\right].
\end{aligned}
\tag{176}
$$

Since $W \geq 1$ we obtain

$$\mathbf{E}_x \left[\tau_{K,n}(t)\right] \leq \frac{1}{c} \left(W(x) - \mathbf{E}_x \left[W(x_{\tau_{K,n}(t)})\right]\right) \leq \frac{1}{c} W(x). \qquad (177)$$

Proceeding as in Theorem 5.9, using Fatou's lemma, we first take the limit $n \to \infty$ and, since $\lim_{n \to \infty} \tau_{K,n}(t) = \tau_K(t)$ we obtain

$$\mathbf{E}_x \left[\tau_K(t)\right] \leq \frac{W(x)}{c}. \qquad (178)$$

Then we take the limit $t \to \infty$ and, since $\lim_{t \to \infty} \tau_K(t) = \tau_K$, we obtain

$$\mathbf{E}_x \left[\tau_K\right] \leq \frac{W(x)}{c}. \qquad (179)$$

We show next the existence of an invariant measure. The construction goes via an embedded (discrete-time) Markov chain. Let us choose a compact set $\tilde{K}$ with K contained in the interior of $\tilde{K}$. We assume they have smooth boundaries which we denote by Γ and $\tilde{\Gamma}$ respectively. We divide now an arbitrary path x_t into cycles in the following way. Let $\tau_0 = 0$, let τ_1' be the first time after τ_0 at which x_t reaches $\tilde{\Gamma}$, τ_1 is the first time after τ_1' at which x_t reaches Γ and so on. It is not difficult to see that, under our assumptions, τ_j and τ_j' are almost surely finite. We define now a discrete-time Markov chain by $X_0 = x \in \Gamma$ and $X_i = x_{\tau_i}$. We denote by $\tilde{P}(x, dy)$ the one-step transition probability of X_n. We note that the Markov chain X_n has a compact phase space and so it possess a stationary distribution $\mu(dx)$ on Γ, by the same argument as the one sketched in Remark (8.2).

We construct now the invariant measure for x_t in the following way. Let $A \subset \mathbf{R}^n$ be a measurable set. We denote σ_A the time spent in A by x_t during the first cycle between 0 and τ_1. We define an unnormalized measure π by

$$\pi(A) = \int_\Gamma \mu(dx) \mathbf{E}_x[\sigma_A]. \qquad (180)$$

Then for any bounded continuous function we have

$$\int_{\mathbf{R}^n} f(x)\pi(dx) = \int_\Gamma \mu(dx) \mathbf{E}_x \left[\int_0^{\tau_1} f(x_s)ds\right]. \qquad (181)$$

In order to show that π is stationary we need to show that, for any bounded continuous f,

$$\int_{\mathbf{R}^n} T_t f(x)\pi(dx) = \int_{\mathbf{R}^n} f(x)\pi(dx), \qquad (182)$$

i.e., using our definition of π

$$\int_\Gamma \mu(dx)\mathbf{E}_x \left[\int_0^{\tau_1} \mathbf{E}_{x_s}[f(x_t)]ds\right] = \int_\Gamma \mu(dx)\mathbf{E}_x \left[\int_0^{\tau_1} f(x(s))ds\right]. \qquad (183)$$

For any measurable continuous function we have

$$\mathbf{E}_x \left[\int_0^{\tau_1} f(x_{t+s})\, ds \right] = \mathbf{E}_x \left[\int_0^\infty \mathbf{E}_x [1_{s<\tau_1} f(x_{t+s})]\, ds \right]$$

$$= \mathbf{E}_x \left[\int_0^\infty 1_{s<\tau_1} \mathbf{E}_{x_s} [f(x(t))]\, ds \right]$$

$$= \mathbf{E}_x \left[\int_0^{\tau_1} \mathbf{E}_{x_s} [f(x_t)]\, ds \right] . \tag{184}$$

Thus we have, using Eqs. (182), (183), and (184),

$$\int_{\mathbf{R}^n} T_t f(x) \pi(dx) = \int_\Gamma \mu(dx) \mathbf{E}_x \left[\int_0^{\tau_1} \mathbf{E}_{x_s} f(x_t)\, ds \right]$$

$$= \int_\Gamma \mu(dx) \mathbf{E}_x \left[\int_0^{\tau_1} f(x_{t+s})ds \right] = \int_\Gamma \mu(dx) \mathbf{E}_x \left[\int_t^{\tau_1+t} f(x_u)\, du \right] \tag{185}$$

$$= \int_\Gamma \mu(dx) \mathbf{E}_x \left[\int_0^{\tau_1} f(x_u)\, du + \int_t^{t+\tau_1} f(x_u)\, du - \int_0^t f(x_u)\, du \right] .$$

Since μ is a stationary distribution for X_n, for any bounded measurable function on Γ

$$\int_\Gamma \mu(dx) \mathbf{E}_x [g(X_1)] = \int_\Gamma \mu(dx) g(X) , \tag{186}$$

and so

$$\int_\Gamma \mu(dx) \mathbf{E}_x \left[\int_t^{t+\tau_1} f(x_u)\, du \right] = \int_\Gamma \mu(dx) \mathbf{E}_x \left[\mathbf{E}_{X_1} \left[\int_0^t f(x_u)\, du \right] \right]$$

$$= \int_\Gamma \mu(dx) \mathbf{E}_x \left[\int_0^t f(x_u)\, du \right] . \tag{187}$$

Combining Eqs. (185) and (187) we obtain that

$$\int_{\mathbf{R}^n} T_t f(x) \pi(dx) = \int_{\mathbf{R}^n} f(x) \pi(dx) . \tag{188}$$

This shows that the measure π satisfies $S_t \pi = \pi$. Finally we note that

$$\pi(\mathbf{R}^n) = \int_\Gamma \mu(dx) \mathbf{E}_x [\tau_K] < \infty . \tag{189}$$

and so π can be normalized to a probability measure. $\square$

Remark 8.6. One can also show (see e.g. [15] or [7]) that that under the same conditions one has convergence to the stationary state: for any x we have

$$\lim_{t\to\infty} \|P_t(x,\cdot) - \pi(\cdot)\| = 0 , \tag{190}$$

where $\|\mu\| = \sup_{\|f\|\le 1} |\int f(x)\mu(dx)|$ is the total variation norm of a signed measure μ. Thus the measure π is also mixing. We don't prove this here, but we will prove an exponential convergence result using a stronger condition than (172).

It is often useful to to quantify the rate of convergence at which an initial distribution converges to the stationary distribution and we prove two such results in that direction which provide exponential convergence. We mention here that polynomial rate of convergence can be also expressed in terms of Liapunov functions, and this makes these functions a particularly versatile tool.

We introduce some notations and definitions. If W is a Liapunov function and μ a signed measure we introduce a weighted total variation norm given by

$$\|\mu\|_W = \sup_{|f| \leq W} | \int f(x)\mu(dx)|, \qquad (191)$$

and a norm on functions $\| \cdot \|_W$ given by

$$\|f\|_W = \sup_{x \in \mathbf{R}^n} \frac{|f(x)|}{W(x)}, \qquad (192)$$

and a corresponding Banach space $\mathcal{H}_W$ given by

$$\mathcal{H}_W = \{f, \|f\|_W < \infty\}. \qquad (193)$$

Theorem 8.7. (Quasicompactness) *Suppose that the conditions* **H1** *and* **H2** *hold. Let K be a compact set and let W be a Liapunov function W. Assume that either of the following conditions hold*

1. *There exists constants $a > 0$ and $b < \infty$ such that*

$$LW(x) \leq -aW(x) + b\mathbf{1}_K(x). \qquad (194)$$

2. *We have $LW \leq cW$ for some $c > 0$ and there exists constants $\kappa < 1$ and $b < \infty$, and a time $t_0 > 0$ such that*

$$T_{t_0}W(x) \leq \kappa W(x) + b\mathbf{1}_K(x). \qquad (195)$$

Then for δ small enough

$$\mathbf{E}_x\left[e^{\delta \tau_K}\right] < \infty, \qquad (196)$$

for all $x \in \mathbf{R}^n$. The Markov process has a stationary distribution x_t and there exists a constants $C > 0$ and $\gamma > 0$ such that

$$\|P_t(x, dy) - \pi(dy)\|_W \leq CW(x)e^{-\gamma t}, \qquad (197)$$

or equivalently

$$\|T_t - \pi\|_W \leq Ce^{-\gamma t}. \qquad (198)$$

Remark 8.8. In Theorem 8.7 one can replace $\mathcal{H}_W$ by any of the space

$$\mathcal{H}_{W,p} = \left\{f, \frac{|f|}{W} \in L^p(dx)\right\}, \qquad (199)$$

with $1 < p < \infty$.

Proof. We will prove here only the part of the argument which is different from the proof of Theorem 8.9. We recall that a bounded operator T acting on a Banach space $\mathcal{B}$ is quasi compact if its spectral radius is M and its essential spectral radius is $\theta < M$. By definition, it means that outside the disk of radius θ, the spectrum of T consists of isolated eigenvalues of finite multiplicity. The following formula for the essential spectral radius θ is proved in [9]

$$\theta = \lim_{n\to\infty} \left(\inf\left\{\|T^n - C\| \mid C \text{ compact}\right\}\right)^{1/n}. \tag{200}$$

Let us assume that Condition 2. holds. If $\|f\| \in \mathcal{H}_W$ then, by definition, we have $|f(x)| \le \|f\|_W W(x)$ and, since T_t is positive, we have

$$|T_t f(x)| \le \|f\|_W T_t W(x). \tag{201}$$

We consider a fixed $t > t_0$. We note that the Liapunov condition implies the bound

$$T_t W(x) \le \kappa W(x) + b. \tag{202}$$

Iterating this bound and using that $T_t 1 = 1$ we have, for all $n \ge 1$, the bounds

$$T_{nt} W(x) \le \kappa^n W(x) + \frac{b}{1 - \kappa}. \tag{203}$$

Let K be any compact set, we have the bound

$$\begin{aligned}
|\mathbf{1}_{K^c}(x) T_{nt} f(x)| &\le W(x) \sup_{y \in K^c} \frac{|T_{nt} f(y)|}{W(y)} \\
&\le W(x)\|f\|_W \sup_{y \in K^c} \frac{T_{nt} W(y)}{W(y)} \\
&\le W(x)\|f\|_W \left(\kappa^n + \frac{b}{1-\kappa} \sup_{y \in K^c} \frac{1}{W(x)}\right). \tag{204}
\end{aligned}$$

Since $\lim_{\|x\|\to\infty} W(x) = \infty$, given $\epsilon > 0$ and $n > 1$ we can choose a compact set K_n such that

$$\|\mathbf{1}_{K_n^c} T_{nt}\|_W \le (\kappa + \epsilon)^n. \tag{205}$$

On the other hand since T_t has a smooth kernel the set, for any compact K, the set

$$\{\mathbf{1}_K(x) T_t f(x) \mid \|f\|_W = 1\}$$

is compact by Arzelà-Ascoli. Therefore we have

$$\inf\left\{\|T_{nt} - C\| \mid C \text{ compact}\right\} \le \|T_{nt} - \mathbf{1}_{K_n} T_{nt}\| \le (\kappa + \epsilon)^n. \tag{206}$$

and therefore the essential spectral radius of T_t is less than κ. In order to obtain the exponential convergence from this one must prove that there is no other eigenvalue than 1 on the unit disk (or our outside the unit disk) and prove that 1 is a simple eigenvalue. We will prove this in Theorem 8.9 and the same argument apply here. Also the assertion on hitting times is proved as in Theorem 8.9. $\square$

Theorem 8.9. (Compactness) *Suppose that the conditions* **H1** *and* **H2** *hold. Let* $\{K_n\}$ *be a sequence of compact sets and let W be a Liapunov function. Assume that either of the following conditions hold*

1. *There exists constants $a_n > 0$ with $\lim_{n\to\infty} a_n = \infty$ and constants $b_n < \infty$ such that*

$$LW(x) \leq -a_n W(x) + b_n \mathbf{1}_{K_n}(x). \tag{207}$$

2. *There exists a constant c such that $LW \leq cW$ and there exists constants $\kappa_n < 1$ with $\lim_{n\to\infty} \kappa_n = 0$, constants $b_n < \infty$, and a time $t_0 > 0$ such that*

$$T_{t_0} W(x) \leq \kappa_n W(x) + b_n \mathbf{1}_{K_n}(x). \tag{208}$$

Then for any (arbitrarily large) δ there exists a compact set $C = C(\delta)$ such that

$$\mathbf{E}_x \left[e^{\delta \tau_C} \right] < \infty, \tag{209}$$

for all $x \in \mathbf{R}^n$. The Markov process x_t has a unique stationary distribution π. The semigroup T_t acting on $\mathcal{H}_W$ is a **compact** *semigroup for $t > t_0$ and there exists a constants $C > 0$ and $\gamma > 0$ such that*

$$\|P_t(x, dy) - \pi(dy)\| \leq CW(x) e^{-\gamma t}. \tag{210}$$

or equivalently

$$\|T_t - \pi\|_W \leq C e^{-\gamma t}. \tag{211}$$

Proof. We will assume that conditions 2. holds. Let us prove the assertion on hitting times. Let X_n be the Markov chain defined by $X_0 =$ and $X_n = x_{nt_0}$ and for a set K let N_K be the least integer such that $X_{N_K} \in K$. We have $N_K \leq \tau_K$ so it is sufficient to prove that $\mathbf{E}_x[e^{\delta N_K}] < \infty$.

Let K_n be the compact set given in Eq. (208). We can assume, by increasing K_n is necessary that K_n is a level set of W, i.e. $K_n = \{W(x) \leq W_n\}$

Using the Liapunov condition and Chebyshev inequality we obtain the following tail estimate

$$\mathbf{P}_x\{N_{K_n} > j\}$$

$$= \mathbf{P}_x\{W(X_j) > W_n, \, X_i \in K_n^c, \, 0 \le i \le j\}$$

$$= \mathbf{P}_x\left\{\prod_{i=i}^{j} \frac{W(X_i)}{W(x_{i-1})} > \frac{W_n}{W(x)}, \, X_i \in K_n^C\right\}$$

$$\le \frac{W(x)}{W_n}\mathbf{E}_x\left[\prod_{i=i}^{j} \frac{W(X_i)}{W(x_{i-1})}, \, X_i \in K_n^C\right]$$

$$\le \frac{W(x)}{W_n}\mathbf{E}_x\left[\prod_{i=i}^{j-1} \frac{W(X_i)}{W(x_{i-1})}\mathbf{E}_{X_{j-1}}\left[\frac{W_{X_j}}{W_{X_{j-1}}}\right], \, X_i \in K_n^C\right]$$

$$\le \frac{W(x)}{W_n}\sup_{y \in K_n^C}\mathbf{E}_y\left[\frac{W(x_1)}{W(y)}\right]\mathbf{E}_x\left[\prod_{i=i}^{j-1} \frac{W(X_i)}{W(x_{i-1})}, \, X_i \in K_n^C\right]$$

$$\le \cdots \le \frac{W(x)}{W_n}\left(\sup_{y \in K_n^C}\mathbf{E}_y\left[\frac{W(x_1)}{W(y)}\right]\right)^j$$

$$\le \frac{W(x)}{W_n}(\kappa_n)^j. \tag{212}$$

We thus have geometric decay of $P_{>j} \equiv P\{N_{k_n} > j\}$ in j. Summing by parts we obtain

$$\mathbf{E}_x\left[e^{\delta N_{k_n}}\right] = \sum_{j=1}^{\infty} e^{\delta j}\mathbf{P}_x\{N_{K_n} = j\}$$

$$= \lim_{M \to \infty}\left(\sum_{j=1}^{M} P_{>j}(e^{\delta(j+1)} - e^{\delta j}) + e^{\delta}P_{>0} - e^{\delta(M+1)}P_{>M}\right)$$

$$\le e^{\delta} + \frac{W(x)}{W_n}(e^{\delta} - 1)\sum_{j=1}^{\infty} \kappa_n^j e^{j\delta}$$

$$\le e^{\delta} + \frac{W(x)}{W_n}(e^{\delta} - 1)\frac{\kappa_n e^{\delta}}{1 - \kappa_n e^{\delta}}, \tag{213}$$

provided $\delta < \ln(\kappa_n^{-1})$. Since we can choose κ_n arbitrarily small, this proves the claim about the hitting time.

If $\|f\| \in \mathcal{H}_W$ then by definition we have $|f(x)| \le \|f\|_W W(x)$ and thus $|T_t f(x)| \le \|f\|_W T_t W(x)$.

The compactness of T_t is a consequence of the following estimate. Using the Liapunov condition we have for $t > t_0$

$$\left|\mathbf{1}_{K_n^c}(x)T_t f(x)\right| \leq W(x) \sup_{y \in K_n^c} \frac{|T_t f(y)|}{W(y)}$$

$$\leq W(x)\|f\|_W \sup_{y \in K_n^c} \frac{T_t W(y)}{W(y)}$$

$$\leq \kappa_n W(x)\|f\|_W, \tag{214}$$

or

$$\|\mathbf{1}_{K_n^c} T_t\|_W \leq \kappa_n. \tag{215}$$

We have thus

$$\lim_{n \to \infty} \|\mathbf{1}_{K_n^c} T_t\|_W = 0, \tag{216}$$

i.e., $\mathbf{1}_{K_n^c} T_t$ converges to 0 in norm. On the other hand since T_t has a smooth kernel and K_n is compact, the operator

$$\mathbf{1}_{K_n} T_t \mathbf{1}_{K_n} \tag{217}$$

is a compact operator by Arzela-Ascoli Theorem. We obtain

$$T_t = \left(\mathbf{1}_{K_n} + \mathbf{1}_{K_n^c}\right) T_{t-\epsilon} \left(\mathbf{1}_{K_n} + \mathbf{1}_{K_n}\right) T_\epsilon$$

$$= \lim_{n \to \infty} \mathbf{1}_{K_n} T_{t-\epsilon} \mathbf{1}_{K_n} T_\epsilon. \tag{218}$$

So T_t is the limit in norm of compact operators, hence it is compact. Its spectrum consists of 0 and eigenvalues with finite multiplicity.

We show that there no eigenvalues of modulus bigger than one. Assume the contrary. Then there is f and λ with $|\lambda| > 1$ such that $T_t f = \lambda f$. Since T_t is positive we have

$$|\lambda||f| = |T_t f| \leq T_t|f|, \tag{219}$$

and therefore

$$T_t|f| - |f| \geq (|\lambda| - 1)|f|. \tag{220}$$

Integrating with the strictly positive stationary distribution we have

$$\int T_t|f|\pi(dx) - \int |f|\pi(dx) \geq (|\lambda| - 1) \int |f|\pi(dx) > 0. \tag{221}$$

This is a contradiction since, by stationarity $\int T_t|f|\pi(dx) = \int |f|\pi(dx)$, and so the r.h.s. of Eq. (221) is 0.

Next we show that 1 is a simple eigenvalue with eigenfunction given by the constant function. Clearly 1 is algebraically simple, if there is another eigenfunction with eigenvalue one, the stationary distribution is not unique. We show that 1 is also geometrically simple. Assume the contrary, then by Jordan decomposition there is a function g such that $T_t g = 1 + g$, so $T_t g - g = 1$. Integrating with respect to the stationary distribution gives a contradiction.

Finally we show that there no other eigenvalues on the unit disk. Assume there exists f, $f \neq 1$ such that $T_t f = \lambda f$ with $|\lambda| = 1$. By the semigroup property T_t

is compact for all $t > t_0$ and $T_t f = e^{i\alpha t} f$, with $\alpha \in \mathbf{R}$, $\alpha \neq 0$. If we choose $t = 2n\pi/\alpha$ we obtain $T_{2n\pi/\alpha} f = f$ and so $f = 1$ which is a contradiction.

Using now the spectral decomposition of compact operators we find that there exists $\gamma > 0$ such that

$$\|T_t - \pi\| \leq C e^{-\gamma t}. \tag{222}$$

This concludes the proof of Theorem 8.9 $\square$

To conclude we show that the correlations in the stationary distribution decay exponentially (exponential mixing).

Corollary 8.10. *Under the assumptions of Theorem 8.7 or 8.9, the stationary distribution π is exponentially mixing: for all f, g such that $f^2, g^2 \in \mathcal{H}_W$ we have*

$$\left| \int f(x) T_t g(x) \pi(dx) - \int f(x) \pi(dx) \int g(x) \pi(dx) \right| C \|f^2\|_W^{1/2} \|g^2\|_W^{1/2} e^{-\gamma t}. \tag{223}$$

Proof. If $f^2 \in \mathcal{H}_W$, then we have

$$|f(x)| \leq \|f^2\|_W^{1/2} W^{1/2}. \tag{224}$$

Further if we have

$$T_{t_0} W(x) \leq \kappa_n W(x) + b_n \mathbf{1}_{K_n}(x), \tag{225}$$

then, using Jensen inequality, the inequality $\sqrt{1+y} \leq 1 + y/2$ and $W \geq 1$ we have

$$\begin{aligned} T_{t_0} \sqrt{W}(x) &\leq \sqrt{T_{t_0} W(x)} \\ &\leq \sqrt{\kappa_n W(x) + b_n \mathbf{1}_{K_n}(x)} \\ &\leq \sqrt{\kappa_n} \sqrt{W(x)} + \frac{b_n \mathbf{1}_{K_n}(x)}{2\kappa_n}. \end{aligned} \tag{226}$$

So we have

$$T_{t_0} \sqrt{W}(x) \leq \kappa_n' \sqrt{W}(x) + b_n' \mathbf{1}_{K_n}(x), \tag{227}$$

with $\kappa_n' = \sqrt{\kappa_n}$ and $b_n' = b_n/2\kappa_n$. Applying Theorem 8.9 or 8.7 with the Liapunov function $\sqrt{W}$ there exist constants $C > 0$ and $\gamma > 0$ such that

$$\left| T_t g(x) - \int g(x) \pi(dx) \right| \leq C \sqrt{W(x)} \|g^2\|_W^{1/2} e^{-\gamma t}. \tag{228}$$

Therefore combining Eqs. (224) and (228) we obtain

$$\begin{aligned} &\left| \int f(x) T_t g(x) \pi(dx) - \int f(x) \pi(dx) \int g(x) \pi(dx) \right| \\ &\leq \int |f(x)| \left| T_t g(x) - \int g(x) \pi(dx) \right| \pi(dx) \\ &\leq C \int W(x) \pi(dx) \|f^2\|^{1/2} \|g^2\|^{1/2} e^{-\gamma t}. \end{aligned} \tag{229}$$

Finally since π is the solution of $S_t\pi = \pi$ we have

$$\int W(x)\pi(dx) \leq \|\pi\|_W < \infty \qquad (230)$$

and this concludes the proof of Corollary 8.10. $\square$

References

1. Hörmander, L.: *The Analysis of linear partial differential operators.* Vol **III**, Berlin: Springer, 1985
2. Has'minskii, R.Z.: *Stochastic stability of differential equations.* Alphen aan den Rijn—Germantown: Sijthoff and Noordhoff, 1980
3. Karatzas, I. and Shreve S.E.: *Brownian Motion and Stochastic Calculus.* Graduate Texts in Mathematics, Berlin: Springer (1997)
4. Kliemann, W.: Recurrence and invariant measures for degenerate diffusions. Ann. of Prob. **15**, 690–702 (1987)
5. Kunita, H.: *Stochastic flows and stochastic differential equations.* Cambridge Studies in Advanced Mathematics, 24. Cambridge: Cambridge University Press (1990)
6. Kunita, H.: Supports of diffusion process and controllability problems. In: *Proc. Inter. Symp. SDE Kyoto 1976.* New-York: Wiley, 1078, pp. 163-185
7. Meyn, S.P. and Tweedie, R.L.: *Markov Chains and Stochastic Stability.* Communication and Control Engineering Series, London: Springer-Verlag London, 1993
8. Norriss, J.: Simplified Malliavin Calculus. In *Séminaire de probabilités XX*, Lectures Note in Math. **1204**, 0 Berlin: Springer, 1986, pp. 101–130
9. Nussbaum R.D.: The radius of the essential spectral radius. Duke Math. J. **37**, 473–478 (1970)
10. Oksendal, B.: *Stochastic differential equations. An introduction with applications.* Fifth edition. Universitext. Berlin: Springer-Verlag, 1998
11. Rey-Bellet L.: Open classical systems. In this volume.
12. Stroock, D.W. and Varadhan, S.R.S.: On the support of diffusion processes with applications to the strong maximum principle. In *Proc. 6-th Berkeley Symp. Math. Stat. Prob.*, Vol **III**, Berkeley: Univ. California Press, 1972, pp. 361–368
13. Varadhan, S.R.S: *Probability Theory.* Courant Institute of Mathematical Sciences - AMS, 2001
14. Varadhan, S.R.S: Lecture notes on Stochastic Processes (Fall 00). http://www.math.nyu.edu/faculty/varadhan/processes.html

Open Classical Systems

Luc Rey-Bellet

Department of Mathematics and Statistics, University of Massachusetts,
Amherst, MA 01003, USA
e-mail :lr7q@math.umass.edu

1	**Introduction**	41
2	**Derivation of the model**	44
	2.1 How to make a heat reservoir	44
	2.2 Markovian Gaussian stochastic processes	48
	2.3 How to make a Markovian reservoir	50
3	**Ergodic properties: the chain**	52
	3.1 Irreducibility	56
	3.2 Strong Feller Property	57
	3.3 Liapunov Function	58
4	**Heat Flow and Entropy Production**	66
	4.1 Positivity of entropy production	69
	4.2 Fluctuation theorem	71
	4.3 Kubo Formula and Central Limit Theorem	75
References		77

1 Introduction

Open systems are usually understood as a small Hamiltonian system (i.e. with a finite number of degrees of freedom) in contact with one or several large reservoirs. There are several ways to model reservoirs and we will take the point of view that the reservoirs are also Hamiltonian systems themselves. It is a convenient physical and mathematical idealization to separate scales and assume that the reservoirs have infinitely many degrees of freedom. We will also assume that, to start with, the reservoirs are in equilibrium, i.e., the initial states of the reservoirs are distributed according to Gibbs distribution with given temperatures. It is also mathematically convenient to assume that the Gibbs measures of the reservoir have very good ergodic properties. This is, in general, a mathematically difficult problem and we will circumvent it by assuming that our reservoirs have a linear dynamics (i.e the Gibbs measures are Gaussian measures).

Our model of a reservoir will be the classical field theory given by a linear wave equation in $\mathbf{R}^d$

$$\partial_t^2 \varphi_t(x) = \Delta \varphi_t(x) \,. \tag{1}$$

This is a Hamiltonian system for the Hamiltonian

$$H(\varphi, \pi) = \int_{\mathbf{R}^d} (|\nabla_x \varphi(x)|^2 + |\pi(x)|^2) \,. \tag{2}$$

If we consider a single particle in a confining potential with Hamiltonian

$$H(p, q) = \frac{p^2}{2} + V(q) \,, \tag{3}$$

we will take as Hamiltonian of the complete system

$$H(\varphi, \pi) + H(p, q) + q \int_{\mathbf{R}^d} \nabla \varphi(x) \rho(x) \, dx \,, \tag{4}$$

which corresponds to a dipole coupling approximation ($\rho(x)$ is a given function which models the coupling of the particle with the field).

If one considers finite-energy solutions for the wave equation, we will say the model is at temperature zero. In this case the physical picture is "radiation damping". The particle energy gets dissipated into the field and relaxes to a stationary point of the Hamiltonian (i.e. $p = 0$, $\nabla V(q) = 0$). This problem is studied in [14] for a slightly different model.

We will say the model is at inverse temperature β if we assume that the initial conditions of the wave equations are distributed according to a Gibbs measure at inverse temperature β. Typical configurations of the field have then infinite energy and thus provide enough energy to let the system "fluctuate". In this case one expects "return to equilibrium", an initial distribution of the system will converge to a stationary state which is given by the Gibbs distribution

$$Z^{-1} e^{-\beta H(p,q)} dp dq \,. \tag{5}$$

The property of return to equilibrium is proved [13] under rather general conditions.

If the small system is coupled to more than one reservoir and the reservoirs have different temperatures (and/or chemical potentials), then one can extract an infinite amount of energy or work from the reservoirs and transmit them through the small system from reservoir to reservoir. Doing this, one can maintain the small systems in a stationary (i.e time independent) nonequilibrium states in which energy and/or matter is flowing. Think e.g. of a bar of metal which is heated at one end and cooled at the other. Contrary to the two previous situations where we know a priori the final state of the system, in nonequilibrium situations, in general, we do not. Even the existence of a stationary state turns out to be a nontrivial mathematical problem and requires a quite detailed understanding of the dynamics. In these lectures we will consider a simple, yet physically realistic model of heat conduction through a lattice of anharmonic oscillators.

Our small system will consist of a chain of anharmonic oscillators with Hamiltonian

$$H(p,q) = \sum_{i=1}^{n} \frac{p_i^2}{2} + V(q_1, \cdots, q_n),$$

$$V(q) = \sum_{i=1}^{n} U^{(1)}(q_i) + \sum_{i=1}^{n-1} U^{(2)}(q_i - q_{i+1}). \tag{6}$$

where $V(q)$ is a confining potential. The number oscillators, n, will be arbitrary (but finite). In a realistic model, the coupling should occur only at the boundary and we will couple the first particle of the chain to one reservoir at inverse temperature β_1 and the n-th particle to another reservoir at inverse temperature β_n. The Hamiltonian of the complete system is

$$H(\varphi_1, \pi_1) + q_1 \int_{\mathbf{R}^d} \nabla \varphi_1(x) \rho_1(x) + H(p_1, \cdots, p_n, q_1, \cdots, q_n)$$

$$+ q_n \int_{\mathbf{R}^d} \nabla \varphi_n(x) \rho_n(x) + H(\varphi_n, \pi_n). \tag{7}$$

Our analysis of this model will establish, that under suitable assumptions on the potential energy V and on the coupling functions ρ_i, we have

1. Existence and uniqueness of stationary states which generalize the Gibbs states of equilibrium.
2. Exponential rate of convergence of initial distribution to the stationary distribution. This is a new result even for equilibrium.
3. Existence of a positive heat flow through the system if the temperatures of the reservoir are different or in other words positivity of entropy production.
4. "Universal" properties of the entropy production. Its (large) fluctuations satisfy a symmetry known as Gallavotti-Cohen fluctuation theorem (large deviation theorem) recently discovered in [8,11]. Its small fluctuations (of central limit theorem type) are governed by Green-Kubo formula which we prove for this model.

See [24, 25] for results on the linear chain and [4–7, 12, 21–23] for results on the nonlinear chain. We follow here mostly [22,23]

An important ingredient in our approach is the use of rather special coupling functions ρ ("rational couplings"). They will allow us to reduce the dynamics of the coupled system to a Markov process in a suitable enlarged phase space and therefore to take advantage of the numerous analytical tools developed for Markov processes (semigroups, PDE's, control theory, see our lecture on Markov processes in these volumes [20]). Physically these couplings are not unreasonable, but it would be nice to go beyond and prove similar results for more general couplings.

We conclude this introduction by mentioning one outstanding problem in mathematical physics: describe quantitatively the transport properties in large systems. One would like, for example, to establish the dependence of the heat flow on the size of the system (here the number n of oscillators). For heat conduction the relevant macroscopic law is the Fourier law of heat conduction. If we denote by J the stationary value of the heat flow and δT the temperature difference between the reservoirs, one would like to prove for example that for large n

$$J \approx \kappa \frac{\delta T}{n}, \tag{8}$$

where the coefficient κ is known as the heat conductivity. A more refined version of this law would to prove that a (local) temperature gradient is established through the system and that the heat flow is proportional to the temperature gradient

$$J \approx -\kappa \nabla T, \tag{9}$$

This is a stationary version of the time dependent macroscopic equation

$$c_v \partial_t T(x,t) = -\nabla J(x,t) = \nabla \kappa \nabla T(x,t), \tag{10}$$

where c_v is the specific heat of the system. This is just the heat equation! In other words the challenge is to derive the heat equation from a microscopic Hamiltonian system. No mechanical model has been shown to obey Fourier law of heat conduction so far (see [1,17] for reviews of this problem and references).

2 Derivation of the model

2.1 How to make a heat reservoir

Our reservoir is modeled by a linear wave equation in $\mathbf{R}$ (the restriction to one-dimension is for simplicity, similar considerations apply to higher dimensions),

$$\partial_t^2 \varphi_t(x) = \partial_x^2 \varphi_t(x), \tag{11}$$

with $t \in \mathbf{R}$ and $x \in \mathbf{R}$. The equation (11) is a second order equation and we rewrite it as a first order equation by introducing a new variable $\pi(x)$

$$\partial_t \varphi_t(x) = \pi_t(x),$$
$$\partial_t \pi_t(x) = \partial_x^2 \varphi_t(x). \tag{12}$$

The system (12) has an Hamiltonian structure. Let us consider the Hamiltonian function

$$H(\varphi, \pi) = \int_{\mathbf{R}} (|\partial_x \varphi(x)|^2 + |\pi(x)|^2) \, dx. \tag{13}$$

then Eqs. (12) are the Hamiltonian equations of motions for the Hamiltonian (13). Let us introduce the notation $\phi = (\varphi, \pi)$ and the norm

$$\|\phi\| = \int_{\mathbf{R}} (|\partial_x(x)|^2 + |\pi|^2)\, dx \,. \tag{14}$$

We have then $H(\phi) = \frac{1}{2}\|\phi\|^2$ and denote $\mathcal{H} = \dot{H}_1(\mathbf{R}) \times L^2(\mathbf{R})$ the corresponding Hilbert space of finite configurations.

In order to study the statistical mechanics of such systems we need to consider the *Gibbs measure* for such systems. We recall that for an Hamiltonian systems with finitely many degrees of freedom with Hamiltonian $H(p,q) = p^2/2 + V(q)$, $p, q \in \mathbf{R}^n$, the Gibbs measure for inverse temperature β is given by

$$\mu_\beta(dpdq) = Z^{-1} e^{-\beta H(p,q)}\, dpdq \,, \tag{15}$$

where $\beta = 1/T$ is the inverse temperature and $Z = \int \exp(-\beta H(p,q))\, dp\, dq$ is a normalization constant which we assume to be finite. One verifies easily that the probability measure μ_β is invariant under the dynamics, i.e., if (p_t, q_t) is a solution of Hamiltonian equations of motion then

$$\int f(p_t, q_t)\mu_\beta(dpdq) \tag{16}$$

is independent of t. (Use conservation of energy and Liouville theorem)

We now construct Gibbs measures for the linear wave equation Eq. (11). If we think of $\{\varphi(x), \pi(x)\}_{x \in \mathbf{R}}$ as the dynamical variables, the Gibbs measure should be, formally, given by

$$\mu_\beta(d\varphi d\pi) = "Z^{-1} \exp(-\beta H(\varphi, \pi)) \prod_{x \in \mathbf{R}} d\varphi(x)d\pi(x)" \,. \tag{17}$$

It turns out that this expression is merely formal: it is a product of three factors which are all infinite, nevertheless the measure can be constructed. We sketch this construction.

We first note that this measure should be a Gaussian measure since H is quadratic in $\phi = (\varphi, \pi)$. A Gaussian measure μ_β on $\mathbf{R}$ with mean 0 and variance β is completely characterized by the fact that its characteristic function (the Fourier transform of the measure) is given by

$$S(\xi) = \int e^{i\langle \xi, x \rangle} \mu_\beta(dx) = e^{-\frac{1}{2\beta}\langle x, x \rangle} \,. \tag{18}$$

is again a Gaussian. Let $\mathcal{H}$ be an infinite dimensional Hilbert space, $\phi \in \mathcal{H}$, $\xi \in \mathcal{H}^* = \mathcal{H}$. Can we construct a measure μ_β on the Hilbert space $\mathcal{H}$, such that

$$S(\xi) = \int e^{i\langle \phi, \xi \rangle} \mu_\beta(d\phi) = e^{-\frac{1}{2\beta}\langle \xi, \xi \rangle} \,? \tag{19}$$

The answer is **NO**.

Proof. By contradiction. Let $\{e_n\}$ be an orthonormal basis of $\mathcal{H}$. We then have $S(e_n) = e^{-1/2\beta}$. For any $\phi \in \mathcal{H}$, $\langle \phi, e_n \rangle \to 0$ as $n \to \infty$. By dominated convergence we have

$$\lim_{n \to \infty} S(e_n) = 1 \neq e^{-\frac{1}{2\beta}}. \tag{20}$$

and this is a contradiction. $\square$

All what this says is that we must give up the requirement that the measure μ be supported on the Hilbert space $\mathcal{H}$ (i.e., on finite energy configurations). The Bochner-Minlos Theorem allows us to construct such measures supported on larger spaces of distributions. Let A be the operator on $\mathcal{H}$ given by

$$A = \begin{pmatrix} (1 - \partial_x^2 + x^2)^{\frac{1}{2}} & 0 \\ 0 & (1 - \partial_x^2 + x^2)^{\frac{1}{2}} \end{pmatrix}. \tag{21}$$

We it leave to the reader to verify that A has compact resolvent and that A^{-s} is Hilbert-Schmidt if $s > 1/2$. For $s > 0$ we define Hilbert spaces

$$K_s = \{ u \in L^2 \,;\, \|u\|_s \equiv \|A^s u\| < \infty \}, \tag{22}$$

and for $s < 0$, $K_s = K_{-s}^*$ where $*$ is the duality in $\mathcal{H}$. We have, for $0 \leq s_1 \leq s_2$,

$$H_{s_2} \subset H_{s_1} \subset L^2 \subset H_{-s_1} \subset H_{-s_2}, \tag{23}$$

with dense inclusions. We set

$$S = \bigcap_s H_s, \quad S' = \bigcup_s H_s. \tag{24}$$

The space S' is simply a space of tempered distributions.

Theorem 2.1. (Bochner-Minlos Theorem) *There is a one-to-one correspondence between measures on S' and functions $S : S \to \mathbf{R}$ which satisfy*

1. S is continuous.
2. $S(0) = 1$.
3. S is of positive type, i.e. $\sum_{i,j=1}^n S(f_i - f_j) \bar{z}_i z_j \geq 0$, for all $n \geq 1$, for all $f_1, \cdots, f_n \in S$, and for all $z \in \mathbf{C}^n$.

The function S is the characteristic function of the measure. The Gaussian Gibbs measures μ_β are then specified by the characteristic function

$$S(\xi) = \int e^{i\langle \phi, \xi \rangle} \mu_\beta(d\phi) = e^{-\frac{1}{2\beta}\langle \xi, \xi \rangle}. \tag{25}$$

where $\langle \phi, \xi \rangle$ denotes now the $S - S'$ duality. If we put $\xi = a_1 \xi_1 + a_2 \xi_2$, the characteristic function allows us to compute the correlation functions (differentiate with respect to a_1, a_2 and compare coefficients):

$$\int_{S'} \langle \phi, \xi \rangle \mu_\beta(d\phi) = 0,$$

$$\int_{S'} \langle \phi, \xi_1 \rangle \langle \phi, \xi_2 \rangle \mu_\beta(d\phi) = \beta^{-1} \langle \xi_1, \xi_2 \rangle. \tag{26}$$

We have then

Lemma 2.2. *If $s > \frac{1}{2}$, then*

$$\int \|A^{-s}\phi\|^2 \, \mu_\beta(d\phi) = \beta^{-1}\mathrm{trace}(A^{-2s}) < \infty. \qquad (27)$$

and thus $\|A^{-s}\phi\|$ is finite μ_β a.s.

Proof. Let λ_j denote the eigenvalues of A and e_j the orthonormal basis of eigenvectors of A. We have $A^{-s}\phi = \sum_j \lambda_j^{-s} e_j \langle \phi, e_j \rangle$ and thus

$$\int \|A^{-s}\phi\|^2 \, \mu_\beta(d\phi) = \sum_j \lambda_j^{-2s} \int (\langle \phi, e_j \rangle)^2 \mu_\beta(d\phi) = \beta^{-1} \sum_j \lambda_j^{-2s}. \quad (28)$$

$\square$

As a consequence we see that for a typical element $\phi = (\varphi, \pi)$ in the support of μ_β, φ has $\frac{1}{2} - \epsilon$ derivatives for all $\epsilon > 0$ and π has $-\frac{1}{2} - \epsilon$ derivatives. Let us compute now the correlations

$$\int \pi(x_1)\pi(x_2)\mu_\beta(d\phi) \quad \text{and} \quad \int \varphi(x_1)\varphi(x_2)\mu_\beta(d\phi). \qquad (29)$$

These expressions have to be interpreted in the distribution sense. Our computations are formal but can be easily justified. If we choose $\xi_1 = (0, \delta(x - x_1))$ and $\xi_2 = (0, \delta(x - x_2))$ we obtain from Eq. (26)

$$\int \pi(x_1)\pi(x_2)\mu_\beta(d\phi) = \beta^{-1}\delta(x_1 - x_2), \qquad (30)$$

i.e., if we think of x as "time" then $\pi(x)$ is a white noise process. On the other hand if we choose $\xi_1 = (\theta(x - t), 0)$ and $\xi_2 = (\theta(x - s), 0)$ and use that $\partial_x \theta(x) = \delta(x)$ we obtain from Eq. (26)

$$\int (\varphi(x_1) - \varphi(x_2))^2 \mu_\beta(d\phi) = \int_{x_1}^{x_2} \int_{x_1}^{x_2} \int \partial_x\varphi(t)\partial_x\varphi(s) \, \mu_\beta(d\phi) \, dt \, ds$$

$$= \beta^{-1} \int_{x_1}^{x_2} \int_{x_1}^{x_2} \delta(t - s) \, dt \, ds$$

$$= \beta^{-1}|x_2 - x_1|. \qquad (31)$$

i.e., if we think of x as "time" then $\varphi(x)$ is a Brownian motion. Note that if we combine this computation with Kolmogorov Continuity Theorem we obtain that the paths of Brownian motion are almost surely Hölder continuous with exponents $\alpha < 1/2$ and almost never Hölder continuous with exponents $\alpha \geq 1/2$.

If we consider the wave equation in $\mathbf{R}^d$, then one obtains similar results (random fields indexed by $\mathbf{R}^d$ instead of "stochastic processes").

2.2 Markovian Gaussian stochastic processes

In this section we describe a few facts on Gaussian stochastic process, in particular we describe a situation when Gaussian stochastic processes are Markovian (see [3]).

Let us consider a one dimensional Gaussian stochastic process x_t. Recall that Gaussian means that for all k and all $t_1 < \cdots < t_k$, the random variable $Z = (x_{t_1}, \cdots, x_{t_k})$ is a normal random variable. Let us assume that x_t has mean 0, $\mathbf{E}[x_t] = 0$, for all t. Then the Gaussian process is uniquely determined by the expectations

$$\mathbf{E}[x_t x_s]. \tag{32}$$

which are called the covariance of x_t. If x_t is stationary, then (32) depends only on $|t - s|$:

$$C(t - s) = \mathbf{E}[x_t x_s]. \tag{33}$$

Note that $C(t - s)$ is positive definite. If C is a continuous function then a special case of Bochner-Minlos theorem (with $\mathcal{S}' = \mathbf{R}$) implies that

$$C(t) = \int_{\mathbf{R}} e^{ikt} d\Delta(k), \tag{34}$$

where $\Delta(k)$ is an odd nondecreasing function with $\lim_{k \to \infty} \Delta(k) < \infty$. If we assume that $\Delta(k)$ has no singular part, then $d\Delta(k) = \Delta'(k)dk$ and the function $\Delta'(k)$ is called the *spectral function* of the Gaussian process x_t. Note that

$$\Delta'(k) \geq 0, \tag{35}$$

since Δ is nondecreasing and that

$$\overline{\Delta'(k)} = \Delta'(-k), \tag{36}$$

since $C(t)$ is real. We will consider here only the special case when $(\Delta')^{-1}$ is a polynomial. By the conditions (35) and (36) there is a polynomial

$$p(k) = \sum_m c_m(-ik)^m, \tag{37}$$

with real coefficients c_m and root in the upper half plane such that

$$\Delta'(k) = \frac{1}{|p(k)|^2}. \tag{38}$$

Under these conditions we have

Proposition 2.3. *If* $p(k) = \sum_m^M c_m(-ik)^m$ *is a polynomial with real coefficients and roots in upper half plane then the Gaussian process with spectral density* $|p(k)|^{-2}$ *is the solution of the stochastic differential equation*

$$\left(p\left(-i\frac{d}{dt}\right)x_t\right) dt = dB_t \tag{39}$$

Proof. The proof follows from the following representation of x_t: let us define a kernel $k(t)$ by

$$k(t) = \frac{1}{\sqrt{2\pi}} \int_{\mathbf{R}} e^{ikt} \frac{1}{p(k)} dk . \tag{40}$$

Since the zeros of p are in the upper half-plane, we have $k(t) = 0$ if $t < 0$. We claim that x_t can be represented as the stochastic integral

$$x_t = \int_{-\infty}^{\infty} k(t-t') dB_{t'} = \int_{-\infty}^{t} k(t-t') dB_{t'} \tag{41}$$

It suffices to compute the variance, we have

$$
\begin{aligned}
\mathbf{E}[x_t x_s] &= \frac{1}{2\pi} \int_{\mathbf{R}} \int_{\mathbf{R}} k(t-t') k(s-s') \mathbf{E}[dB_{t'} dB_{s'}] \\
&= \frac{1}{2\pi} \int_{\mathbf{R}} \int_{\mathbf{R}} k(t-t') k(s-s') \delta(t'-s') \, dt' \, ds' \\
&= \frac{1}{2\pi} \int_{\mathbf{R}} k(t-s') k(s-s') \, ds' \\
&= \frac{1}{2\pi} \int_{\mathbf{R}} \int_{\mathbf{R}} \int_{\mathbf{R}} e^{ik(t-s')} e^{ik'(s-s')} \frac{1}{p(k)p(k')} \, ds' \, dk \, dk' \\
&= \int_{\mathbf{R}} \int_{\mathbf{R}} e^{ikt} e^{ik's} \frac{1}{p(k)p(k')} \delta(k+k') \, dk \, dk' \\
&= \int_{\mathbf{R}} e^{i(t-s)k} \frac{1}{p(k)p(-k)} dk \tag{42}
\end{aligned}
$$

and this proves the claim. From Eq. (41) we obtain

$$
\begin{aligned}
p\left(-i\frac{d}{dt}\right) x_t &= \int_{-\infty}^{t} \int_{\mathbf{R}} p\left(-i\frac{d}{dt}\right) e^{ik(t-t')} \frac{1}{p(k)} \, dk \, dB_{t'} \\
&= \int_{-\infty}^{t} \int_{\mathbf{R}} e^{ik(t-t')} \, dk \, dB_{t'} = \frac{dB}{dt} . \tag{43}
\end{aligned}
$$

and this concludes the proof of the proposition. $\square$

For example let us take

$$\Delta'(k) = \frac{\gamma}{\pi} \frac{1}{k^2 + \gamma^2} \tag{44}$$

and so $p(k) \propto (ik + \gamma)$ Then

$$C(t) = \frac{\gamma}{\pi} \int e^{ikt} \frac{1}{k^2 + \gamma^2} dk = e^{-\gamma|t|}, \tag{45}$$

and

$$k(t) = \sqrt{2\gamma}e^{-\gamma t}, \quad t \geq 0.$$

$$x(t) = \sqrt{2\gamma}\int_{-\infty}^{t} e^{-\gamma(t-t')}dB_{t'} \tag{46}$$

and we obtain

$$dx_t = -\gamma x_t + \sqrt{2\gamma}dB_t . \tag{47}$$

This is the Ornstein-Uhlenbeck process.

It is good exercise to compute the covariances $C(t)$ and derive the corresponding stochastic differential equations for the spectral densities with $p(k) \propto (ik + iu + \gamma)(ik - iu + \gamma)$ and $p(k) \propto (ik + \gamma)^2$.

2.3 How to make a Markovian reservoir

We derive effective equations for the small system. In spirit we are close to [21], although we are deriving different equations. Let us consider first a model of one single particle with Hamiltonian $H_S(p, q) = p^2/2 + V(q)$, where $(p, q) \in \mathbf{R} \times \mathbf{R}$, coupled to a single reservoir. The total Hamiltonian is, using the notation (14)

$$H(\phi, p, q) = \frac{1}{2}\|\phi\|^2 + p^2 + V(q) + q\int \partial_x \varphi(x)\rho(x)\,dx \tag{48}$$

$$= H_B(\phi) + H_S(p, q) + q\langle \phi, \alpha \rangle,$$

where, in Fourier space, $\hat{\alpha}(k) = (-ik\hat{\rho}(k)/k^2, 0)$. Let $\mathcal{L}$ be the linear operator given by

$$\mathcal{L} = \begin{pmatrix} 0 & 1 \\ \partial_x^2 & 0 \end{pmatrix}. \tag{49}$$

In Fourier space the semigroup $e^{t\mathcal{L}}$ is given by

$$e^{t\mathcal{L}} = \begin{pmatrix} \cos(kt) & k^{-1}\sin(kt) \\ -k\sin(kt) & \cos(kt) \end{pmatrix}. \tag{50}$$

Let us introduce the covariance function $C(t) = \langle \exp(\mathcal{L}t)\alpha, \alpha \rangle$. We have

$$C(t) = \int k^2 \frac{ik}{k^2}\overline{\hat{\rho}(k)}\cos(kt)\frac{-ik}{k^2}\hat{\rho}(k)\,dk = \int |\hat{\rho}(k)|^2 e^{ikt}\,dk, \tag{51}$$

and thus $C(t)$ is the covariance function of a Gaussian process with spectral density $|\rho(k)|^2$. We also define a coupling constant λ by setting

$$\lambda^2 = C(0) = \int dk|\rho(k)|^2. \tag{52}$$

The equations of motion of the coupled system particle and reservoir are

$$\dot{q}_t = p_t,$$
$$\dot{p}_t = -\partial_q V(q_t) - \langle \phi, \alpha \rangle, \tag{53}$$
$$\dot{\phi}_t(k) = \mathcal{L}(\phi_t(k) + q_t\alpha(k)).$$

Integrating the last equation of (53) we have

$$\phi_t(k) = e^{\mathcal{L}t}\phi_0(k) + \int_0^t e^{\mathcal{L}(t-s)} \mathcal{L}\alpha(k) q_s \, ds \,. \tag{54}$$

Inserting into the second equation of (53) we obtain

$$\dot{q}_t = p_t \,,$$
$$\dot{p}_t = -\partial_q V(q_t) - \int_0^t D(t-s) q_s \, ds - \langle \phi_0, e^{-\mathcal{L}t}\alpha \rangle \,. \tag{55}$$

where

$$D(t) = \langle e^{\mathcal{L}t} \mathcal{L}\alpha, \alpha \rangle = \dot{C}(t) \,. \tag{56}$$

Let us assume that the initial conditions of the reservoir ϕ_0 are distributed according to the Gibbs measure μ_β defined in Section 2.1. Then

$$y_t = \langle \psi_0 e^{-\mathcal{L}t}\alpha \rangle \,, \tag{57}$$

is a Gaussian process with covariance

$$\mathbf{E}[y_t y_s] = \int \langle \phi_0, e^{-\mathcal{L}t}\alpha \rangle \langle \phi_0, e^{-\mathcal{L}s}\alpha \rangle \mu_\beta(d\phi) = \beta^{-1} C(t-s) \,. \tag{58}$$

The equation (55) is a random integro-differential equation, since it contains memory terms both deterministic and random. The relation between the kernel D in the deterministic memory term and the covariance of the random term goes under the general name of Fluctuation-Dissipation Theorem. The solution of (55) is a random process. Note that the randomness in this equation comes from our choice on initial conditions of the reservoir. Let us choose the coupling function ρ, as in Section 2.2, such that

$$|\rho(k)|^2 = \frac{1}{|p(k)|^2} \,, \tag{59}$$

where p is a real polynomial in ik with roots in the lower half-plane. For simplicity we choose $p(k) \propto (ik+\gamma)$ This assumption together with the fluctuation-dissipation relation permits, by extending the phase space with one auxiliary variable, to rewrite the integro-differential equations (55) as a Markov process. We have then $C(t) = \lambda^2 e^{-\gamma|t|}$. It is convenient to introduce the variable r which is defined defined by

$$\lambda r_t = \lambda^2 q_t + \int_0^t D(t-s) q_s \, ds + y_t \,, \tag{60}$$

and we obtain from Eqs.(55) the set of Markovian differential equations:

$$dq_t = dp_t \, dt \,,$$
$$dp_t = (-\partial_q V_{\text{eff}}(q_t) - \lambda r_t) \,,$$
$$dr_t = (-\gamma r_t + \lambda p_t) \, dt + \sqrt{2\beta^{-1}\gamma} \, dB_t \,. \tag{61}$$

where $V_{\text{eff}}(q) = V(q) - \lambda^2 q^2 / 2$. The potential V is renormalized by the coupling to the reservoir. This is an artifact of the dipole approximation we have been using. Namely if we start with a translation invariant coupling of the form

$$\int \phi(x) \rho(x - q) \, dx \,, \qquad (62)$$

the dipole expansion leads to terms of the form

$$q \int \partial_x \phi(x) \rho(x) \, dx + \frac{q^2}{2} \int |\rho(x)|^2 \, dx \,, \qquad (63)$$

and the second term exactly compensates the normalization of the potential. We will ignore this renormalization in the sequel.

If one chooses other polynomials, similar equations can be derived. One should add one auxiliary variable for each pole of the polynomial $p(k)$. It is a good exercise to derive the SDE's for a particle coupled to a wave equation if we choose $p(k) \propto (ik + iu - \gamma)(ik - iu - \gamma)$ or $p(k) \propto (ik - \gamma)^2$.

One can recover the Langevin equation,

$$dq_t = dp_t \, dt \,,$$
$$dp_t = (-\partial V(q_t) \, dt - \kappa p_t) \, dt + \sqrt{2\beta^{-1}\kappa} \, dB_t \,.$$

but only in a suitable limit. Formally one would obtain these equations if $C(t - s) \propto \delta(t - s)$ (this corresponds to choosing $\rho(k) = 1$ which is not square integrable). But then the coupling constant $\lambda^2 = C(0)$ becomes infinite. Rather one should consider a suitable sequence of covariance which tends to a delta function and simultaneously rescale the coupling constant.

3 Ergodic properties: the chain

We consider here a model of non-equilibrium statistical mechanics: a one-dimensional "crystal" coupled at each end to reservoirs at different temperatures.

Let us consider a chain of n anharmonic oscillators given by the Hamiltonian

$$H_S(p, q) = \sum_{i=1}^{n} \frac{p_i^2}{2} + V(q_1, \cdots, q_n) \,,$$

$$V(q) = \sum_{i=1}^{n} U^{(1)}(q_i) + \sum_{i=1}^{n-1} U^{(2)}(q_i - q_{i+1}) \,.$$

where $(p_i, q_i) \in \mathbf{R} \times \mathbf{R}$. Our assumptions on the potential $V(q)$ are

(H1) Growth at infinity: The potentials $U^{(1)}(x)$ and $U^{(2)}(x)$ are $\mathcal{C}^\infty$ and grow at infinity like $|x|^{k_1}$ and $|x|^{k_2}$: There exist constants $C_i, D_i, i = 1, 2$ such that

$$\lim_{\lambda \to \infty} \lambda^{-k_i} U^{(i)}(\lambda x) = a^{(i)} |x|^{k_i}, \tag{64}$$

$$\lim_{\lambda \to \infty} \lambda^{-k_i+1} \partial_x U^{(i)}(\lambda x) = a^{(i)} k_i |x|^{k_i-1}, \tag{65}$$

$$|\partial_x^2 U^{(i)}(x)| \leq (C_i + D_i U^{(i)}(x))^{1-\frac{2}{k_i}}. \tag{66}$$

Moreover we will assume that

$$k_2 \geq k_1 \geq 2, \tag{67}$$

so that, for large $|x|$ the interaction potential $U^{(2)}$ is "stiffer" than the one-body potential $U^{(1)}$.

(H2) Non-degeneracy: The coupling potential between nearest neighbors $U^{(2)}$ is non-degenerate in the following sense: For any $q \in \mathbf{R}$, there exists $m = m(q) \geq 2$ such that $\partial^m U^{(2)}(q) \neq 0$. This means that $U^{(2)}$ has no flat pieces nor infinitely degenerate critical points. Note that we require this condition for $U^{(2)}$ only and not for $U^{(1)}$.

For example if $U^{(1)}$ and $U^{(2)}$ are polynomials of even degree, with a positive coefficients for the monomial of highest degree and $\deg U^{(2)} \geq \deg U^{(1)} \geq 2$, then both conditions **H1** and **H2** are satisfied.

We couple the first and the nth particle to reservoirs at inverse temperatures β_1 and β_n, respectively. We assume that the couplings to be as in Section 2.3 so that, by introducing two auxiliary variables r_1 and r_n, we obtain the set of stochastic differential equations equations

$$
\begin{aligned}
dq_{1t} &= dp_{1t}\, dt\,, \\
dp_{1t} &= (-\partial_{q_1} V(q_t) - \lambda r_{1t}) dt\,, \\
dr_{1t} &= (-\gamma r_{1t} + \lambda p_{1t})\, dt + (2\beta_1^{-1}\gamma)^{1/2} dB_{1t}\,, \\
dq_{jt} &= dp_{jt}\, dt\,, & j = 2, \ldots, n-1\,, \\
dp_{jt} &= -\partial_{q_j} V(q_t)\, dt\,, & j = 2, \ldots, n-1\,, \\
dq_{nt} &= p_{nt}\, dt\,, \\
dp_{nt} &= (-\partial_{q_{nt}} V(q_t) - \lambda r_{nt})\, dt\,, \\
dr_{nt} &= (-\gamma r_{nt} + \lambda p_{nt})\, dt + (2\beta_n^{-1}\gamma)^{1/2} dB_{nt}\,.
\end{aligned}
\tag{68}
$$

It will be useful to introduce the following notation. We define the linear maps $\Lambda : \mathbf{R}^n \to \mathbf{R}^2$ by $\Lambda(x_1, \ldots, x_n) = (\lambda x_1, \lambda x_n)$ and $T : \mathbf{R}^2 \to \mathbf{R}^2$ by $T(x, y) = (\beta_1^{-1} x, \beta_n^{-1} y)$. We can rewrite Eq.(68) in the compact form

$$
\begin{aligned}
dq_t &= p_t\, dt\,, \\
dp_t &= (-\nabla_q V(q_t) - \Lambda^* r)\, dt\,, \\
dr_t &= (-\gamma r_t + \Lambda p_t)\, dt + (2\gamma T)^{1/2} dB_t\,.
\end{aligned}
\tag{69}
$$

where $B = (B_1, B_n)$ is a two dimensional Brownian motion. The solution $x_t = (p_t, q_t, r_t) \in \mathbf{R}^{2n+2}$ of Eq.(69) is a Markov process. We denote T_t the associated semigroup and $P_t(x, dy)$ the transition probabilities

$$T_t f(x) = \mathbf{E}_x[f(x_t)] = \int_{\mathbf{R}^n} P_t(x, dy) f(y) . \tag{70}$$

The generator of T_t is given by

$$L = \gamma \left(\nabla_r T \nabla_r - r \nabla_r \right) + \left(\Lambda p \nabla_r - r \Lambda \nabla_p \right) + \left(p \nabla_q - (\nabla_q V(q)) \nabla_p \right) , \tag{71}$$

and the adjoint of L (Fokker-Planck operator) is given by

$$L^* = \gamma \left(\nabla_r T \nabla_r + \nabla_r r \right) - \left(\Lambda p \nabla_r - r \Lambda \nabla_p \right) - \left(p \nabla_q - (\nabla_q V(q)) \nabla_p \right) . \tag{72}$$

There is a natural energy function which is associated to Eq.(69), given by

$$G(p, q, r) = \frac{r^2}{2} + H(p, q) . \tag{73}$$

Since we assumed that $H(p, q)$ is a smooth function, we have local solutions for the SDE (69). A straightforward computation shows that we have

$$LG(p, q, r) = \gamma(\operatorname{tr}(T) - r^2) \leq \gamma \operatorname{tr}(T) . \tag{74}$$

Therefore we obtain global existence for the solutions of (69) (see Theorem 5.9 in [20]). Also a straightforward computation shows that in the special case of equilibrium, i.e., if $\beta_1 = \beta_n = \beta$ we have

$$L^* e^{-\beta G(p,q,r)} = 0 , \tag{75}$$

and therefore $Z^{-1} e^{-\beta G(p,q,r)}$ is, in that special case, the density of a stationary distribution for the Markov process $x(t)$. In the sequel we will refer to G as the energy of the system.

We are going to construct a Liapunov function (see Section 8 of [20]) for this system and it is quite natural to try functions of the energy G: let us denote

$$W_\theta = \exp(\theta G) . \tag{76}$$

A computation shows that

$$LW_\theta = \gamma \theta W_\theta \left(\operatorname{Tr}(T) - r(1 - \theta T)r \right) \tag{77}$$

This not quite a Liapunov function, but nearly so. The r.h.s. of Eq. (77) is negative provided $\theta/\beta_i < 1$ which we will always assume in the sequel and provided r is not too close to 0. Our proof is based on the following idea: at times r will be small, this corresponds to the situation where there is no dissipation of energy into the reservoir. But we will show that over small time interval, if we start the system at sufficiently large energy E, then with very large probability r^2 will be of order E^α

where $\alpha = 2/k_2$ is related to the growth exponent of the interaction energy in the chain (this where we use that $k_2 \geq k_1$). So if we integrate the equation of a small time interval $[0, t]$ we will show that if $G(x) > E$ and E is large enough

$$T_t W_\theta(x) \leq \kappa(E) W_\theta(x) \tag{78}$$

where $\kappa(E) \sim \exp(-E^\alpha)$.

We denote as $|\cdot|_\theta$ the weighted total variation norm given by

$$\|\pi\|_\theta = \sup_{|f| \leq W_\theta} \left| \int f \, d\pi \right|, \tag{79}$$

for any (signed) measure π. We introduce norms $\|\cdot\|_\theta$ and Banach spaces $\mathcal{H}_\theta$ given by

$$\|f\|_\theta = \sup_{x \in X} \frac{|f(x)|}{W_\theta(x)}, \quad \mathcal{H}_\theta = \{f : \|f\|_\theta < \infty\}, \tag{80}$$

and write $\|K\|_\theta$ for the norm of an operator $K : \mathcal{H}_\theta \to \mathcal{H}_\theta$.

Our results on the ergodic properties of Eqs. (69) are summarized in

Theorem 3.1. : Ergodic properties   *Let us assume condition* **H1** *and* **H2**.

(a) The Markov process $x(t)$ has a unique stationary distribution μ and μ has a C^∞ everywhere positive density.

(b) For any θ with $0 < \theta < \beta_{\min} = \min(\beta_1, \beta_n)$ the semigroup $T_t : \mathcal{H}_\theta \to \mathcal{H}_\theta$ is compact for all $t > 0$. In particular the process $x(t)$ converges exponentially fast to its stationary state μ: there exist constants $\gamma = \gamma(\theta) > 0$ and $R = R(\theta) < \infty$ such that

$$|P_t(x, \cdot) - \mu|_\theta \leq R e^{-\gamma t} W_\theta(x), \tag{81}$$

for all $x \in X$ or equivalently

$$\|T_t - \mu\|_\theta \leq R e^{-\gamma t}. \tag{82}$$

(c) The Markov process x_t is ergodic: For any $f \in L^1(\mu)$

$$\lim_{t \to \infty} \frac{1}{t} \int_0^t f(x_s) \, ds = \int f(x) \mu(dx), \tag{83}$$

for all initial condition x and for almost all realizations of the noise B_t. The Markov process is exponentially mixing: for all functions f, g with $f^2, g^2 \in \mathcal{H}_\theta$ and all $t > 0$ we have

$$\left| \int g T_t f \, d\mu - \int f \, d\mu \int g \, d\mu \right| \leq R e^{-\gamma t} \|f^2\|_\theta^{1/2} \|g^2\|_\theta^{1/2}. \tag{84}$$

With the tools we have developed in our lecture on Markov process citeRB, in order to prove Theorem 3.1 it will suffice to prove the following properties:

1. **Strong-Feller property**. The transition probabilities have a density $p_t(x, y)$ which is C^∞ in (t, x, y).
2. **Irreducibility**. For all $t > 0$, and all x supp $P_t(x, \cdot) = X$.
3. **Liapunov function**. For any $t > 0$, $\theta < \beta_{\min}$, and $E > 0$ there exists functions $\kappa(E) = \kappa(E, \theta, t)$ and $b(E) = b(E, \theta, t)$ with $\lim_{E \to \infty} \kappa(E) = 0$ such that

$$T_t W_\theta(x) \leq \kappa(E) W_\theta(x) + b(E) 1_{G \leq E}(x). \tag{85}$$

3.1 Irreducibility

Using the results of Section 6 in [20] we consider the control system

$$\begin{aligned}
\dot{q}_t &= p_t, \\
\dot{p}_t &= -\nabla_q V(q_t) - \Lambda^* r_t, \\
\dot{r}_t &= -\gamma r_t + \Lambda p_t + u_t.
\end{aligned} \tag{86}$$

where $t \mapsto u_t \in \mathbf{R}^2$ is a piecewise smooth control. One shows that for this system the set of accessible points from x in time t

$$\overline{A_t(x)} = \mathbf{R}^{2n+2}, \tag{87}$$

for any $x \in \mathbf{R}^{2n+2}$ and any $t > 0$.

We will illustrate here how this can be done sketching the proof by for the simpler problem of two oscillators coupled to a single reservoir and by assuming that $\partial_q U^{(2)}(q)$ is a diffeomorphism. Our assumption **H2** only ensures that the map $\partial_q U^{(2)}(q)$ is surjective and that we can find a piecewise smooth right inverse. This is enough to generalize the following argument, but one has to be careful if the initial or final points are one of the points where the right inverse of $\partial_q U^{(2)}(q)$ is not smooth. Let us consider the control system

$$\begin{aligned}
\dot{r}_t &= -\gamma r_t + \lambda p_{1t} + u_t, \\
\ddot{q}_{1t} &= -\partial_{q_1} U^{(1)}(q_{1t}) - \partial_{q_1} U^{(2)}(q_{1t} - q_{2t}) - \lambda r_t, \\
\ddot{q}_{2t} &= -\partial_{q_2} U^{(1)}(q_{2t}) - \partial_{q_2} U^{(2)}(q_{1t} - q_{2t}).
\end{aligned} \tag{88}$$

and let us choose arbitrary initial and final conditions

$$\begin{aligned}
x_0 &= (q_{10}, p_{10}, q_{20}, p_{20}, r_0) \\
x_1 &= (q_{1t}, p_{1t}, q_{2t}, p_{2t}, r_t).
\end{aligned} \tag{89}$$

Since the map $\partial_q U^{(2)}(q)$ is a diffeomorphism we first rewrite Eq. (88) as

$$\begin{aligned}
u_t &= f_1(r_t, \dot{r}_t, \dot{q}_{1t}), \\
r_t &= f_2(q_{1t}, q_{2t}, \ddot{q}_{1t}), \\
q_{1t} &= f_3(q_{2t}, \ddot{q}_{2t}),
\end{aligned} \tag{90}$$

for some smooth function f_i. Then there exists a function F such that

$$u_t = F(q_{2t}, \dot{q}_{2t}, \cdots, q_{2t}^{(5)}).$$ (91)

On the other hand by differentiating repeatedly the equation of motion we find function smooth function g_k such that

$$q_{2t}^{(k)} = g_k(q_{1t}, \dot{q}_{1t}, q_{2t}, \dot{q}_{2t}, r_t).$$ (92)

for $k = 0, 1, 2, 3, 4$. Let us choose now any curve q_{2t} which satisfies the boundary conditions

$$q_{20}^{(k)} = g_k(q_{10}, \dot{q}_{10}, q_{20}, \dot{q}_{20}, r_0),$$
$$q_{2t}^{(k)} = g_k(q_{1t}, \dot{q}_{1t}, q_{2t}, \dot{q}_{2t}, r_t).$$ (93)

We then define the desired control u_t by

$$u_t = F(q_{2t}, \dot{q}_{2t}, \cdots, q_{2t}^{(5)}).$$ (94)

which drives the system from x_0 to x_1 in time t. Since x_0 and x_1, and t are arbitrary this proves (87).

3.2 Strong Feller Property

We apply Hörmander's Theorem (see Section 7 of [20]) to show that the transition probabilities have a smooth density.

The generator of the Markov process $x(t)$ can be written in the form

$$L = \sum_{i=1}^{2} X_i^2 + X_0.$$ (95)

with $X_1 = \partial_{r_1}$ $X_2 = \partial_{r_n}$ and

$$x_0 = -\gamma r \nabla_r + (\Lambda p \nabla_r - r \Lambda \nabla_p) + (p \nabla_q - (\nabla_q V(q)) \nabla_p),$$ (96)

Let us verify that Hörmander condition is satisfied.

The vector fields X_i, $i = 1, 2$ are, up to a constant, ∂_{r_i}, $i = 1, n$. We have

$$[\partial_{r_1}, X_0] = -\gamma \partial_{r_1} - \lambda \partial_{p_1},$$
$$[\partial_{p_1}, X_0] = \lambda \partial_{r_1} + \partial_{q_1},$$

and so we can express the vector fields ∂_{p_1} and ∂_{q_1} as linear combinations of X_1, $[X_1, X_0]$, $[[X_1, X_0]X_0]$. Furthermore

$$[\partial_{q_1}, X_0] = (\partial^2 U^{(1)}(q_1) + \partial^2 U^{(1)}(q_1 - q_2))\partial_{q_1} - \partial^2 U^{(2)}(q_1 - q_2)\partial_{p_2}.$$ (97)

If $U^{(2)}$ is strictly convex, $\partial^2 U^{(2)}(q_1 - q_2)$ is positive and this gives ∂_{p_2} as a linear combination X_1, $[X_1, X_0]$, $[[X_1, X_0]X_0]$, and $[[[X_1, X_0]X_0]X_0]$. In general case we use Condition **H2**: for any q, there exists $m > 2$ such that $\partial^m U^{(2)}(q) \neq 0$ and we consider the commutators

$$\left[\partial_{q_1}, \left[\cdots, \left[\partial_{q_1}, \partial^2 U^{(2)}(q_1 - q_2)\partial_{p_2}\right]\right]\right]$$
$$= \partial^m U^{(2)}(q_1 - q_2)\partial_{p_2}.$$

and therefore we can express, at a given point q, ∂_{p_2} as a linear combination of commutators.

Proceeding by induction, we obtain, see Corollary 7.2 of [20]

Proposition 3.2. *If Condition* **H2** *is satisfied then the Lie algebra*

$$\{X_i\}_{i=1}^2, \quad \{[X_i, X_i]\}_{i,j=0}^2, \quad \{[[X_i, X_j], X_k]\}_{i,j,k=0}^2, \quad \cdots \tag{98}$$

has rank $\mathbf{R}^{2n+2}$ *at every point* x. *The transition probabilities* $P_t(x, y)$ *have a density* $p_t(x, y)$ *which is* C^∞ *in* (t, x, y).

3.3 Liapunov Function

We first consider the question of energy dissipation for the following deterministic equations

$$\begin{aligned}
\dot{q}_t &= p_t, \\
\dot{p}_t &= -\nabla_q V(q_t) - \Lambda^* r_t, \\
\dot{r}_t &= -\gamma r_t + \Lambda p_t,
\end{aligned} \tag{99}$$

obtained from Eq.(69) by setting $\beta_1 = \beta_n = \infty$. This corresponds to an initial condition 0 for the reservoirs. A simple computation shows that the energy $G(p, q, r)$ is non-increasing along the flow $x_t = (p_t, q_t, r_t)$ given by Eq.(99):

$$\frac{d}{dt}G(p_t, q_t, r_t) = -\gamma r_t^2 \leq 0. \tag{100}$$

We now show by a scaling argument that for any initial condition with sufficiently high energy, after a small time, a substantial amount of energy is dissipated.

At high energy, the two-body interaction $U^{(2)}$ in the potential dominates the term $U^{(1)}$ since $k_2 \geq k_1$ and so for an initial condition with energy $G(x) = E$, the natural time scale – essentially the period of a single one-dimensional oscillator in the potential $|q|^{k_2}$ – is $E^{1/k_2 - 1/2}$. We scale a solution of Eq.(99) with initial energy E as follows

$$\begin{aligned}
\tilde{p}_t &= E^{-\frac{1}{2}} p_{E^{\frac{1}{k_2}-\frac{1}{2}}t}, \\
\tilde{q}_t &= E^{-\frac{1}{k_2}} q_{E^{\frac{1}{k_2}-\frac{1}{2}}t}, \\
\tilde{r}_t &= E^{-\frac{1}{k_2}} r_{E^{\frac{1}{k_2}-\frac{1}{2}}t}.
\end{aligned} \tag{101}$$

Accordingly the energy scales as $G(p, q, r) = E\tilde{G}_E(\tilde{p}, \tilde{q}, \tilde{r})$, where

$$\tilde{G}_E(\tilde{p}, \tilde{q}, \tilde{r}) = E^{\frac{2}{k_2}-1}\frac{\tilde{r}^2}{2} + \frac{\tilde{p}^2}{2} + \tilde{V}_E(\tilde{q}),$$

$$\tilde{V}_E(\tilde{q}) = \sum_{i=1}^{n}\tilde{U}^{(1)}(\tilde{q}_i) + \sum_{i=1}^{n-1}\tilde{U}^{(2)}(\tilde{q}_i - \tilde{q}_{i+1}),$$

$$\tilde{U}^{(i)}(\tilde{x}) = E^{-1}\tilde{U}^{(i)}(E^{\frac{1}{k_2}}x), \quad i = 1, 2.$$

The equations of motion for the rescaled variables are

$$\dot{\tilde{q}}_t = \tilde{p}_t,$$
$$\dot{\tilde{p}}_t = -\nabla_{\tilde{q}}\tilde{V}_E(\tilde{q}_t) - E^{\frac{2}{k_2}-1}\Lambda^* r_t,$$
$$\dot{\tilde{r}}_t = -E^{\frac{1}{k_2}-\frac{1}{2}}\gamma\tilde{r}_t + \Lambda\tilde{p}_t. \qquad (102)$$

By assumption **H1**, as $E \to \infty$ the rescaled energy becomes

$$\tilde{G}_\infty(\tilde{p}, \tilde{q}, \tilde{r}) \equiv \lim_{E\to\infty}\tilde{G}_E(\tilde{p}, \tilde{q}, \tilde{r})$$

$$= \begin{cases} \tilde{p}^2/2 + \tilde{V}_\infty(\tilde{q}) & k_1 = k_2 > 2 \text{ or } k_2 > k_1 \geq 2 \\ \tilde{r}^2/2 + \tilde{p}^2/2 + \tilde{V}_\infty(\tilde{q}) & k_1 = k_2 = 2 \end{cases},$$

where

$$V_\infty(\tilde{q}) = \begin{cases} \sum a^{(1)}|\tilde{q}_i|^{k_2} + \sum a^{(2)}|\tilde{q}_i - \tilde{q}_{i+1}|^{k_2} & k_1 = k_2 \geq 2 \\ \sum a^{(2)}|\tilde{q}_i - \tilde{q}_{i+1}|^{k_2} & k_2 > k_1 \geq 2 \end{cases}. \qquad (103)$$

The equations of motion scale in this limit to

$$\dot{\tilde{q}}_t = \tilde{p}_t,$$
$$\dot{\tilde{p}}_t = -\nabla_{\tilde{q}}\tilde{V}_\infty(\tilde{q}_t),$$
$$\dot{\tilde{r}}_t = \Lambda\tilde{p}_t, \qquad (104)$$

in the case $k_2 > 2$, while they scale to

$$\dot{\tilde{q}}_t = \tilde{p}_t,$$
$$\dot{\tilde{p}}_t = -\nabla_{\tilde{q}}\tilde{V}_\infty(\tilde{q}_t) - \Lambda^*\tilde{r}_t,$$
$$\dot{\tilde{r}}_t = -\gamma\tilde{r}_t + \Lambda\tilde{p}_t, \qquad (105)$$

in the case $k_1 = k_2 = 2$.

Remark 3.3. Had we supposed, instead of **H1**, that $k_1 > k_2$, then the natural time scale at high energy would be $E^{1/k_1 - 1/2}$. Scaling the variables (with k_2 replaced by k_1) would yield the limiting Hamiltonian $\tilde{p}^2/2 + \sum a^{(1)}|\tilde{q}_i|^{k_1}$, i.e., the Hamiltonian of n *uncoupled* oscillators. So in this case, at high energy, essentially no energy is transmitted through the chain. While this does not necessary preclude the existence of an invariant measure, we expect in this case the convergence to a stationary state to be much slower. In any case even the existence of the stationary state in this case remains an open problem.

Theorem 3.4. *Given $\tau > 0$ there are constants $c > 0$ and $E_0 < \infty$ such that for any x with $G(x) = E > E_0$ and any solution $x(t)$ of Eq.(99) with $x(0) = x$ we have the estimate, for $t_E = E^{1/k_2 - 1/2}\tau$,*

$$G(x_{t_E}) - E \leq -cE^{\frac{3}{k_2} - \frac{1}{2}}. \tag{106}$$

Remark 3.5. In view of Eqs. (106) and (100), this shows that r is at least typically $O(E^{1/k_2})$ on the time interval $[0, E^{1/k_2 - 1/2}\tau]$.

Proof. Given a solution of Eq.(99) with initial condition x of energy $G(x) = E$, we use the scaling given by Eq.(101) and we obtain

$$G(x(t_E)) - E = -\gamma \int_0^{t_E} dt\, r_t^2 = -\gamma E^{\frac{3}{k_2} - \frac{1}{2}} \int_0^{\tau} dt\, \tilde{r}_t^2, \tag{107}$$

where $\tilde{r}_t$ is the solution of Eq.(102) with initial condition $\tilde{x}$ of (rescaled) energy $\tilde{G}_E(\tilde{x}) = 1$. By Assumption **H2** we may choose E_0 so large that for $E > E_0$ the critical points of $\tilde{G}_E$ are contained in, say, the set $\{\tilde{G}_E \leq 1/2\}$.

For a fixed E and x with $G(x) = E$, we show that there is a constant $c_{x,E} > 0$ such that

$$\int_0^{\tau} dt\, \tilde{r}_t^2 \geq c_{\tilde{x},E}. \tag{108}$$

The proof is by contradiction. Suppose that $\int_0^{\tau} dt\, \tilde{r}_t^2 = 0$, then we have $\tilde{r}_t = 0$, for all $t \in [0, \tau]$. From the third equation in (102) we conclude that $\tilde{p}_{1t} = \tilde{p}_{nt} = 0$ for all $t \in [0, \tau]$, and so from the first equation in (102) we see that $\tilde{q}_{1t}$ and $\tilde{q}_{nt}$ are constant on $[0, \tau]$. The second equation in (102) gives then

$$0 = \dot{\tilde{p}}_1(t) = -\partial_{q_1}\tilde{V}(\tilde{q}_t) = -\partial_{q_1}\tilde{U}^{(1)}(\tilde{q}_{1t}) - \partial_{q_1}\tilde{U}^{(2)}(\tilde{q}_{1t} - \tilde{q}_{2t}), \tag{109}$$

together with a similar equation for $\dot{\tilde{p}}_n$. By our assumption **H1** the map $\nabla\tilde{U}^{(2)}$ has a right inverse g which is piecewise smooth thus we obtain

$$\tilde{q}_{2t} = \tilde{q}_{1t} - g(\tilde{U}^{(1)}(\tilde{q}_{1t})). \tag{110}$$

Since $\tilde{q}_1$ is constant, this implies that $\tilde{q}_2$ is also constant on $[0, \tau]$. Similarly we see that $\tilde{q}_{n-1}$ is constant on $[0, \tau]$. Using again the first equation in (102) we obtain now $\tilde{p}_{2t} = \tilde{p}_{n-1t} = 0$ for all $t \in [0, \tau]$. Inductively one concludes that $\tilde{r}_t = 0$ implies $\tilde{p}_t = 0$ and $\nabla_{\tilde{q}}\tilde{V} = 0$ and thus the initial condition $\tilde{x}$ is a critical point of $\tilde{G}_E$. This contradicts our assumption and Eq. (108) follows.

Now for given E, the energy surface $\tilde{G}_E$ is compact. Using the continuity of the solutions of O.D.E with respect to initial conditions we conclude that there is a constant $c_E > 0$ such that

$$\inf_{\tilde{x} \in \{\tilde{G}_E = 1\}} \int_0^{\tau} dt\, \tilde{r}_t^2 \geq c_E. \tag{111}$$

Finally we investigate the dependence on E of c_E. We note that for $E = \infty$, $\tilde{G}_\infty$ has a well-defined limit given by Eq.(103) and the rescaled equations of motion, in the limit $E \rightarrow \infty$, are given by Eqs. (104) in the case $k_2 > 2$ and by Eq. (105) in the case $k_1 = k_2 = 2$. Except in the case $k_1 = k_2 = 2$ the energy surface $\{\tilde{G}_\infty = 1\}$ is *not* compact. However, in the case $k_1 = k_2 > 2$, the Hamiltonian $\tilde{G}_\infty$ and the equation of motion are invariant under the translation $r \mapsto r + a$, for any $a \in \mathbf{R}^2$. And in the case $k_2 > k_1 > 2$ the Hamiltonian $\tilde{G}_\infty$ and the equation of motion are invariant under the translation $r \mapsto r + a \ q \mapsto q + b$, for any $a \in \mathbf{R}^2$ and $b \in \mathbf{R}^n$. The quotient of the energy surface $\{\tilde{G}_\infty = 1\}$ by these translation, is compact.

Note that for a given $\tilde{x} \in \{\tilde{G}_\infty = 1\}$ a similar argument as above show that $\int_0^\tau dt (\tilde{r} + a)^2 > 0$, for any $a > 0$ and since this integral clearly goes to ∞ as $a \rightarrow \infty$ there exists a constant $c_\infty > 0$ such that

$$\inf_{\tilde{x} \in \{\tilde{G}_\infty = 1\}} \int_0^\tau \tilde{r}_t^2 \, dt > c_\infty . \tag{112}$$

Using again that the solution of O.D.E depends smoothly on its parameters, we obtain

$$\inf_{E > E_0} \inf_{\tilde{x} \in \{\tilde{G}_E = 1\}} \int_0^\tau dt \, \tilde{r}_t^2 > c . \tag{113}$$

This estimate, together with Eq. (107) gives the conclusion of Theorem 3.4. □

Next we show, that at sufficiently high energies, the overwhelming majority of the random paths $x_t = x_t(\omega)$ solving Eqs.(69) follows very closely the deterministic paths x_t^{det} solving Eqs.(99). As a consequence, for most random paths the same amount of energy is dissipated into the reservoirs as for the corresponding deterministic ones. We need the following *a priori* "no-runaway" bound on the growth of $G(x_t)$.

Lemma 3.6. *Let $\theta \leq (\max\{T_1, T_n\})^{-1}$. Then $\mathbf{E}_x[\exp(\theta G(x_t))]$ satisfies the bound*

$$\mathbf{E}_x[\exp(\theta G(x_t))] \leq \exp(\gamma \mathrm{Tr}(T)\theta t) \exp(\theta G(x)) . \tag{114}$$

Moreover for any x with $G(x) = E$ and any $\delta > 0$ we have the estimate

$$\mathbf{P}_x \left\{ \sup_{0 \leq s \leq t} G(x_s)) \geq (1 + \delta)E \right\} \leq \exp(\gamma \mathrm{Tr}(T)\theta t) \exp(-\delta \theta E) . \tag{115}$$

Remark 3.7. The lemma shows that for E sufficiently large, with very high probability, $G(x_t) = O(E)$ if $G(x) = E$. The assumption on θ here arises naturally in the proof, where we need $(1 - \theta T) \geq 0$, cf. Eq. (116).

Proof. For $\theta \leq (\max\{T_1, T_n\})^{-1}$ we have the bound (the generator L is given by Eq. (11))

$$L \exp(\theta G(x)) = \gamma \theta \exp(\theta G(x)) (\mathrm{Tr}(T) - r(1 - \theta T)r)$$

$$\leq \gamma \theta \mathrm{Tr}(T) \exp(\theta G(x)) , \tag{116}$$

Then we apply Theorem 5.4 of [20]. □

We have the following "tracking" estimates to the effect that the random path closely follows the deterministic one at least up to time t_E for a set of paths which have nearly full measure. We set $\Delta x_t \equiv x_t(\omega) - x_t^{\text{det}} = (\Delta r_t, \Delta p_t, \Delta q_t)$ with both $x_t(\omega)$ and x_t^{det} having initial condition x. Consider the event

$$S(x, E, t) = \{x.(\omega)\,;\, G(x) = E \text{ and } \sup_{0 \le s \le t} G(x_s) < 2E\}. \qquad (117)$$

By Lemma 3.6, $\mathbf{P}\{S(x, E, t)\} \ge 1 - \exp\left(\gamma\theta\mathrm{Tr}(T)t - \theta E\right)$.

Proposition 3.8. *There exist constants $E_0 < \infty$ and $c > 0$ such that for paths $x_t(\omega) \in S(x, E, t_E)$ with $t_E = E^{1/k_2/-1/2}\tau$ and $E > E_0$ we have*

$$\sup_{0 \le t \le t_E} \begin{pmatrix} \|\Delta q_t\| \\ \|\Delta p_t\| \\ \|\Delta r_t\| \end{pmatrix} \le c \sup_{0 \le t \le t_E} \|\sqrt{2\gamma T}B_t(\omega)\| \begin{pmatrix} E^{\frac{2}{k_2}-1} \\ E^{\frac{1}{k_2}-\frac{1}{2}} \\ 1 \end{pmatrix}. \qquad (118)$$

Proof. We write differential equations for Δx_t again assuming both the random and deterministic paths start at the same point x with energy $G(x) = E$. These equations can be written in the somewhat symbolic form:

$$d\Delta q_t = \Delta p_t dt\,,$$
$$d\Delta p_t = \left(O(E^{1-2/k_2})\Delta q_t - \Lambda^*\Delta r_t\right) dt\,,$$
$$d\Delta r_t = (-\gamma\Delta r_t + \Lambda\Delta p_t)\,dt + \sqrt{2\gamma T}dB_t \qquad (119)$$

The $O(E^{1-2/k_2})$ coefficient refers to the difference between forces, $-\nabla_q V(\cdot)$ evaluated at $x_t(\omega)$ and x_t^{det}; we have that $G(x_t) \le 2E$, so that $\nabla_q V(q_t(\omega)) - \nabla_q V(q_t^{\text{det}}) = O(\partial^2 V)\Delta q_t = O(E^{1-2/k_2})\Delta q_t$. For later purposes we pick a constant c' so large that

$$\rho = \rho(x) = c'E^{1-\frac{2}{k_2}} \ge \sup_i \sum_j \sup_{\{q:V(q)\le 2E\}} \left|\frac{\partial^2 V(q)}{\partial q_i \partial q_j}\right| \qquad (120)$$

for all sufficiently large E.

In order to estimate the solutions of Eqs. (119), we consider the 3×3 matrix which bounds the coefficients in this system, and which is given by

$$M = \begin{pmatrix} 0 & 1 & 0 \\ \rho & 0 & \lambda \\ 0 & \lambda & \gamma \end{pmatrix} \qquad (121)$$

We have the following estimate on powers of M; For $\Delta X^{(0)} = (0, 0, 1)^T$, we set $\Delta X^{(m)} \equiv M^m \Delta X^{(0)}$. For $\alpha = \max(1, \gamma + \lambda)$, we obtain $\Delta X^{(1)} \le \alpha(0, 1, 1)^T$, $\Delta X^{(2)} \le \alpha^2(1, 1, 1)^T$, and, for $m \ge 3$,

$$\Delta X^{(m)} \equiv \begin{pmatrix} u^{(m)} \\ v^{(m)} \\ w^{(m)} \end{pmatrix} \le \alpha^m 2^{m-2} \begin{pmatrix} \rho^{\frac{m-2}{2}} \\ \rho^{\frac{m-1}{2}} \\ \rho^{\frac{m-2}{2}} \end{pmatrix},$$

where the inequalities are componentwise. From this we obtain the bound

$$e^{tM} \begin{pmatrix} 0 \\ 0 \\ 1 \end{pmatrix} \leq \begin{pmatrix} \frac{1}{2}(\alpha t)^2 e^{\sqrt{\rho}2\alpha t} \\ \alpha t e^{\sqrt{\rho}2\alpha t} \\ 1 + \alpha t + \frac{1}{2}(\alpha t)^2 e^{\sqrt{\rho}2\alpha t} \end{pmatrix}. \tag{122}$$

If $0 \leq t \leq t_E$ we have $\sqrt{\rho}t < \sqrt{c'}$. Then the exponentials in the above equation are bounded, and

$$e^{tM} \begin{pmatrix} 0 \\ 0 \\ 1 \end{pmatrix} \leq c \begin{pmatrix} 1/\rho \\ 1/\sqrt{\rho} \\ 1 \end{pmatrix}, \tag{123}$$

for some constant c.

Returning now to the original differential equation system Eq.(119), we write this equation in the usual integral equation form:

$$\begin{pmatrix} \Delta q_t \\ \Delta p_t \\ \Delta r_t \end{pmatrix} = \int_0^t \begin{pmatrix} \Delta p_s \\ -\nabla_q V(q_s(\omega)) \, ds + \nabla_q V(q_s^{\text{det}}) - \Gamma^* \Delta r_s \\ -\gamma \Delta r_s + \Lambda \Delta p_s \end{pmatrix} ds$$

$$+ \begin{pmatrix} 0 \\ 0 \\ \sqrt{2\gamma T} B_t \end{pmatrix}. \tag{124}$$

From this we obtain the bound

$$\begin{pmatrix} \|\Delta q_t\| \\ \|\Delta p_t\| \\ \|\Delta r_t\| \end{pmatrix} \leq \int_0^t M \begin{pmatrix} \|\Delta q_t\| \\ \|\Delta p_t\| \\ \|\Delta r_t\| \end{pmatrix} ds + \begin{pmatrix} 0 \\ 0 \\ B_{\max} \end{pmatrix}, \tag{125}$$

where M is given by Eq.(121), and $B_{\max} = \sup_{t \leq t_E} \|\sqrt{2\gamma T} B_t\|$. Note that the solution of the integral equation

$$\Delta X_t = \int_0^t ds \, M \Delta X_s + \begin{pmatrix} 0 \\ 0 \\ B_{\max} \end{pmatrix}, \tag{126}$$

is $\Delta X_t = \exp(tM)(0, 0, B_{\max})^T$. We can solve both Eq. (124) and Eq. (126) by iteration. Let Δx_{ms}, ΔX_{ms} denote the respective m^{th} iterates (with $\Delta x_{0s} = (0, 0, \sqrt{2\gamma T} B_s)^T$, and $\Delta X_{0s} = (0, 0, B_{\max})^T$, $0 \leq s \leq t_E$). The ΔX_m's are monotone increasing in m. Then it is easy to see that

$$\begin{pmatrix} \|\Delta q_{mt}\| \\ \|\Delta p_{mt}\| \\ \|\Delta r_{mt}\| \end{pmatrix} \leq \Delta X_{mt} \leq \Delta X_t, \tag{127}$$

for each iterate. By Eqs.(122), (123), and the definition of ρ the conclusion Eq. (118) follows. □

As a consequence of Theorem 3.4 and Proposition 3.8 we obtain

Corollary 3.9. Let $\Omega(E) = E^\alpha$ with $\alpha < 1/k_2$ and assume that B_t is such that $\sup_{0 \le t \le t_E} \|\sqrt{2\gamma T} B_t\| \le \Omega(E)$ and $x.(\omega) \in S(x, E, t_E)$. Then there are constants $c > 0$ and $E_0 < \infty$ such that all paths $x_t(\omega)$ with initial condition x with $G(x) = E > E_0$ satisfy the bound

$$\int_0^{t_E} r_s^2 ds \ge c E^{\frac{3}{k_2} - \frac{1}{2}} . \tag{128}$$

Remark 3.10. For large energy E, paths *not* satisfying the hypotheses of the corollary have measure bounded by

$$\mathbf{P}_x \left\{ \sup_{0 \le s \le t_E} \|\sqrt{2\gamma T} B_s\| > \Omega(E) \right\} + \mathbf{P} \left\{ S(x, E, t_E)^C \right\}$$

$$\le \frac{a}{2} \exp \left(-\frac{\Omega(E)^2}{b \gamma T_{\max} t_E} \right) + \exp \left(\theta(\gamma \mathrm{Tr}(T) t_E - E) \right)$$

$$\le a \exp \left(-\frac{\Omega(E)^2}{b \gamma T_{\max} t_E} \right) , \tag{129}$$

where a and b are constants which depend only on the dimension of ω. Here we have used the reflection principle to estimate the first probability and Eq. (115) and the definition of S to estimate the second probability. For E large enough, the second term is small relative to the first.

Proof: It is convenient to introduce the L^2-norm on functions on $[0, t]$, $\|f\|_t \equiv \left(\int_0^t \|f_s\|^2 ds \right)^{1/2}$. By Theorem 3.4, there are constants E_1 and c_1 such that for $E > E_1$ the deterministic paths x_s^{det} satisfy the bound

$$\|r^{\mathrm{det}}\|_{t_E}^2 = \int_0^{t_E} (r_s^{\mathrm{det}})^2 ds \ge c_1 E^{\frac{3}{k_2} - \frac{1}{2}} . \tag{130}$$

By Proposition 3.8, there are constants E_2 and c_2 such that $\|\Delta r_s\| \le c_2 \Omega(E)$, uniformly in s, $0 \le s \le t_E$, and uniformly in x with $G(x) > E_2$. So we have

$$\|r\|_{t_E} \ge \|r^{\mathrm{det}}\|_{t_E} - \|\Delta r\|_{t_E} \ge \left(c_1 E^{\frac{3}{k_2} - \frac{1}{2}} \right)^{1/2} - c_2 \Omega(E) \left(E^{\frac{1}{k_2} - \frac{1}{2}} \right)^{1/2} . \tag{131}$$

But the last term is $O(E^{\alpha - 1/4 + 1/2k_2})$, which is of lower order than the first since $\alpha < 1/k_2$, so the corollary follows, for an appropriate constant c and E sufficiently large. $\quad\square$

With these estimates we now prove the existence of a Liapunov function.

Theorem 3.11. Let $t > 0$ and $\theta < \beta_{\min}$. Then there are functions $\kappa(E) = \kappa(E, t, \theta) < 1$ and $b(E) = b(E, t, \theta) < \infty$ such that

$$T_t W_\theta(x) \le \kappa(E) W_\theta(x) + b(E) \mathbf{1}_{\{G \le E\}}(x) . \tag{132}$$

The function $\kappa(E)$ satisfies the bound

$$\kappa(E) \leq A \exp(-BE^{2/k_2}), \tag{133}$$

for some constants A and B.

Proof. For any compact set U and for any t, $T^t \exp(\theta G)(x)$ is a bounded function on U, uniformly on $[0, t]$. So, in order to prove Eq.(132), we only have to prove that there exist a compact set U and $\kappa < 1$ such that

$$\sup_{x \in U^C} \mathbf{E}_x \left[\exp\left(\theta(G(x_t) - G(x))\right) \right] \leq \kappa < 1. \tag{134}$$

Using Ito's Formula to compute $G(x_t) - G(x)$ in terms of a stochastic integral we obtain

$$\mathbf{E}_x \left[\exp\left(\theta(G(x_t) - G(x))\right) \right]$$
$$= \exp\left(\theta\gamma\mathrm{tr}(T)t\right)\mathbf{E}_x \left[\exp\left(-\theta \int_0^t \gamma r_s^2 \, ds + \theta \int_0^t \sqrt{2\gamma T} r_s dB_s \right) \right]. \tag{135}$$

For any $\theta < \beta_{\min}$, we choose $p > 1$ such that $\theta p < \beta_{\min}$. Using Hölder inequality we obtain,

$$\mathbf{E}_x \left[\exp\left(-\theta \int_0^t \gamma r_s^2 \, ds + \theta \int_0^t \sqrt{2\gamma T} r_s dB_s \right) \right]$$
$$= \mathbf{E}_x \left[\exp\left(-\theta \int_0^t \gamma r_s^2 \, ds + \frac{p\theta^2}{2} \int_0^t (\sqrt{2\gamma T} r_s)^2 \, ds \right) \times \right.$$
$$\left. \times \exp\left(-\frac{p\theta^2}{2} \int_0^t (\sqrt{2\gamma T} r_s)^2 \, ds + \theta \int_0^t \sqrt{2\gamma T} r_s dB_s \right) \right]$$
$$\leq \mathbf{E}_x \left[\exp\left(-q\theta \int_0^t \gamma r_s^2 \, ds + \frac{qp\theta^2}{2} \int_0^t (\sqrt{2\gamma T} r_s)^2 \, ds \right) \right]^{1/q} \times$$
$$\times \mathbf{E}_x \left[\exp\left(-\frac{p^2\theta^2}{2} \int_0^t (\sqrt{2\gamma T} r_s)^2 \, ds + \theta p \int_0^t \sqrt{2\gamma T} r_s dB_s \right) \right]^{1/p}$$
$$= \mathbf{E}_x \left[\exp\left(-q\theta \int_0^t \gamma r_s^2 \, ds + \frac{qp\theta^2}{2} \int_0^t (\sqrt{2\gamma T} r_s)^2 \, ds \right) \right]^{1/q}.$$

Here, in the next to last line, we have used Girsanov theorem and so the second expectation is equal to 1. Finally we obtain the bound

$$\mathbf{E}_x \left[\exp\left(\theta(G(x_t) - G(x))\right) \right]$$
$$\leq \exp\left(\theta\gamma\mathrm{tr}(T)t\right)\mathbf{E}_x \left[\exp\left(-q\theta(1 - p\theta T_{\max}) \int_0^t \gamma r_s^2 \, ds \right) \right]^{1/q}. \tag{136}$$

In order to proceed we need to distinguish two cases according if $3/k_2 - 1/2 > 0$ or $3/k_2 - 1/2 \leq 0$ (see Corollary 3.9). In the first case we let E_0 be defined by

$t = E_0^{1/k_2 - 1/2}\tau$. For $E > E_0$ we break the expectation Eq. (136) into two parts according to whether the paths satisfy the hypotheses of Corollary 3.9 or not. For the first part we use Corollary 3.9 and that $\int_0^t r_s^2 ds \geq \int_0^{t_E} r_s^2 \geq cE^{3/k_2 - 1/2}$; for the second part we use estimate (129) in Remark 3.10 on the probability of unlikely paths together with the fact that the exponential under the expectation in Eq. (136) is bounded by 1. We obtain for all x with $G(x) = E > E_0$ the bound

$$\mathbf{E}_x\left[\exp\left(\theta(G(x_t) - G(x))\right)\right] \leq \exp\left(\theta\gamma\mathrm{tr}(T)t_{E_0}\right) \times$$

$$\times \left[\exp\left(-q\theta(1 - p\theta T_{\max})cE^{\frac{3}{k_2}-\frac{1}{2}}\right) + a\exp\left(-\frac{\Omega(E)^2\theta_0}{b\gamma t_E}\right)\right]^{1/q}. (137)$$

Choosing the set $U = \{x\,;\, G(x) \leq E_1\}$ with E_1 large enough we can make the term in Eq. (137) as small as we want.

If $3/k_2 - 1/2 \leq 0$, for a given t and a given x with $G(x) = E$ we split the time interval $[0, t]$ into $E^{1/2 - 1/k_2}$ pieces $[t_j, t_{j+1}]$, each one of size of order $E^{1/k_2 - 1/2}t$. For the "good" paths, i.e., for the paths x_t which satisfy the hypotheses of Corollary 3.9 on each time interval $[t_j, t_{j+1}]$, the tracking estimates of Proposition 3.8 imply that $G(x_t) = O(E)$ for t in each interval. Applying Corollary 3.9 and using that $G(x_{t_j}) = O(E)$ we conclude that $\int_0^t r_s^2 ds$ is at least of order $E^{3/k_2 - 1/2} \times E^{1/2 - 1/k_2} = E^{2/k_2}$. The probability of the remaining paths can be estimated, using Eq. (129), not to exceed

$$1 - \left(1 - a\exp\left(-\frac{\Omega_{\max}^2\theta_0}{b\gamma t_E}\right)\right)^{E^{\frac{1}{2}-\frac{1}{k_2}}}. \qquad (138)$$

The remainder of the argument is essentially as above, Eq. (137) and this concludes the proof of Theorem 3.11. $\quad\square$

4 Heat Flow and Entropy Production

In this section we study some thermodynamical properties of the stationary distribution. Most interesting is the case where the temperatures of the two reservoirs are different, we expect then to have heat (i.e., energy) flowing through the system from the hot reservoir into the cold one. Very little is known about the properties of systems in a nonequilibrium stationary state. The Kubo formula and Onsager reciprocity relations are such properties which are known to hold near equilibrium (i.e., if the temperatures of the reservoirs are close). In the recent years a new general fact about nonequilibrium has been discovered, the so-called Gallavotti-Cohen fluctuation Theorem. It asserts that the fluctuation of the ergodic mean of the entropy production has a certain symmetry. This symmetry can be seen as a generalization of Kubo formula and Onsager reciprocity relations to situations far from equilibrium. It has been discovered in numerical experiments in [8]. As a theorem it has been proved for Anosov maps [11], these deterministic systems are used to model nonequilibrium systems with reservoirs described by non-Hamiltonian deterministic forces (the so-called Gaussian thermostat). The fluctuation theorem has been formulated and extended to Markov process in [15, 16, 18] and proved for simple systems like Markov chains with a finite state space or non-degenerate diffusions.

We will prove this fluctuation theorem for our model. Both the degeneracy of the Markov process and the non-compactness of the phase space are the technical difficulties which have to be overcome. Our model is the first model which is completely derived from first principles (it is Hamiltonian to start with) and for which the fluctuation theorem can be proved.

To define the heat flow and the entropy production we write the energy of the chain H as a sum of local energies $H = \sum_{i=1}^{n} H_i$ where

$$
\begin{aligned}
H_1 &= \frac{p_1^2}{2} + U^{(1)}(q_1) + \frac{1}{2} U^{(2)}(q_1 - q_1) , \\
H_i &= \frac{p_i^2}{2} + U^{(1)}(q_i) + \frac{1}{2} \left(U^{(2)}(q_{i-1} - q_i) + U^{(2)}(q_i - q_{i+1}) \right) , \quad (139) \\
H_n &= \frac{p_n^2}{2} + U^{(1)}(q_n) + \frac{1}{2} U^{(2)}(q_n - q_{n-1}) .
\end{aligned}
$$

Using Ito's Formula one finds

$$
dH_i(x_t) = (\Phi_{i-1}(x_t) - \Phi_i(x_t)) \, dt , \tag{140}
$$

where

$$
\begin{aligned}
\Phi_0 &= -\lambda r_1 p_1 , \\
\Phi_i &= \frac{(p_i + p_{i+1})}{2} \partial_q U^{(2)}(q_i - q_{i+1}) , \quad (141) \\
\Phi_n &= \lambda r_n p_n .
\end{aligned}
$$

It is natural interpret Φ_i, $i = 1, \cdots, n-1$ as the heat flow from the ith to the $(i+1)^{th}$ particle, Φ_0 as the flow from the left reservoir into the chain, and Φ_n as the flow from the chain into the right reservoir. We define corresponding entropy productions by

$$
\sigma_i = (\beta_n - \beta_1)\Phi_i . \tag{142}
$$

There are other possible definitions of heat flows and corresponding entropy production that one might want to consider. One might, for example, consider the flows at the boundary of the chains, and define $\sigma_b = \beta_1 \Phi_0 - \beta_n \Phi_n$. Also our choice of local energy is somewhat arbitrary, other choices are possible but this does not change the subsequent analysis. Our results on the heat flow are summarized in

Theorem 4.1. : Entropy production

(a) **Positivity of entropy production.** *The expectation of the entropy production σ_j in the stationary state is independent of j and nonnegative*

$$\int \sigma_j d\mu \geq 0, \tag{143}$$

and it is positive away from equilibrium

$$\int \sigma_j d\mu = 0 \qquad \text{if and only if} \qquad \beta_1 = \beta_n. \tag{144}$$

(b) **Large deviations and fluctuation theorem.** *The ergodic averages*

$$\bar{\sigma}_j{}^t \equiv \frac{1}{t} \int_0^t \sigma_j(x_s) \tag{145}$$

satisfy the large deviation principle: There exist a neighborhood O of the interval $[-\int \sigma_j d\mu, \int \sigma_j d\mu]$ and a rate function $e(w)$ (both are independent of j) such that for all intervals $[a, b] \subset O$ we have

$$\lim_{t \to \infty} -\frac{1}{t} \log \mathbf{P}_x \{\bar{\sigma}_j{}^t \in [a, b]\} = \inf_{w \in [a,b]} e(w). \tag{146}$$

The rate function $e(w)$ satisfy the relation

$$e(w) - e(-w) = -w, \tag{147}$$

i.e., the odd part of e is linear with slope $-1/2$.

(c) **Kubo formula and central limit theorem.** *Let us introduce the parameters $\beta = (\beta_1 + \beta_n)/2$ and $\eta = \beta_n - \beta_1$. We have*

$$\frac{\partial}{\partial \eta} \left(\int \phi_j d\mu \right) \bigg|_{\eta=0} = \int_0^\infty \left(\int (T_t^{\text{eq}} \phi_j) \phi_j d\mu^{\text{eq}} \right) ds, \tag{148}$$

where μ^{eq} is the Gibbs stationary distribution at equilibrum (see Eq. (75)) and T_t^{eq} is the semigroup at equilibrium. Moreover, if we consider the fluctuations of the heat flow at equilibrium, they satisfy a central limit theorem

$$\mathbf{P}_x \left\{ a < \frac{1}{\sqrt{\kappa^2 t}} \int_0^t \Phi_j(x_s)\, ds < b \right\} \longrightarrow \frac{1}{\sqrt{2\pi}} \int_a^b \exp(-\frac{y^2}{2})\, dy \tag{149}$$

as $t \to \infty$, the constant κ^2 is positive, independent of j, and is given by

$$\kappa^2 = \int_0^\infty \left(\int \Phi_j(x) T_s^{\text{eq}} \Phi_j(x) \mu(dx) \right) ds. \tag{150}$$

Loosely speaking the fluctuation theorem has the following interpretation,

$$\frac{\mathbf{P}_x \{\bar{\sigma}_j{}^t \approx a\}}{\mathbf{P}_x \{\bar{\sigma}_j{}^t \approx -a\}} \approx e^{ta}, \tag{151}$$

in other words this gives a bound on the probability to observe a fluctuation of the entropy production which would give rise to a energy flow from the cold reservoir to the hot reservoir (i.e., a "violation" of the second law of thermodynamics). As we will see the Kubo formula is a consequence of the fluctuation theorem and thus we can also view the fluctuation theorem as a generalization of Kubo formula to large fields. We will elaborate on this interpretation later.

4.1 Positivity of entropy production

Let us consider the functions R_j given by

$$R_j = \beta_1 \left(\frac{r_1^2}{2} + \sum_{k=1}^{j} H_k(p,q) \right) + \beta_n \left(\sum_{k=j+1}^{n} H_k(p,q) + \frac{r_n^2}{2} \right), \qquad (152)$$

so that $\exp(-R_j)$ is a kind of "two-temperatures" Gibbs state. We also denote by J the time reversal operator which changes the sign of the momenta of all particles $Jf(p,q,r) = f(-p,q,r)$.

The following identities can be regarded as operator identities on $\mathcal{C}^\infty$ functions. That the left and right side of Eq. (154) actually generate semigroups for some non trivial domain of α is a non trivial result which we will discuss later.

Lemma 4.2. *Let us consider e^{R_i} and e^{-R_i} as multiplication operators. Then we have the operator identities*

$$e^{R_i} J L^* J e^{-R_i} = L - \sigma_i, \qquad (153)$$

and also for any constant α

$$e^{R_i} J (L^* - \alpha \sigma_i) J e^{-R_i} = L - (1-\alpha)\sigma_i. \qquad (154)$$

Proof. We write the generator L as $L = L_0 + L_1$ with

$$L_0 = \gamma (\nabla_r T \nabla_r - r \nabla_r) \qquad (155)$$

$$L_1 = (\Lambda p \nabla_r - r \Lambda \nabla_p) + (p \nabla_q - (\nabla_q V(q)) \nabla_p). \qquad (156)$$

Since L_1 is a first order differential operator we have

$$e^{-R_i} L_1 e^{R_i} = L_1 + L_1 R_i = L_1 + \sigma_i. \qquad (157)$$

Using that $\nabla_r R_i = T^{-1} r$ we obtain

$$e^{-R_i} L_0 e^{R_i} = e^{-R_i} \gamma (\nabla_r - T^{-1} r) T \nabla_r e^{R_i}$$

$$= \gamma \nabla_r T (\nabla_r + T^{-1} r) = L_0^*.$$

This gives

$$e^{-R_i} L e^{R_i} = L_0^* + L_1 + \sigma_i = J L^* J + \sigma_i, \qquad (158)$$

which is Eq. (153). Since $J \sigma_i J = -\sigma_i$, Eq. (154) follows immediately from Eq. (153). $\square$

Proof of Theorem 4.1 (a): We write the positive density $\rho(x)$ of $\mu(dx) = \rho(x)dx$ as

$$\rho = J e^{-R_j} e^{-F_j}. \qquad (159)$$

Let $L^\dagger$ denote the adjoint of L on $L^2(\mu)$, it is given by $L^\dagger = \rho^{-1}L^*\rho$ and using Eq. (153) a simple computation shows that

$$JL^dagger = e^{F_j}(L - \sigma_j)e^{-F_j}$$
$$= L - \sigma_j - (LF_j) - 2(T\nabla_r F_j)\nabla_r + |T^{1/2}\nabla_r F_j|^2). \quad (160)$$

It is easy to see that the operator $JL^\dagger J$ satisfies $JL^\dagger J1 = 0$ and so applying the Eq. (160) to the constant function we find

$$\sigma_j = |T^{1/2}\nabla F_j|^2 - LF_i. \quad (161)$$

The first term is obviously positive while the expectation of the second term in the stationary state vanishes and so we obtain Eq. (143).

In order to prove positivity of the entropy production, we will make a proof by contradiction. Let us suppose that $\beta_1 \neq \beta_n$ and that $\int \sigma_i(x)\mu(dx) = 0$. Since all σ_i have the same stationary value, it is enough to consider one of them and we choose $\sigma_0 = (\beta_1 - \beta_n)\lambda p_1 r_1$. The assumption implies that $\int |T^{1/2}\nabla_r F_0|^2\mu(dx) = 0$. Since ρ is positive, this means that $\nabla_r F_0 = 0$, and therefore F_0 does not depend on the r variables. From Eq. (161) we obtain

$$\sigma_0 = -LF_0. \quad (162)$$

Using the definition of L and σ_0 and the fact that F_0 does not depend on r, we obtain the equation

$$0 = (p \cdot \nabla_q \varphi - (\nabla_q V) \cdot \nabla_p) F_0 + \lambda r_1 \partial_{p_1} F_0 + \lambda r_n \partial_{p_n} F_0 = (\beta_n - \beta_1)\lambda r_1 p_1.$$

Since F_0 does not depend on r we get the sytem of equations

$$(p\nabla_q - (\nabla_q V)\nabla_p)F_0 = 0,$$
$$\partial_{p_1} F_0 = (\beta_n - \beta_1)p_1,$$
$$\partial_{p_n} F_0 = 0. \quad (163)$$

We will show that this system of linear equations has no solution unless $\beta_1 = \beta_n$. To see this we consider the system of equations

$$(p\nabla_q - (\nabla_q V)\nabla_p)F_0 = 0,$$
$$\partial_{p_1} F_0 = (\beta_n - \beta_1)p_1. \quad (164)$$

This system has a solution which is given by $(\beta_n - \beta_1)H(q, p)$. We claim that this the unique solution (up to an additive constant) of Eq. (164). If this holds true, then the only solution of Eq. (163) is given by $(\beta_n - \beta_1)H(q, p)$ and this is incompatible with the third equation in (163) when $\beta_1 \neq \beta_n$.

Since Eq. (164) is a linear inhomogeneous equation, it is enough to show that the only solutions of the homogeneous equation

$$(p\nabla_q - (\nabla_q V)\nabla_p)F_0 = 0,$$
$$\partial_{p_1} F_0 = 0. \quad (165)$$

are the constant functions. Since $\partial_{p_1} F_0 = 0$, F_0 does not depend on p_1, we conclude that the first equation in (165) reads

$$p_1 \partial_{q_1} F_0 + f_1(q_1, \ldots, q_n, p_2, \ldots p_n) = 0, \qquad (166)$$

where f_1 does not depend on the variable p_1. Thus we see that $\partial_{q_1} F_0 = 0$ and therefore F_0 does not depend on the variable q_1 either. By the first equation in (165) we now get

$$(\partial_{q_1} U^{(2)}(q_1 - q_2)) \partial_{p_2} F_0 + f_2(q_2, \ldots, q_n, p_2, \ldots, p_n) = 0, \qquad (167)$$

where f_2 does not depend on p_1 and q_1. By condition **H2** we see that $\partial_{p_2}\varphi = 0$ and hence f does not depend on p_2. Iterating the above procedure we find that the only solutions of (165) are the constant functions.

As a consequence, the stationary state $\mu = \mu_{\beta_1, \beta_n}$ sustains a non-vanishing heat flow in the direction from the hotter to the colder reservoir. Of course if $\beta_1 = \beta_n$ the heat flow vanishes since Φ_j is an odd function of p and the density of the stationary distribution is even in p. □

4.2 Fluctuation theorem

Let us consider now the part (b) of Theorem 4.1. Let us first give an outline of the proof. To study the large deviations of $t^{-1} \int_0^t \sigma_i(x_s) ds$ one considers moment generating function

$$\Gamma_x^j(t, \alpha) = \mathbf{E}_x \left[e^{-\alpha \int_0^t \sigma_j(x_s) ds} \right]. \qquad (168)$$

A formal application of Feynman-Kac formula gives

$$\frac{d}{dt} \mathbf{E}_x \left[e^{-\alpha \int_0^t \sigma_j(x_s) ds} f(x_t) \right] = (L - \alpha \sigma_j) \left[e^{-\alpha \int_0^t \sigma_j(x_s) ds} f(x_t) \right], \qquad (169)$$

but since is σ_j is not a bounded function, it is not clear that the expectation $\Gamma_x^j(t, \alpha)$ is even well defined. We will show below that there exists a neighborhood O of the interval $[0, 1]$ such that $\Gamma_x^j(t, \alpha)$ is well defined if $\alpha \in O$. We denote then $T_t^{(\alpha)}$ the semigroup with generator $(L - \alpha \sigma_j)$. We then have

$$\Gamma_x^j(t, \alpha) = \mathbf{E}_x \left[e^{-\alpha \int_0^t \sigma_j(x_s) ds} \right] = T_t^{(\alpha)} 1(x) \qquad (170)$$

Next one shows that the following limit

$$e(\alpha) \equiv \lim_{t \to \infty} -\frac{1}{t} \log \Gamma_x^j(t, \alpha), \qquad (171)$$

exists, is independent of x and j, and is a C^1 function of α. We will do this by a Perron-Frobenius like argument and identify $\exp(-te(\alpha))$ as the (real) eigenvalue of $T_t^{(\alpha)}$ with biggest modulus (on a suitable function space).

Then a standard and general argument of the theory of large deviations [2] (the Gärtner-Ellis Theorem) gives a large deviation principle for the ergodic average $t^{-1} \int_0^t \sigma_i(x_s) ds$ with a large deviation functional $e(w)$ which is given by the Legendre transform of the function $e(\alpha)$.

Formally, from Eq. (154) we see that $T_t^{(\alpha)}$ is conjugated to $(T_t^{(1-\alpha)})^*$, but since $T_t^{(\alpha)}$ has the same spectrum as $(T_t^{(\alpha)})^*$ we conclude that

$$e(\alpha) = e(1 - \alpha). \tag{172}$$

Taking now a Legendre transform we have

$$I(w) = \sup_{\alpha} \{e(\alpha) - \alpha w\} = \sup_{\alpha} \{e(1 - \alpha) - \alpha w\}$$
$$= \sup_{\beta} \{e(\beta) - (1 - \beta)w\} = I(-w) - w.$$

and this gives the part (b) of Theorem 4.1.

Let us explain how to make this argument rigorous, by making yet another conjugation.

Lemma 4.3. *We have the identity*

$$L - \alpha\sigma_j = e^{\alpha R_j} \overline{L}_\alpha e^{-\alpha R_j}, \tag{173}$$

where

$$\overline{L}_\alpha = \tilde{L}_\alpha - ((\alpha - \alpha^2)\gamma r T^{-1} r - \alpha \mathrm{tr}(\gamma I)) \tag{174}$$

and

$$\tilde{L}_\alpha = L + 2\alpha\gamma r \nabla_r. \tag{175}$$

Proof. As in Lemma 4.2 we write the generator L as $L = L_0 + L_1$, see Eqs.(156) and (155). Since L_1 is a first order differential operator we have

$$e^{-\alpha R_j} L_1 e^{\alpha R_j} = L_1 + \alpha(L_1 R_j) = L_1 + \alpha\sigma_j. \tag{176}$$

Using that $\nabla_r R_j = T^{-1} r$ is independent of j we find that

$$e^{-\alpha R_j} L_0 e^{\alpha R_j} = \gamma \left((\nabla_r + \alpha T^{-1} r) T (\nabla_r + \alpha T^{-1} r) - r(\nabla_r + \alpha T^{-1} r) \right)$$
$$= L_0 + \alpha\gamma(r\nabla_r + \nabla_r r) + (\alpha^2 - \alpha)\gamma r T^{-1} r$$
$$= L_0 + 2\alpha\gamma r \nabla_r + (\alpha^2 - \alpha)\gamma r T^{-1} r + \alpha \mathrm{tr} \gamma I. \tag{177}$$

Combining Eqs. (176) and (177) gives the desired result. □

The point of this computation is that it shows that $L - \alpha\sigma_i$ is conjugated to the operator $\overline{L}_\alpha$ which is independent of i. Furthermore $\overline{L}_\alpha$ has the form L plus terms which are quadratic in r and ∇_r. Combining Feynman-Kac and Girsanov formulas we can analyze the spectral properties of this operator by the same methods as the operator L. The basic identity here is as in Section 3.3.

$$\overline{L}_\alpha \exp \theta G(x) =$$

$$= \exp \theta G(x) \gamma \left[\operatorname{tr}(\theta T + \alpha I) + r(\theta^2 T - (1 - 2\alpha)\theta - \alpha(1 - \alpha)T^{-1})r \right]$$

$$\leq C \exp \theta G(x), \tag{178}$$

provided α and T_i, $i = 1, n$ satisfy the inequality

$$\theta^2 T_i - (1 - 2\alpha)\theta - \alpha(1 - \alpha)T_i^{-1} \leq 0, \tag{179}$$

or

$$-\alpha < \theta T_i < 1 - \alpha. \tag{180}$$

In particular we see that the semigroup $\overline{T}_t^{(\alpha)}$ defined by

$$\overline{T}_t^{(\alpha)} = e^{-\alpha R_j} T_t^{(\alpha)} e^{\alpha R_j} \tag{181}$$

and with generator $\overline{L}_\alpha$ is well defined on the Banach space $\mathcal{H}_\theta$ if $-\alpha < \theta T_i < 1 - \alpha$. Furthermore it has the following properties

1. **Strong-Feller property.** The semigroup $\overline{T}_t^{(\alpha)}$ has a kernel $p_t^{(\alpha)}(x, y)$ which is C^∞ in (t, x, y).
2. **Irreducibility.** For all $t > 0$, and all nonnegative f, $\overline{T}_t^{(\alpha)} f$ is positive.
3. **Liapunov function.** For any $t > 0$ and θ such that $-\alpha < \theta T_i < 1 - \alpha$, there exists functions $\kappa(E) = \kappa(E, \theta, t)$ and $b(E) = b(E, \theta, t)$ with $\lim_{E \to \infty} \kappa(E) = 0$ such that
$$\overline{T}_t^{(\alpha)} W_\theta(x) \leq \kappa(E) W_\theta(x) + b(E) 1_{G \leq E}(x). \tag{182}$$

These properties are proved exactly as in for the operator L, using in addition Girsanov and Feynman-Kac formula (see [22] for details).

As a consequence, by Theorem 8.9 of [20], we obtain that on $\mathcal{H}_\theta$, with $-\alpha < \theta T_i < 1 - \alpha$ the semigroup $\overline{T}_t^{(\alpha)}$ is a compact semigroup, it has exactly one eigenvalue with maximal modulus which, in addition is real. In particular $\overline{T}_t^{(\alpha)}$ has a spectral gap. We then obtain

Theorem 4.4. *If*

$$\alpha \in \left(-\frac{\beta_{\max}}{\beta_{\min} - \beta_{\max}}, 1 + \frac{\beta_{\max}}{\beta_{\min} - \beta_{\max}} \right), \tag{183}$$

then

$$e(\alpha) = \lim_{t \to \infty} -\frac{1}{t} \log \Gamma_x^j(t, \alpha) \tag{184}$$

exists, is finite and independent both of j and x.

Proof. The semigroup $\overline{T}_t^{(\alpha)}$ is well defined on $\mathcal{H}_\theta$ if

$$-\alpha < \theta T_i < 1 - \alpha. \tag{185}$$

A simple computation shows that for given α, β_1, and β_n the set of θ we can choose is non-empty provided if

$$\alpha \in \left(-\frac{\beta_{\max}}{\beta_{\min} - \beta_{\max}}, 1 + \frac{\beta_{\max}}{\beta_{\min} - \beta_{\max}} \right), \tag{186}$$

Using the definition of R_i, Eq. (152), $e^{-\alpha R_i} \in \mathcal{H}_\theta$ since $-\alpha + \theta T_i < 0$. Using now Lemma 2.7, we see that $\Gamma_x^i(t, \alpha)$ exists and is given by

$$\Gamma_x^i(t, \alpha) = T_t^{(\alpha)} 1(x) = e^{\alpha R_i} \overline{T}_t^{(\alpha)} e^{-\alpha R_i}(x). \tag{187}$$

From the spectral properties of $\overline{T}_t^{(\alpha)}$ we infer the existence of a one-dimensional projector P_α such that

1. $P_\alpha f > 0$ if $f \geq 0$
2. We have

$$\overline{T}_t^{(\alpha)} = e^{-te(\alpha)} P_\alpha + \overline{T}_t^{(\alpha)} (1 - P_\alpha), \tag{188}$$

and there exists a constants $d(\alpha) > e(\alpha)$ and C such that

$$\|\overline{T}_t^{(\alpha)}(1 - P_\alpha)\| \leq Ce^{-td(\alpha)}, \tag{189}$$

or, in other words,

$$|\overline{T}_t^{(\alpha)}(1 - P_\alpha)g| \leq Ce^{-td(\alpha)} \|g\|_\theta W_\theta(x). \tag{190}$$

From Lemma 4.3 and Eq. (190) we obtain, for all x, that

$$\lim_{t \to \infty} -\frac{1}{t} \log \Gamma_x^j(t, \alpha)$$

$$= \lim_{t \to \infty} -\frac{1}{t} \log e^{\alpha R_j} \overline{T}_t^{(\alpha)} e^{-\alpha R_j}(x)$$

$$= \lim_{t \to \infty} -\frac{1}{t} \log e^{\alpha R_j} e^{-te(\alpha) R_j} \left(P_\alpha e^{-\alpha R_j} + e^{te(\alpha)} \overline{T}_t^{(\alpha)} (1 - P_\alpha) e^{-\alpha R_j}(x) \right)$$

$$= \lim_{t \to \infty} -\frac{1}{t} (\alpha R_j(x) - te(\alpha) + $$

$$\log \left(P_\alpha e^{-\alpha R_i}(x) + e^{te(\alpha)} \overline{T}_t^{(\alpha)} (1 - P_\alpha) e^{-\alpha R_i}(x) \right) \right)$$

$$= e(\alpha).$$

This concludes the proof of Theorem 4.4. $\quad\square$

It is straightforward now to obtain the symmetry of the Gallavotti-Cohen fluctuation theorem

Theorem 4.5. *If*

$$\alpha \in \left(-\frac{T_{\min}}{T_{\max} - T_{\min}}, 1 + \frac{T_{\min}}{T_{\max} - T_{\min}} \right), \tag{191}$$

then

$$e(\alpha) = e(1 - \alpha). \tag{192}$$

Proof. Let us consider the dual semigroup $(\overline{T}_t^{(\alpha)})^*$ acting on $\mathcal{H}_\theta^*$. Since $\overline{T}_t^{(\alpha)}$ has a smooth kernel, $(\overline{T}_t^{(\alpha)})^*\nu$ is a measure with a smooth density, and we denote by $(\overline{T}_t^{(\alpha)})^*$ its action on densities

$$(\overline{T}_t^{(\alpha)})^*\nu(dx) = \left((\overline{T}_t^{(\alpha)})^*\rho(x)\right) dx. \tag{193}$$

Combining Lemmas 4.2 and 4.3 we have

$$\begin{aligned}
\overline{L}_\alpha &= e^{-\alpha R_j}(L - \alpha\sigma_j)e^{\alpha R_j} \\
&= e^{-(1-\alpha)R_j}J(L - (1-\alpha)\sigma_j)^*Je^{(1-\alpha)R_j} \\
&= J\left(e^{(1-\alpha)R_j}(L - (1-\alpha)\sigma_j)e^{-(1-\alpha)R_j}\right)^* J \\
&= J\overline{L}_{1-\alpha}{}^* J
\end{aligned} \tag{194}$$

or

$$\overline{T}_t^{(\alpha)} = J(\overline{T}_t^{(1-\alpha)})^* J. \tag{195}$$

The spectral radius formula concludes the proof of Theorem 4.5. □

Combining this fact with the formal argument given above, we obtain the proof of part (b) of Theorem 4.1.

4.3 Kubo Formula and Central Limit Theorem

One can derive the Kubo formula of linear response theory from the fluctuation theorem. Here the external "field" driving the system out of equilibrium is the inverse temperature difference $\eta = (\beta_n - \beta_1)$ and we have $\sigma_j = \eta\phi_j$. Instead of the function $e(\alpha)$, we consider a the function $f(a, \eta)$ given by

$$f(a, \eta) \equiv \lim_{t\to\infty} -\frac{1}{t}\log \mathbf{E}_\mu\left[e^{-a\int_0^t \phi_i(x(s))\,ds}\right], \tag{196}$$

where $a = \alpha\eta$ and the second variable in f indicates the dependence of the dynamics and of the stationary state μ on η. From our compactness results for the semigroup, one can show that $f(a, \eta)$ is a real-analytic function of both variables a and F. The relation $e(\alpha) = e(1 - \alpha)$ now reads

$$f(a, \eta) = f(\eta - a, \eta). \tag{197}$$

Differentiating this relation one finds

$$\frac{\partial^2 f}{\partial a\partial\eta}(0,0) = -\frac{\partial^2 f}{\partial a\partial\eta}(0,0) - \frac{\partial^2 f}{\partial a^2}(0,0). \tag{198}$$

and thus

$$\frac{\partial^2 f}{\partial a\partial\eta}(0,0) = -\frac{1}{2}\frac{\partial^2 f}{\partial a^2}(0,0). \tag{199}$$

This relation is indeed Kubo formula, although in a disguised form. Differentiating and using the stationarity we find

$$\frac{\partial f}{\partial a}(0, \eta) = \mathbf{E}_\mu \left[\frac{1}{t} \int_0^t \phi_j(x_s) ds \right] = \int \phi_j d\mu, \qquad (200)$$

and therefore

$$\frac{\partial^2 f}{\partial a \partial \eta}(0,0) = \frac{\partial}{\partial \eta} \left(\int \phi_j d\mu \right) \Big|_{\Delta\beta=0} \qquad (201)$$

is the derivative of the heat flow at equilibrium. On the other hand

$$\frac{\partial^2 f}{\partial a^2}(0, \eta)$$

$$= \lim_{t\to\infty} \mathbf{E}_\mu \left[\frac{1}{t} \int_0^t \phi_j(x_s) ds \right]^2 - \mathbf{E}_\mu \left[\frac{1}{t} \int_0^t \phi_j(x_s) ds \int_0^t \phi_j(x_u) du \right] \quad (202)$$

At equilibrium, $\eta = 0$, the first term vanishes since there is no heat flow at equilibrium. For the second term, we obtain, using stationarity, and changing variables

$$\frac{1}{t} \int_0^t ds \int_0^t du \mathbf{E}_\mu \left[\phi_j(x_s)\phi_j(x_u) \right]$$

$$= 2\frac{1}{t} \int_0^t ds \int_s^t du \mathbf{E}_\mu \left[\phi_j(x_s)\phi_j(x_u) \right]$$

$$= 2\frac{1}{t} \int_0^t ds \int_s^t du \mathbf{E}_\mu \left[\phi_j(x_0)\phi_j(x_{u-s}) \right]$$

$$= 2\frac{1}{t} \int_0^t ds \int_0^{t-s} du \mathbf{E}_\mu \left[\phi_j(x_0)\phi_j(x_u) \right]$$

$$= 2\frac{1}{t} \int_0^t ds \int_0^s du \mathbf{E}_\mu \left[\phi_j(x_0)\phi_j(x_u) \right] \qquad (203)$$

By Theorem 3.1 we obtain

$$\mathbf{E}_\mu \left[\phi_j(x_0)\phi_j(x_u) \right] = \int \phi_j(x) T_u \phi_j(x) \mu(dx) \leq C e^{-u\gamma} \| \phi_j^2 \|_{W\infty} \qquad (204)$$

and thus it is an integrable function of u. We then obtain

$$\frac{\partial^2 f}{\partial a^2}(0,0) = \lim_{t\to\infty} 2\frac{1}{t} \int_0^t ds \int_0^s du \mathbf{E}_\mu \left[\phi_j(x_0)\phi_j(x_u) \right]$$

$$= 2 \int_0^\infty ds \int \phi_j(x) T_u \phi_j(x) \mu(dx), \qquad (205)$$

is the integral of the flow autocorrelation function. Combining Eqs. (199), (201), and (205) we obtain

$$\frac{\partial}{\partial \eta} \left(\int \phi_j d\mu \right) \Bigg|_{\eta=0} = \int_0^\infty \left(\int (T_t \phi_j) \phi_j d\mu \right) ds, \qquad (206)$$

and this is the familiar Kubo formula. Note that this formula involves only the equilibrium dynamics and the equilibrium stationary distribution.

The appearance of an autocorrelation function is not fortuitous and can be interpreted in terms of the central limit theorem. With the strong ergodic properties we have established in Theorem 3.1, one can prove [19] a central limit theorem for any function f such that $f^2 \in \mathcal{H}_\theta$ (see the condition for exponential mixing in Theorem 3.1). For any such function we have that

$$\mathbf{P}_x \left\{ a < \frac{1}{\sqrt{\kappa^2 t}} \int_0^t \left(f(x_t) - \int f(x)\mu(dx) \right) ds < b \right\} \longrightarrow \frac{1}{\sqrt{2\pi}} \int_a^b e^{-\frac{y^2}{2}} dy \qquad (207)$$

provided the variance

$$\kappa^2 = \int_0^\infty \left(\int g(x) T_t g(x) \mu(dx) - (\int g(x)\mu(dx))^2 \right) \qquad (208)$$

does not vanish. In our case $f = \phi_j$, it follows from (206) and from the positivity of entropy production that κ^2 is positive.

References

1. Bonetto, F., Lebowitz J.L., and Rey-Bellet, L.: Fourier Law: A challenge to Theorists. In: *Mathematical Physics 2000*, Imp. Coll. Press, London 2000, pp. 128–150
2. Dembo, A. and Zeitouni, O.: *Large deviations techniques and applications*. Applications of Mathematics **38**. New-York: Springer-Verlag 1998
3. Dym H. and McKean, H.P.: *Gaussian processes, function theory, and the inverse spectral problem*. Probability and Mathematical Statistics, Vol. **31**, New York–London: Academic Press, 1976
4. Eckmann, J.-P. and Hairer, M.: Non-equilibrium statistical mechanics of strongly anharmonic chains of oscillators. Commun. Math. Phys. **212**, 105–164 (2000)
5. Eckmann, J.-P. and Hairer, M.: Spectral properties of hypoelliptic operators. Commun. Math. Phys. **235**, 233–253 (2003)
6. Eckmann, J.-P., Pillet C.-A., and Rey-Bellet, L.: Non-equilibrium statistical mechanics of anharmonic chains coupled to two heat baths at different temperatures. Commun. Math. Phys. **201**, 657–697 (1999)
7. Eckmann, J.-P., Pillet, C.-A., and Rey-Bellet, L.: Entropy production in non-linear, thermally driven Hamiltonian systems. J. Stat. Phys. **95**, 305–331 (1999)
8. Evans, D.J., Cohen, E.G.D., and Morriss, G.P.: Probability of second law violation in shearing steady flows. Phys. Rev. Lett. **71**, 2401–2404 (1993)
9. Ford, G.W., Kac, M. and Mazur, P.: Statistical mechanics of assemblies of coupled oscillators. J. Math. Phys. **6**, 504–515 (1965)
10. Gallavotti, G.: Chaotic hypothesis: Onsager reciprocity and fluctuation-dissipation theorem. J. Stat. Phys. **84**, 899–925 (1996)

11. Gallavotti, G. and Cohen E.G.D.: Dynamical ensembles in stationary states. J. Stat. Phys. **80**, 931–970 (1995)
12. Hérau, F. and Nier, F.: Isotropic hypoellipticity and trend to equilibrium for Fokker-Planck equation with high degree potential. Arch. Ration. Mech. Anal. **171**, 151–218 (2004)
13. Jakšić, V. and Pillet, C.-A.: Ergodic properties of classical dissipative systems. I. Acta Math. **181**, 245–282 (1998)
14. Komech, A., Kunze, M., and Spohn, H.: Long-time asymptotics for a classical particle interacting with a scalar wave field. Comm. Partial Differential Equations **22**, 307–335 (1997)
15. Kurchan, J: Fluctuation theorem for stochastic dynamics. J. Phys.**A 31**, 3719–3729 (1998)
16. Lebowitz, J.L. and Spohn, H.: A Gallavotti-Cohen-type symmetry in the large deviation functional for stochastic dynamics. J. Stat. Phys. **95**, 333-365 (1999)
17. Lepri, S., Livi, R., and Politi, A.: Thermal conduction in classical low-dimensional lattices. Phys. Rep. **377**, 1–80 (2003)
18. Maes, C.: The fluctuation theorem as a Gibbs property. J. Stat. Phys. **95**, 367–392 (1999)
19. Meyn, S.P. and Tweedie, R.L.: *Markov Chains and Stochastic Stability.* Communication and Control Engineering Series, London: Springer-Verlag London, 1993
20. Rey-Bellet, L.: Lecture notes on Ergodic properties of Markov processes. This volume.
21. Rey-Bellet, L. and Thomas, L.E.: Asymptotic behavior of thermal non-equilibrium steady states for a driven chain of anharmonic oscillators. Commun. Math. Phys. **215**, 1–24 (2000)
22. Rey-Bellet, L. and Thomas, L.E.: Exponential convergence to non-equilibrium stationary states in classical statistical mechanics. Commun. Math. Phys. **225**, 305–329 (2002)
23. Rey-Bellet, L. and Thomas, L.E.: Fluctuations of the entropy production in anharmonic chains. Ann. H. Poinc. **3**, 483–502 (2002)
24. Rieder, Z., Lebowitz, J.L., and Lieb, E.: Properties of a harmonic crystal in a stationary non-equilibrium state. J. Math. Phys. **8**, 1073–1085 (1967)
25. Spohn, H. and Lebowitz, J.L.: Stationary non-equilibrium states of infinite harmonic systems. Commun. Math. Phys. **54**, 97–120 (1977)

Quantum Noises

Stéphane Attal

Institut Camille Jordan, Université Claude Bernard Lyon 1,
21 av Claude Bernard, 69622 Villeurbanne Cedex, France
e-mail: attal@math.univ-lyon1.fr
URL: http://math.univ-lyon1.fr/~attal

1	**Introduction**	80
2	**Discrete time**	81
	2.1 Repeated quantum interactions	81
	2.2 The Toy Fock space	83
	2.3 Higher multiplicities	89
3	**Itô calculus on Fock space**	93
	3.1 The continuous version of the spin chain: heuristics	93
	3.2 The Guichardet space	94
	3.3 Abstract Itô calculus on Fock space	97
	3.4 Probabilistic interpretations of Fock space	105
4	**Quantum stochastic calculus**	110
	4.1 An heuristic approach to quantum noise	110
	4.2 Quantum stochastic integrals	113
	4.3 Back to probabilistic interpretations	122
5	**The algebra of regular quantum semimartingales**	123
	5.1 Everywhere defined quantum stochastic integrals	124
	5.2 The algebra of regular quantum semimartingales	127
6	**Approximation by the toy Fock space**	130
	6.1 Embedding the toy Fock space into the Fock space	130
	6.2 Projections on the toy Fock space	132
	6.3 Approximations	136
	6.4 Probabilistic interpretations	138
	6.5 The Itô tables	139
7	**Back to repeated interactions**	139

  7.1 Unitary dilations of completely positive semigroups 140
  7.2 Convergence to Quantum Stochastic Differential Equations 142

8   Bibliographical comments 145

References 145

1 Introduction

In the Markovian approach of quantum open systems, the environment acting on a simple quantum system is unknown, or is not being given a model. The only effective data that the physicists deal with is the evolution of the simple quantum system. This evolution shows up the fact that the system is not isolated and is dissipating.

One of the question one may ask then is wether one can give a model for the environment and its action that gives an accounct of this effective evolution. One way to answer that question is to describe the exterior system as a noise, a quantum random effect of the environment which perturbs the Hamiltonian evolution of the small system.

This approach is a quantum version of what has been exposed in L. Rey-Bellet's courses: a dissipative dynamical evolution on some system is represented as resulting of the evolution of a closed but larger system, in which part of the action is represented by a noise, a Brownian motion for example.

This is the aim of R. Rebolledo's course and F. Fagnola's course, following this one in this volume, to show up how the dissipative quantum systems can be dilated into a closed evolution driven by quantum noises. But before hands, the mathematical theory of these quantum noises needs to be developed. This theory is not an obvious extension of the classical theory of stochastic processes and stochastic integration. It needs its own developments, where the fact that we are dealing with unbounded operators calls for being very careful with domain constraints.

On the other hand, the quantum theory of noise is somehow easier than the classical one, it can be described in a very natural way, it contains very natural physical interpretations, it is deeply connected to the classical theory of noises. This is the aim of this course to develop the theory of quantum noises and quantum stochastic integration, to connect it with its classical counterpart, while trying to keep it connected with some physical intuition.

The intuitive construction of this theory and its final rules, such as the quantum Itô formula, are not very difficult to understand, but the whole precise mathematical theory is really much more difficult and subtle, it needs quite long and careful developments. We have tried to be as precise as possible in this course, the most important proofs are there, but we have tried to keep it reasonable in size and to always preserves the intuition all along the constructions, without getting lost in long expositions of technical details.

The theory of quantum noises and quantum stochastic integration was started in quantum physics with the notion of quantum Langevin equations (see for example [1], [21], [22]). They have been given many different meanings in terms of several definitions of quantum noises or quantum Brownian motions (for example [23], [25], [24]). One of the most developed and useful mathematical languages developed for that purpose is the quantum stochastic calculus of Hudson and Parthasarathy and their quantum stochastic differential equations ([25]). The quantum Langevin equations they allow to consider have been used very often to model typical situations of quantum open systems: continual quantum measurement ([12], [14]), quantum optics ([19], [20], [13]), electronic transport ([16]), thermalization ([8], [28], [27]), repeated quantum interactions ([8], [11]). This theory can be found much more developed in the books [2], [30] and [29].

The theory of quantum noises and quantum stochastic integration we present in this course is rather different from the original approach of Hudson and Parthasarathy. It is an extention of it, essentially developed by the author, which presents several advantages: it gives a maximal definition of quantum stochastic integrals in terms of domains, it admits a very intuitive approach in terms of discrete approximations with spin chains, it gives a natural language for connecting this quantum theory of noises to the classical one. This is the point of view we adopt all along this course, the main reference we follow here is [2].

2 Discrete time

2.1 Repeated quantum interactions

We first motivate the theory of quantum noises and quantum stochastic differential equations through a family of physical examples: the continuous time limit of repeated quantum interactions. This physical context is sufficiently wide to be of real interest in many applications, but it is far from being the only motivation for the introduction of quantum noises. We present it here for it appears to be an illuminating application in the context of these volumes. The approach presented in this section has been first developed in [11].

We consider a *small quantum system* $\mathcal{H}_0$ (a finite dimensional Hilbert space in this course, but the infinite dimensional case can also be handled) and another quantum system $\mathcal{H}$ which represents a piece of the environment: a measurement apparatus, an incoming photon, a particle ... or any other system which is going to interact with the small system. We assume that these two systems are coupled and interact during a small interval of time of length h. That is, on the space $\mathcal{H}_0 \otimes \mathcal{H}$ we have an Hamiltonian H which describes the interaction, the evolution is driven by the unitary operator $U = e^{ihH}$. An initial state $\rho \otimes \omega$ for the system is thus transformed into

$$U^*(\rho \otimes \omega)U.$$

After this time h the two systems are separated and another copy of $\mathcal{H}$ is presented before $\mathcal{H}_0$ in order to interact with it, following the same unitary operator U. And so on, for an arbitrary number of interactions. One can think of several sets of examples where this situation arises: in repeated quantum measurement, where a family of identical measurement devices is repeatedly presented before the system $\mathcal{H}_0$ (or one single device which is refreshened after every use); in quantum optics, where a sequence of independent atoms arrive one after the other to interact with $\mathcal{H}_0$ (a cavity with a strong electromagnetic field) for a short time; a particle is having a succession of chocs with a gas of other particles ...

In order to describe the first two interactions we need to consider the space $\mathcal{H}_0 \otimes \mathcal{H} \otimes \mathcal{H}$. We put U_1 to be the operator acting as U on the tensor product of $\mathcal{H}_0$ with the first copy of $\mathcal{H}$ and which acts as the identity on the second copy of $\mathcal{H}$. We put U_2 to be the operator acting as U on the tensor product of $\mathcal{H}_0$ with the second copy of $\mathcal{H}$ and which acts as the identity on the first copy of $\mathcal{H}$.

For an initial state $\rho \otimes \omega \otimes \omega$, say, the state after the first interaction is

$$U_1(\rho \otimes \omega \otimes \omega)U_1^*$$

and after the second interaction is

$$U_2U_1(\rho \otimes \omega \otimes \omega)U_1^*U_2^*.$$

It is now easy to figure out what the setup should be for an indefinite number of repeated interactions: we consider the state space

$$\mathcal{H}_0 \otimes \bigotimes_{I\!N} \mathcal{H},$$

(this countable tensor product will be made more precise later on). For every $n \in I\!N$, the operator U_n is the copy of the operator U but acting on the tensor product of $\mathcal{H}_0$ with the n-th copy of $\mathcal{H}$, it acts as the identity on the other copies of $\mathcal{H}$. Let

$$V_n = U_n \ldots U_2U_1,$$

then the result of the n-th measurement on the initial state

$$\rho \otimes \bigotimes_{n\in I\!N} \omega$$

is given by the state

$$V_n \left(\rho \otimes \bigotimes_{n\in I\!N} \omega \right) V_n^*.$$

Note that the V_n are solution of

$$V_{n+1} = U_{n+1}V_n,$$

with $V_0 = I$. This way, the V_n's describe the Hamiltonian evolution of the repeated quantum interactions. It is more exactly a time-dependent Hamiltonian evolution, it can also be seen as a Hamiltonian evolution in interaction picture.

We wish to pass to the limit $h \to 0$, that is, to pass to the limit from repeated interactions to continuous interactions. Our model of repeated interactions can be considered as a toy model for the interaction with a quantum field, we now want to pass to a more realistic model: a continuous quantum field.

We will not obtain a non trivial limit if no assumption is made on the Hamiltonian H. Clearly, it will need to satisfy some normalization properties with respect to the parameter h. As we will see later, this situation is somehow like for the central limit theorem: if one considers a Bernoulli random walk with time step h and if one tries to pass to the limit $h \to 0$ then one obtains 0; the only scale of normalization of the walk which gives a non trivial limit (namely the Brownian motion) is obtained when scaling the random walk by $\sqrt{h}$. Here, in our context we can wonder what are the scaling properties that the Hamiltonian should satisfy and what type of limit evolution we shall get for V_n.

Note that the evolution $(V_n)_{n \in \mathbb{N}}$ is purely Hamiltonian, in particular it is completely deterministic, the only ingredient here being the Hamiltonian operator H which drives everything in this setup.

At the end of this course, we will be able to give a surprising result: under some renormalization conditions on H, in the continuous limit, we obtain a limit evolution equation for $(V_t)_{t \in \mathbb{R}^+}$ which is a Schrödinger evolution perturbed by quantum noise terms, a *quantum Langevin equation*.

The point with that result is that it shows that these quantum noise terms are spontaneously produced by the limit equation and do not arise by an assumption or a model made on the interaction with the field. The limit quantum Langevin equation is really the effective continuous limit of the Hamiltonian description of the repeated quantum interactions.

We shall illustrate this theory with a very basic example. Assume $\mathcal{H}_0 = \mathcal{H} = \mathbb{C}^2$ that is, both are two-level systems with basis states Ω (the fundamental state) and X (the excited state). Their interaction is described as follows: if the states of the two systems are the same (both fondamental or both excited) then nothing happens, if they are different (one fundamental and the other one excited) then they can either be exchanged or stay as they are. Following this description, in the basis $\{\Omega \otimes \Omega, \Omega \otimes X, X \otimes \Omega, X \otimes X\}$ we take the unitary operator U to be of the form

$$\begin{pmatrix} 1 & 0 & 0 & 0 \\ 0 & \cos \alpha & -\sin \alpha & 0 \\ 0 & \sin \alpha & \cos \alpha & 0 \\ 0 & 0 & 0 & 1 \end{pmatrix}.$$

2.2 The Toy Fock space

The spin chain structure

We start with a description of the structure of the chain $\otimes_{\mathbb{N}} \mathcal{H}$ in the case where $\mathcal{H} = \mathbb{C}^2$. This is the simplest case, but it contains all the ideas. We shall later indicate how the theory is to be changed when $\mathcal{H}$ is larger (even infinite dimensional).

In every copy of $\mathbb{C}^2$ we choose the same orthonormal basis $\{\Omega, X\}$, representing fundamental or excited states. An orthonormal basis of the space $T\Phi = \otimes_{\mathbb{N}} \mathbb{C}^2$ is given by the set

$$\{X_A; A \in \mathcal{P}_{\mathbb{N}}\}$$

where $\mathcal{P}_{\mathbb{N}}$ is the set of finite subsets of $\mathbb{N}$ and X_A denotes the tensor product

$$X_{i_1} \otimes \ldots \otimes X_{i_n}$$

where $A = \{i_1, \ldots, i_n\}$ and the above vector means we took tensor products of X in each of copies number i_k with Ω in all the other copies. If $A = \emptyset$, we put $X_\emptyset = \Omega$, that is, the tensor product of Ω in each copy of $\mathbb{C}^2$. This is to say that the countable tensor product above has been constructed as associated to the stabilizing sequence $(\Omega)_{n \in \mathbb{N}}$.

Note that any element f of $T\Phi$ is of the form

$$f = \sum_{A \in \mathcal{P}_{\mathbb{N}}} f(A) X_A$$

with

$$\|f\|^2 = \sum_{A \in \mathcal{P}_{\mathbb{N}}} |f(A)|^2 < \infty.$$

The space $T\Phi$ defined this way is called the *Toy Fock space* .

This particular choice of a basis gives $T\Phi$ a particular structure. If we denote by $T\Phi_{i]}$ the space generated by the X_A such that $A \subset \{0, \ldots, i\}$ and by $T\Phi_{[j}$ the one generated by the X_A such that $A \subset \{j, j+1, \ldots\}$, we get an obvious natural isomorphism between $T\Phi$ and $T\Phi_{i-1]} \otimes T\Phi_{[i}$ given by

$$[f \otimes g](A) = f(A \cap \{0, \ldots, i-1\}) \, g(A \cap \{i, \ldots\}).$$

Operators on the spin chain

We consider the following basis of matrices on $\mathbb{C}^2$:

$$a^\times = \begin{pmatrix} 1 & 0 \\ 0 & 0 \end{pmatrix}$$

$$a^+ = \begin{pmatrix} 0 & 0 \\ 1 & 0 \end{pmatrix}$$

$$a^- = \begin{pmatrix} 0 & 1 \\ 0 & 0 \end{pmatrix}$$

$$a^\circ = \begin{pmatrix} 0 & 0 \\ 0 & 1 \end{pmatrix}.$$

For every $\varepsilon = \{\times, +, -, \circ\}$, we denote by a_n^ε the operator which acts on $T\Phi$ as a^ε on the copy number n of $\mathbb{C}^2$ and the identity elsewhere. On the basis X_A this gives

$$a_n^{\times} X_A = X_A \, \mathbb{1}_{n \notin A}$$

$$a_n^+ X_A = X_{A \cup \{n\}} \, \mathbb{1}_{n \notin A}$$

$$a_n^- X_A = X_{A \setminus \{n\}} \, \mathbb{1}_{n \in A}$$

$$a_n^{\circ} X_A = X_A \, \mathbb{1}_{n \in A}.$$

Note that the von Neumann algebra generated by all the operators a_n^{ε} is the whole of $\mathcal{B}(\mathbb{T\Phi})$, for there is no non-trivial subspace of $\mathbb{T\Phi}$ which is invariant under this algebra. But this kind of theorem does not help much to give an explicit representation of a given bounded operator H on $\mathbb{T\Phi}$ in terms of the operators a_n^{ε}. There are two concrete ways of representing an (eventually unbounded) operator on $\mathbb{T\Phi}$ in terms of these basic operators. The first one is a representation as a kernel

$$H = \sum_{\mathcal{P}^3} k(A, B, C) \, a_A^+ \, a_B^{\circ} \, a_C^-$$

where $a_A^{\varepsilon} = a_{i_1}^{\varepsilon} \ldots a_{i_n}^{\varepsilon}$ if $A = \{i_1 \ldots i_n\}$.

Note that the term $a^{\times}$ does not appear in the above kernel. The reason is that $a^{\times} + a^{\circ}$ is the identity operator and introducing the operator $a^{\times}$ in the above representation will make us lose the uniqueness of the above representation. Note that a° is not necessary either for it is equal to $a^+ a^-$. But if we impose the sets A, B, C to be two by two disjoint then the above representation is unique.

We shall not discuss much this kind of representation here, but better a different kind of representation (which one can derive from the above kernel representation by grouping the terms in 3 packets depending on which set A, B or C contains $\max A \cup B \cup C$). This is the so-called *integral representation*:

$$H = \sum_{\varepsilon = +, \circ, -} \sum_{i \in \mathbb{N}} H_i^{\varepsilon} \, a_i^{\varepsilon} \tag{1}$$

where the H_i^{ε} are operators acting on $\mathbb{T\Phi}_{i-1]}$ only (and as the identity on $\mathbb{T\Phi}_{[i}$). This kind of representation will be of great interest for us in the sequel.

For the existence of such a representation we have very mild conditions, even for unbounded H ([32]).

Theorem 2.1. *If the orthonormal basis $\{X_A, A \in \mathcal{P}_{\mathbb{N}}\}$ belongs to* $\mathrm{Dom}\, H \cap \mathrm{Dom}\, H^*$ *then there exists a unique integral representation of H of the form (1).*
□

One important point needs to be understood at that stage. The integral representation of a single operator H as in (1) makes use of only 3 of the four matrices a_i^{ε}. The reason is the same as for the kernel representation above: the sum $a_i^{\circ} + a_i^{\times}$ is the identity operator I, if we allow $a_i^{\times}$ to appear in the representation, we lose uniqueness. But, very often one has to consider processes of operators, that is, families $(H_i)_{i \in \mathbb{N}}$ of operators on $\mathbb{T\Phi}_{i]}$ respectively. In that case, the fourth family is necessary and we get representations of the form

$$H_i = \sum_{\varepsilon=+,\circ,-,\times} \sum_{j \leq i} H_j^\varepsilon a_j^\varepsilon. \qquad (2)$$

One interesting point with the integral representations is that they are stable under composition. The integral representation of a composition of integral representations is given by the *discrete quantum Itô formula*, which is almost straightforward if we forget about details on the domain of operators.

Theorem 2.2 (Discrete quantum Itô formula). *If*

$$H_i = \sum_{\varepsilon=+,\circ,-,\times} \sum_{j \leq i} H_j^\varepsilon a_j^\varepsilon$$

and

$$K_i = \sum_{\varepsilon=+,\circ,-,\times} \sum_{j \leq i} K_j^\varepsilon a_j^\varepsilon$$

are operators on $\mathbb{TP}_{i]}$, *indexed by* $i \in \mathbb{N}$, *then we have the following "integration by part formula":*

$$H_i K_i = \sum_{\varepsilon=+,\circ,-,\times} \sum_{j \leq i} H_{j-1} K_j^\varepsilon a_j^\varepsilon + \sum_{\varepsilon=+,\circ,-,\times} \sum_{j \leq i} H_j^\varepsilon K_{j-1} a_j^\varepsilon +$$
$$+ \sum_{\varepsilon,\nu=+,\circ,-,\times} \sum_{j \leq i} H_j^\varepsilon K_j^\nu a_j^\varepsilon a_j^\nu$$

where the products $a^\varepsilon a^\nu$ *are given by the following table*

	a^+	a^-	a°	$a^\times$
a^+	0	a°	0	a^+
a^-	$a^\times$	0	a^-	0
a°	a^+	0	a°	0
$a^\times$	0	a^-	0	$a^\times$

Note the following two particular cases of the above formula, which will be of many consequences for the probabilistic interpretations of quantum noises.

The Pauli matrix

$$\sigma_x = \begin{pmatrix} 0 & 1 \\ 1 & 0 \end{pmatrix} = a^+ + a^-$$

satisfies

$$\sigma_x^2 = 1. \qquad (3)$$

The matrix

$$X_\lambda = \sigma_x + \lambda a^\circ$$

satisfies

$$X_\lambda^2 = I + \lambda X_\lambda. \tag{4}$$

These two very simple matrix relations are actually the discrete version of the famous relations

$$(dW_t)^2 = dt$$

and

$$(dX_t)^2 = dt + \lambda \, dX_t$$

which characterise the Brownian motion $(W_t)_{t \geq 0}$ and the compensated Poisson process $(X_t)_{t \geq 0}$, with intensity λ. We shall give a rigourous meaning to these affirmations in section 6.4.

Probabilistic interpretations

In this section, we describe the *probabilistic interpretations* of the space $T\Phi$ and of its basic operators.

We realize a Bernoulli random walk on its canonical space. Let $\Omega = \{0,1\}^{I\!\!N}$ and $\mathcal{F}$ be the σ-field generated by finite cylinders. One denotes by ν_n the coordinate mapping : $\nu_n(\omega) = \omega_n$, for all $n \in I\!\!N$. Let $p \in \,]0,1[$ and $q = 1-p$. Let μ_p be the probability measure on $(\Omega, \mathcal{F})$ which makes the sequence $(\nu_n)_{n \in I\!\!N}$ a sequence of independent, identically distributed Bernoulli random variables with law $p\delta_1 + q\delta_0$. Let $I\!\!E_p[\,\cdot\,]$ denote the expectation with respect to μ_p. We have $I\!\!E_p[\nu_n] = I\!\!E_p[\nu_n^2] = p$. Thus the random variables

$$X_n = \frac{\nu_n - p}{\sqrt{pq}},$$

satisfy the following:

i) they are independent,

ii) they take the value $\sqrt{q/p}$ with probability p and $-\sqrt{p/q}$ with probability q,

iii) $I\!\!E_p[X_n] = 0$ and $I\!\!E_p[X_n^2] = 1$.

Let $T\Phi_p$ denote the space $L^2(\Omega, \mathcal{F}, \mu_p)$. We define particular elements of $T\Phi_p$ by

$$\begin{cases} X_\emptyset = 1\!\!1, & \text{in the sense } X_\emptyset(\omega) = 1 \text{ for all } \omega \in \Omega \\ X_A = X_{i_1} \cdots X_{i_n}, & \text{if } A = \{i_1 \ldots i_n\} \text{ is any finite subset of } I\!\!N. \end{cases}$$

Recall that $\mathcal{P}_{I\!\!N}$ denotes the set of finite subsets of $I\!\!N$. From *i)* and *iii)* above it is clear that $\{X_A \,;\, A \in \mathcal{P}_{I\!\!N}\}$ is an orthonormal set of vectors in $T\Phi_p$.

Proposition 2.3. *The family* $\{X_A; A \in \mathcal{P}_{I\!\!N}\}$ *is an orthonormal basis of* $T\Phi_p$.

Proof. We only have to prove that $\{X_A, A \in \mathcal{P}_{I\!\!N}\}$ forms a total set in $T\Phi_p$. In the same way as for the X_A, define

$$\begin{cases} \nu_\emptyset = 1\!\!1 \\ \nu_A = \nu_{i_1} \cdots \nu_{i_n} & \text{for } A = \{i_1 \ldots i_n\}. \end{cases}$$

It is sufficient to prove that the set $\{\nu_A\ ;\ A{\in}\mathcal{P}_{I\!N}\}$ is total. The space $(\Omega, \mathcal{F}, \mu_p)$ can be identified to $([0,1], \mathcal{B}([0,1]), \widetilde{\mu}_p)$ for some probability measure $\widetilde{\mu}_p$, via the base 2 decomposition of real numbers. Note that

$$\nu_n(\omega) = \omega_n = \begin{cases} 1 & \text{if } \omega_n = 1 \\ 0 & \text{if } \omega_n = 0 \end{cases}$$

thus $\nu_n(\omega) = 1\!\!1_{\omega_n=1}$. As a consequence $\nu_A(\omega) = 1\!\!1_{\omega_{i_1}=1} \cdots 1\!\!1_{\omega_{i_n}=1}$. Now let $f{\in}\mathbb{T}\!\Phi_p$ be such that $\langle f, \nu_A \rangle = 0$ for all $A{\in}\mathcal{P}_{I\!N}$. Let $I = \left[k2^{-n}, (k+1)2^{-n}\right]$ be a dyadic interval with $k < 2^n$. The base 2 decomposition of $k2^{-n}$ is of the form $(\alpha_1 \ldots \alpha_n, 0, 0, \ldots)$. Thus

$$\int_I f(\omega)\, d\widetilde{\mu}_p(\omega) = \int_{[0,1]} f(\omega) 1\!\!1_{\omega_1=\alpha_1} \cdots 1\!\!1_{\omega_n=\alpha_n}\, d\widetilde{\mu}_p(\omega)\,.$$

The function $1\!\!1_{\omega_1=\alpha_1} \cdots 1\!\!1_{\omega_n=\alpha_n}$ can be clearly written as a linear combination of the ν_A. Thus $\int_I f\, d\widetilde{\mu}_p = 0$. The integral of f vanishes on every dyadic interval, thus on all intervals. It is now easy to conclude that $f \equiv 0$. $\square$

We have proved that every element $f \in \mathbb{T}\!\Phi_p$ admits a unique decomposition

$$f = \sum_{A\in\mathcal{P}_{I\!N}} f(A) X_A \tag{5}$$

with

$$\|f\|^2 = \sum_{A\in\mathcal{P}_{I\!N}} |f(A)|^2 < \infty\,. \tag{6}$$

This means that there exists a natural isomorphism between $\mathbb{T}\!\Phi$ and $\mathbb{T}\!\Phi_p$ which consists in identifying the natural orthonormal basis $\{X_A; A \in \mathcal{P}_{I\!N}\}$ of both space. For each $p \in]0,1[$, the space $\mathbb{T}\!\Phi_p$ is called the *p-probabilistic interpretation* of $\mathbb{T}\!\Phi$. That is, it gives an interpretation of $\mathbb{T}\!\Phi$ in terms of a probabilistic space: it is the canonical space associated to the Bernoulli random walk with parameter p.

Identifying the basis element $X_{\{n\}}$ of $\mathbb{T}\!\Phi$ with the random variable $X_n \in \mathbb{T}\!\Phi_p$, as elements of some Hilbert spaces, does not give much information on the probabilistic nature of X_n. One cannot read this way the distribution of X_n or its independence with respect to other X_m's, ... The only way to represent the random variable $X_n \in \mathbb{T}\!\Phi_p$ with all its probabilistic structure, inside the structure of $\mathbb{T}\!\Phi$, is to consider the operator of *multiplication* by X_n acting on $\mathbb{T}\!\Phi_p$ and to represent it as a self-adjoint operator in $\mathbb{T}\!\Phi$ through the above natural isomorphism. When knowing the multiplication operator by X_n one knows all the probabilistic information on the random variable X_n. One cannot make the difference between the multiplication operator by X_n pushed on $\mathbb{T}\!\Phi$ and the "true" random variable X_n in $\mathbb{T}\!\Phi_p$.

Let us compute this multiplication operator by X_n. The way we have chosen the basis of $\mathbb{T}\!\Phi_p$ makes the product being determined by the value of X_n^2, $n{\in}I\!N$. Indeed, if $n \notin A$ then $X_n X_A = X_{A\cup\{n\}}$.

Proposition 2.4. *In* $\mathbb{T}\Phi_p$ *we have*

$$X_n^2 = 1 + c_p X_n$$

where $c_p = (q - p)/\sqrt{pq}$. *Furthermore* $p \mapsto c_p$ *is a one to one application from* $]0,1[$ *to* $\mathbb{R}$.

Proof.

$$X_n^2 = \frac{1}{pq}(\nu_n^2 + p^2 - 2p\nu_n) = \frac{1}{pq}\left(p^2 + (1-2p)\nu_n\right)$$

$$= \frac{1}{pq}\left(p^2 + (q-p)\nu_n\right) = 1 + \frac{p^2-qp}{qp} + \frac{q-p}{qp}\nu_n$$

$$= 1 - \frac{pc_p}{\sqrt{pq}} + \frac{c_p}{\sqrt{pq}}\nu_n = 1 + c_p\frac{\nu_n - p}{\sqrt{pq}}. \qquad \square$$

The above formula determines an associative product on $\mathbb{T}\Phi$ which is called the *p-product*. The operator of p-multiplication by X_n in $\mathbb{T}\Phi$ is the exact representation of the random variable X_n in the p-probabilistic interpretation. By means of all these p-multiplication operators we are able to put in a single structure a whole continuum of probabilistic situations that had no relation whatsoever: the canonical Bernoulli random walks with parameter p, for every $p \in]0,1[$. What's more we get a very simple represention of these multiplication operators.

Proposition 2.5. *The operator* $M_{X_n}^p$ *of p-multiplication by X_n on* $\mathbb{T}\Phi$ *is given by*

$$M_{X_n}^p = a_n^+ + a_n^- + c_p a_n^\circ.$$

Proof.

$$X_n X_A = X_{A \cup \{n\}} \mathbb{1}_{n \notin A} + X_{A \setminus \{n\}}(1 + c_p X_n) \mathbb{1}_{n \in A}$$

$$= a_n^+ X_A + a_n^- X_A + c_p a_n^\circ X_A. \qquad \square$$

This result is amazing in the sense that the whole continuum of different probabilistic situations, namely $\mathbb{T}\Phi_p$, $p \in]0,1[$, can be represented in $\mathbb{T}\Phi$ by means of very simple linear combinations of only 3 differents operators!

2.3 Higher multiplicities

In the case where $\mathcal{H}$ is not $\mathbb{C}^2$ but $\mathbb{C}^{N+1}$, or any separable Hilbert space, the above presentation is changed as follows. Let us consider the case $\mathbb{C}^{N+1}$ (the infinite dimensional case can be easily derived from it).

Each copy of $\mathbb{C}^{N+1}$ is considered with the same fixed orthonormal basis $\{\Omega, X_1, \dots,$ We shall sometimes write $X_0 = \Omega$. The space

$$\mathbb{T}\Phi = \bigotimes_{k \in \mathbb{N}} \mathbb{C}^{N+1}$$

has a natural orthonormal basis X_A indexed by the subsets

$$A = \{(n_1, i_1), \ldots, (n_k, i_k)\}$$

of $\mathbb{N} \otimes \{1, \ldots, N\}$, such that the n_j's are different. This is the so-called *Toy Fock space with multiplicity N*.

The basis for the matrices on $\mathbb{C}^{N+1}$ is the usual one:

$$a_j^i X_k = \delta_{ik} X_j$$

for all $i, j, k = 0, \ldots, N$. We also have their natural ampliations to $\mathbb{TФ}$: $a_j^i(k)$, $k \in \mathbb{N}$.

We now develop the probabilistic interpretations of the space $\mathbb{TФ}$ in the case of multiplicity higher than 1. Their structure is very rich and interesting, but it is not used in the rest of this course. The reader is advised to skip that part at first reading.

Let X be a random variable in $\mathbb{R}^N$ which takes exactly $N+1$ different values $v_1, \ldots, v_{N+1}$ with respective probability $\alpha_1, \ldots, \alpha_{N+1}$ (all different from 0 by hypothesis). We assume, for simplicity, that X is defined on its canonical space $(A, \mathcal{A}, P)$, that is, $A = \{1, \ldots, N+1\}$, $\mathcal{A}$ is the σ-field of subsets of A, the probability measure P is given by $P(\{i\}) = \alpha_i$ and X is given by $X(i) = v_i$, for all $i = 1, \ldots, N+1$.

Such a random variable X is called *centered and normalized* if $\mathbb{E}[X] = 0$ and $\text{Cov}(X) = I$.

A family of elements $v_1, \ldots, v_{N+1}$ of $\mathbb{R}^N$ is called an *obtuse system* if

$$< v_i, v_j > = -1$$

for all $i \neq j$.

We consider the coordinates $X_1, \ldots, X_N$ of X in the canonical basis of $\mathbb{R}^N$, together with the random variable Ω on $(A, \mathcal{A}, P)$ which is deterministic always equal to 1. We put $\widetilde{X}_i$ to be the random variable $\widetilde{X}_i(j) = \sqrt{\alpha_j} X_i(j)$ and $\widetilde{\Omega}(j) = \sqrt{\alpha_j}$. For any element $v = (a_1, \ldots, a_N)$ of $\mathbb{R}^N$ we put $\widehat{v} = (1, a_1, \ldots, a_N) \in \mathbb{R}^{N+1}$. The following proposition is rather straightforward and left to the reader.

Proposition 2.6. *The following assertions are equivalent.*

i) X is centered and normalized.

ii) The $(N+1) \times (N+1)$-matrix $(\widetilde{\Omega}, \widetilde{X}_1, \ldots, \widetilde{X}_N)$ is unitary.

iii) The $(N+1) \times (N+1)$-matrix $(\sqrt{\alpha_1}\,\widehat{v}_1, \ldots, \sqrt{\alpha_{N+1}}\,\widehat{v}_{N+1})$ is unitary.

iv) The family $v_1, \ldots, v_{N+1}$ is an obtuse system of $\mathbb{R}^N$ and

$$\alpha_i = \frac{1}{1 + ||v_i||^2}.$$

Let T be a 3-tensor in $\mathbb{R}^N$, that is (at least, this is the way we interpret them here), a linear mapping from $\mathbb{R}^N$ to $M_N(\mathbb{R})$. We denote by T_k^{ij} the coefficients of T in the canonical basis of $\mathbb{R}^N$, that is,

$$(T(x))_{i,j} = \sum_{k=1}^{N} T_k^{ij} x_k.$$

Such a 3-tensor T is called *sesqui-symmetric* if

i) $(i, j, k) \longmapsto T_k^{ij}$ is symmetric

ii) $(i, j, l, m) \longmapsto \sum_k T_k^{ij} T_k^{lm} + \delta_{ij}\delta_{lm}$ is symmetric.

Theorem 2.7. *If X is a centered and normalized random variable in $\mathbb{R}^N$, taking exactly $N+1$ values, then there exists a unique sesqui-symmetric 3-tensor T such that*

$$X \otimes X = I + T(X). \tag{7}$$

Proof. By Proposition 2.6, the matrix $(\sqrt{\alpha_1}\,\widehat{v}_1, \ldots, \sqrt{\alpha_{N+1}}\,\widehat{v}_{N+1})$ is unitary. In particular the matrix $(\widehat{v}_1, \ldots, \widehat{v}_{N+1})$ is invertible. But the lines of this matrix are the values of the random variables $\Omega, X_1, \ldots, X_N$. As a consequence, these $N+1$ random variables are linearly independent. They thus form a basis of $L^2(A, \mathcal{A}, P)$ which is a $N+1$ dimensional space.

The random variable $X_i X_j$ belongs to $L^2(A, \mathcal{A}, P)$ and can thus be written uniquely as

$$X_i X_j = \sum_{k=0}^{N} T_k^{ij} X_k$$

where X_0 denotes Ω and for some real coefficients T_k^{ij}, $k = 0, \ldots, N$, $i, j = 1, \ldots N$. The fact that $\mathbb{E}[X_k] = 0$ and $\mathbb{E}[X_i X_j] = \delta_{ij}$ implies $T_0^{ij} = \delta_{ij}$. This gives the representation (7).

The fact that the 3-tensor T associated to the above coefficients T_k^{ij}, $i, j, k = 1, \ldots N$, is sesqui-symmetric is an easy consequence of the fact that the expressions $X_i X_j$ are symmetric in i, j and $X_i(X_j X_m) = (X_i X_j) X_m$ for all i, j, m. We leave this to the reader. $\square$

The following theorem is an interesting characterization of the sesqui-symmetric tensors. The proof of this result is far from obvious, but as we shall not need it we omit the proof and convey the interested reader to read the proof in [6], Theorem 2, p. 268-272.

Theorem 2.8. *The formulas*

$$S = \{x \in \mathbb{R}^N ; x \otimes x = I + T(x)\}.$$

and

$$T(y) = \sum_{x \in S} p_x <x\,,\,y> x \otimes x,$$

where $p_x = 1/(1 + ||x||^2)$, *define a bijection between the set of sesqui-symmetric 3-tensor T on $\mathbb{R}^N$ and the set of obtuse systems S in $\mathbb{R}^N$.*

Now we wish to consider the random walks (or more exactly the sequences of independent copies of induced by obtuse random variables). That is, on the probability space $(A^{\mathbb{N}}, \mathcal{A}^{\otimes \mathbb{N}}, P^{\otimes \mathbb{N}})$, we consider a sequence $(X(n))_{n \in \mathbb{N}^*}$ of independent random variables with the same law as a given obtuse random variable X (once again, the use of the terminology "random walk" is not correct here in the sense that it usually refers to the *sum* of these independent random variables, but we shall anyway use it here as it is shorter and essentially means the same).

For any $A \in \mathcal{P}_n$ we define the random variable

$$X_A = \prod_{(n,i) \in A} X_i(n)$$

with the convention

$$X_\emptyset = \mathbb{1}.$$

Proposition 2.9. *The family $\{X_A; A \in \mathcal{P}_{\mathbb{N}}\}$ forms an orthonormal basis of the space $L^2(A^{\mathbb{N}}, \mathcal{A}^{\otimes \mathbb{N}}, P^{\otimes \mathbb{N}})$.*

Proof. For any $A, B \in \mathcal{P}_n$ we have

$$< X_A, X_B > = \mathbb{E}[X_A X_B] = \mathbb{E}[X_{A \Delta B}] \mathbb{E}[X^2_{A \cap B}]$$

by the independence of the $X(n)$. For the same reason, the first term $\mathbb{E}[X_{A \Delta B}]$ gives 0 unless $A \Delta B = \emptyset$, that is $A = B$. The second term $\mathbb{E}[X^2_{A \cap B}]$ is then equal to $\prod_{(n,i) \in A} \mathbb{E}[X_i(n)^2] = 1$. This proves the orthonormal character of the family $\{X_A; A \in \mathcal{P}_n\}$.

Let us now prove that it generates a dense subspace of $L^2(A^{\mathbb{N}}, \mathcal{A}^{\otimes \mathbb{N}}, P^{\otimes \mathbb{N}})$. Had we considered random walks indexed by $\{0, \ldots, M\}$ instead of $\mathbb{N}$, it would be clear that the $X_A, A \subset \{0, \ldots, M\}$ form an orthonormal basis of $L^2(A^M, \mathcal{A}^{\otimes M}, P^{\otimes M})$, for the dimensions coincide. Now a general element f of $L^2(A^{\mathbb{N}}, \mathcal{A}^{\otimes \mathbb{N}}, P^{\otimes \mathbb{N}})$ can be easily approximated by a sequence $(f_M)_M$ such that $f_M \in L^2(A^M, \mathcal{A}^{\otimes M}, P^{\otimes M})$, for all M, by taking conditional expectations on the trajectories of X up to time M. $\square$

For every obtuse random variable X, we thus obtain a Hilbert space $\mathbb{T}\Phi(X) = L^2(A^{\mathbb{N}}, \mathcal{A}^{\otimes \mathbb{N}}, P^{\otimes \mathbb{N}})$, with a natural orthonormal basis $\{X_A; A \in \mathcal{P}_{\mathbb{N}}\}$ which emphasizes the independence of the $X(n)$'s. In particular there is a natural isomorphism between all the spaces $\mathbb{T}\Phi(X)$ which consists in identifying the associated bases. In the same way, all these canonical spaces $\mathbb{T}\Phi(X)$ of obtuse random walks are naturally isomorphic to the Toy Fock space $\mathbb{T}\Phi$ with multiplicity N (again by identifying their natural orthonormal bases).

In the same way as in multiplicity 1 we compute the representation of the multiplication operator by $X_i(k)$ in $\mathbb{T}\Phi$.

Theorem 2.10. Let X be an obtuse random variable, let $(X(k))_{k \in \mathbb{N}}$ be the associated random walk on the canonical space $T\Phi(X)$. Let T be the sesqui-symmetric 3-tensor associated to X by Theorem 2.7. Let U be the natural unitary isomorphism from $T\Phi(X)$ to $T\Phi$. Then, for all $k \in \mathbb{N}, i = \{1, \dots, n\}$ we have

$$U \mathcal{M}_{X_i(k)} U^* = a_i^0(k) + a_0^i(k) + \sum_{j,l} T_i^{jl} a_l^j(k).$$

Proof. It suffices to compute the action of $X_i(k)$ on the basis elements X_A, $A \in \mathcal{P}_n$. We get

$$X_i(k)X_A = \mathbb{1}_{(k,\cdot) \notin A} X_i(k)X_A + \sum_j \mathbb{1}_{(k,j) \in A} X_i(k)X_A$$

$$= \mathbb{1}_{(k,\cdot) \notin A} X_{A \cup \{(k,i)\}} + \sum_j \mathbb{1}_{(k,j) \in A} X_i(k)X_j(k)X_{A \setminus \{(k,j)\}}$$

$$= \mathbb{1}_{(k,\cdot) \notin A} X_{A \cup \{(k,i)\}} + \sum_j \mathbb{1}_{(k,j) \in A} (\delta_{ij} + \sum_l T_l^{ij} X_l(k)) X_{A \setminus \{(k,j)\}}$$

$$= \mathbb{1}_{(k,\cdot) \notin A} X_{A \cup \{(k,i)\}} + \mathbb{1}_{(k,i) \in A} X_{A \setminus (k,i)} +$$
$$+ \sum_j \sum_l \mathbb{1}_{(k,j) \in A} T_l^{ij} X_{A \setminus \{(k,j)\} \cup \{(k,i)\}}$$

and we recognize the formula for

$$a_i^0(k)X_A + a_0^i(k)X_A + \sum_{k,l} T_l^{ij} a_l^j(k)X_A. \qquad \square$$

This ends the section on the discrete time setting for quantum noises. We shall come back to it later when using it to approximate the Fock space structure.

3 Itô calculus on Fock space

3.1 The continuous version of the spin chain: heuristics

We now present the structure of the continuous version of $T\Phi$. By a continuous version of the spin chain we mean a Hilbert space which should be of the form

$$\Phi = \bigotimes_{\mathbb{R}^+} \mathbb{C}^2.$$

We first start with a heuristical discussion in order to make out an idea of how this space should be defined. We mimick, in a continuous time version, the structure of $T\Phi$.

The countable orthonormal basis $X_A, A \in \mathcal{P}_{\mathbb{N}}$ is replaced by a continuous orthonormal basis $d\chi_\sigma, \sigma \in \mathcal{P}$, where $\mathcal{P}$ is the set of finite subsets of $\mathbb{R}^+$. With the same idea as for $T\!\Phi$, this means that each copy of $\mathbb{C}^2$ is equipped with an orthonormal basis $\Omega, d\chi_t$ (where t is the parameter attached to the copy we are looking at). The orthonormal basis $d\chi_\sigma$ is the one obtained by specifying a finite number of sites $t_1, \ldots, t_n$ which are going to be excited, the other ones being in the fundamental state Ω.

The representation of an element f of $T\!\Phi$:

$$f = \sum_{A \in \mathcal{P}_N} f(A)\, X_A$$

$$\|f\|^2 = \sum_{A \in \mathcal{P}_N} |f(A)|^2$$

is replaced by an integral version of it in Φ:

$$f = \int_{\mathcal{P}} f(\sigma)\, d\chi_\sigma,$$

$$\|f\|^2 = \int_{\mathcal{P}} |f(\sigma)|^2\, d\sigma,$$

where, in the last integral, the measure $d\sigma$ is a "Lebesgue measure" on $\mathcal{P}$, that we shall explain later.

A good basis of operators acting on Φ can be obtained by mimicking the operators a_n^ε of $T\!\Phi$. Here we have a set of infinitesimal operators da_t^ε acting on the copy t of $\mathbb{C}^2$ by

$$da_t^\times \Omega = dt\, \Omega \qquad \text{and} \qquad da_t^\times d\chi_t = 0,$$

$$da_t^+ \Omega = d\chi_t \qquad \text{and} \qquad da_t^+ d\chi_t = 0,$$

$$da_t^- \Omega = 0 \qquad \text{and} \qquad da_t^- d\chi_t = dt\, \Omega,$$

$$da_t^\circ \Omega = 0 \qquad \text{and} \qquad da_t^\circ d\chi_t = d\chi_t.$$

In the basis $d\chi_\sigma$, this means

$$da_t^\times d\chi_\sigma = d\chi_\sigma\, dt\, \mathbb{1}_{t \notin \sigma}$$

$$da_t^+ d\chi_\sigma = d\chi_{\sigma \cup \{t\}}\, \mathbb{1}_{t \notin \sigma}$$

$$da_t^- d\chi_\sigma = d\chi_{\sigma \setminus \{t\}}\, dt\, \mathbb{1}_{t \in \sigma}$$

$$da_t^\circ d\chi_\sigma = d\chi_\sigma\, \mathbb{1}_{t \in \sigma}.$$

3.2 The Guichardet space

We now describe a setting in which the above heuristic discussion is made rigorous.

Notations

Let $\mathcal{P}$ denote the set of finite subsets of $\mathbb{R}^+$. That is, $\mathcal{P} = \cup_n \mathcal{P}_n$ where $\mathcal{P}_0 = \{\emptyset\}$ and $\mathcal{P}_n$ is the set of n elements subsets of $\mathbb{R}^+$, $n \geq 1$. By ordering elements of a $\sigma = \{t_1, t_2 \ldots t_n\} \in \mathcal{P}_n$ we identify $\mathcal{P}_n$ with $\Sigma_n = \{0 < t_1 < t_2 < \cdots < t_n\} \subset (\mathbb{R}^+)^n$. This way $\mathcal{P}_n$ inherits the measured space structure of $(\mathbb{R}^+)^n$. By putting the Dirac measure $\delta_\emptyset$ on $\mathcal{P}_0$, we define a σ-finite measured space structure on $\mathcal{P}$ (which coincides with the n-dimensional Lebesgue measure on each $\mathcal{P}_n$) whose only atom is $\{\emptyset\}$. The elements of $\mathcal{P}$ are denoted with small greek letters $\sigma, \omega, \tau, \ldots$ the associated measure is denoted $d\sigma, d\omega, d\tau \ldots$ (with in mind that $\sigma = \{t_1 < t_2 < \cdots < t_n\}$ and $d\sigma = dt_1 dt_2 \cdots dt_n$).

The space $L^2(\mathcal{P})$ defined this way is naturally isomorphic to the symmetric Fock space $\Phi = \Gamma_s(L^2(\mathbb{R}^+))$. Indeed, $L^2(\mathcal{P}) = \bigoplus_n L^2(\mathcal{P}_n)$ is isomorphic to $\bigoplus_n L^2(\Sigma_n)$ (with $\Sigma_0 = \{\emptyset\}$) that is Φ by identifying the space $L^2(\Sigma_n)$ to the space of symmetric functions in $L^2((\mathbb{R}^+)^n)$. In order to be really clear, the isomorphism between Φ and $L^2(\mathcal{P})$ can be explicitly written as:

$$V : \Phi \longrightarrow L^2(\mathcal{P})$$
$$f \longmapsto Vf$$

where $f = \sum_n f_n$ and

$$[Vf](\sigma) = \begin{cases} f_0 & \text{if } \sigma = \emptyset \\ f_n(t_1 \ldots t_n) & \text{if } \sigma = \{t_1 < \cdots < t_n\}. \end{cases}$$

Let us fix some notations on $\mathcal{P}$. If $\sigma \neq \emptyset$ we put $\vee\sigma = \max \sigma$, $\sigma- = \sigma \setminus \{\vee\sigma\}$. If $t \in \sigma$ then $\sigma \setminus t$ denotes $\sigma \setminus \{t\}$. If $\{t \notin \sigma\}$ then $\sigma \cup t$ denotes $\sigma \cup \{t\}$. If $0 \leq s \leq t$ then

$$\sigma_{s)} = \sigma \cap [0, s[$$
$$\sigma_{(s,t)} = \sigma \cap]s, t[$$
$$\sigma_{(t} = \sigma \cap]t, +\infty[.$$

We also put

$$\mathbb{1}_{\sigma \leq t} = \begin{cases} 1 & \text{if } \sigma \subset [0, t] \\ 0 & \text{otherwise.} \end{cases}$$

If $0 \leq s \leq t$ then

$$\mathcal{P}^{s)} = \{\sigma \in \mathcal{P}; \sigma \subset [0, s[\}$$
$$\mathcal{P}^{(s,t)} = \{\sigma \in \mathcal{P}; \sigma \subset]s, t[\}$$
$$\mathcal{P}^{(t} = \{\sigma \in \mathcal{P}; \sigma \subset]t, +\infty[\} .$$

Finally, $\#\sigma$ denotes the cardinal of σ.

If we put $\Phi_{t]} = \Gamma_s(L^2([0,t]))$, $\Phi_{[t} = \Gamma_s(L^2([t,+\infty[)$ and so on ... we clearly have

$$\Phi_{s]} \simeq L^2(\mathcal{P}^{s)})$$
$$\Phi_{[s,t]} \simeq L^2(\mathcal{P}^{(s,t)})$$
$$\Phi_{[t} \simeq L^2(\mathcal{P}^{(t)}) .$$

In the following we make several identifications:
 • Φ is not distinguished from $L^2(\mathcal{P})$ (and the same holds for $\Phi_{s]}$ and $L^2(\mathcal{P}^{s)})$, etc...)
 • $L^2(\mathcal{P}^{s)})$, $L^2(\mathcal{P}^{(s,t)})$ and $L^2(\mathcal{P}^{(t)})$ are seen as subspaces of $L^2(\mathcal{P})$: the subspace of $f \in L^2(\mathcal{P})$ such that $f(\sigma) = 0$ for all σ such that $\sigma \not\subset [0,s]$ (resp. $\sigma \not\subset [s,t]$, resp. $\sigma \not\subset [t,+\infty[$).

A particular family of elements of Φ is of great use: the space of *coherent vectors*. For every $h \in L^2(\mathbb{R}^+)$, consider the element $\varepsilon(h)$ of Φ defined by

$$[\varepsilon(h)](\sigma) = \prod_{s \in \sigma} h(s)$$

with the convention that the empty product is equal to 1. They satisfy the relation

$$\langle \varepsilon(h) , \varepsilon(k) \rangle = e^{\langle h , k \rangle}.$$

The linear space $\mathcal{E}$ generated by these vectors is dense in Φ and any finite family of distinct coherent vectors is linearly free.

If $\mathcal{M}$ is any dense subset of $L^2(\mathbb{R}^+)$ then $\mathcal{E}(\mathcal{M})$ denotes the linear space spaned by the vectors $\varepsilon(h)$ such that $h \in \mathcal{M}$. This forms a dense subspace of Φ.

The *vacuum* element of Φ is the element Ω given by

$$\Omega(\sigma) = \mathbb{1}_{\sigma=\emptyset}.$$

The Σ-Lemma

The following lemma is a very important and useful combinatoric result that we shall use quite often in the sequel.

Theorem 3.1 (Σ-Lemma). *Let f be a measurable positive (resp. integrable) function on $\mathcal{P} \times \mathcal{P}$. Define a function g on $\mathcal{P}$ by*

$$g(\sigma) = \sum_{\alpha \subset \sigma} f(\alpha, \sigma \setminus \alpha) .$$

Then g is measurable positive (resp. integrable) and

$$\int_{\mathcal{P}} g(\sigma) \, d\sigma = \int_{\mathcal{P} \times \mathcal{P}} f(\alpha, \beta) \, d\alpha \, d\beta .$$

Proof. By density arguments one can restrict ourselves to the case where $f(\alpha, \beta) = h(\alpha)k(\beta)$ and where $h = \varepsilon(u)$ and $k = \varepsilon(v)$ are coherent vectors. In this case one has

$$\int_{\mathcal{P}\times\mathcal{P}} f(\alpha,\beta)\, d\alpha\, d\beta = \int_{\mathcal{P}} \varepsilon(u)(\alpha)\, d\alpha \int_{\mathcal{P}} \varepsilon(v)(\beta)\, d\beta$$

$$= e^{\int_0^\infty u(s)\, ds} e^{\int_0^\infty v(s)\, ds} \quad \text{(take } u, v \in L^1 \cap L^2(\mathbb{R}^+))$$

and

$$\int_{\mathcal{P}} \sum_{\alpha \subset \sigma} f(\alpha, \sigma \setminus \alpha)\, d\sigma = \int_{\mathcal{P}} \sum_{\alpha \subset \sigma} \prod_{s \in \alpha} u(s) \prod_{s \in \sigma \setminus \alpha} v(s)\, d\sigma$$

$$= \int_{\mathcal{P}} \prod_{s \in \sigma}(u(s) + v(s))\, d\sigma = e^{\int_0^\infty u(s)+v(s)\, ds}. \qquad \square$$

In the same way as for the Toy Fock space we have a natural isomorphism between Φ and $\Phi_{t]} \otimes \Phi_{[t}$.

Theorem 3.2. *The mapping:*

$$\Phi_{t]} \otimes \Phi_{[t} \longrightarrow \Phi$$
$$f \otimes g \longmapsto h$$

where $h(\sigma) = f(\sigma_{t)})g(\sigma_{(t)})$ *defines an isomorphism between* $\Phi_{t]} \otimes \Phi_{[t}$ *and* Φ.

Proof.

$$\int_{\mathcal{P}} |h(\sigma)|^2\, d\sigma = \int_{\mathcal{P}} |f(\sigma_{t)})|^2 |g(\sigma_{(t)})|^2\, d\sigma$$

$$= \int_{\mathcal{P}} \sum_{\alpha \subset \sigma} \mathbb{1}_{\alpha \subset [0,t]} \mathbb{1}_{\sigma \setminus \alpha \subset [t,+\infty[} |f(\alpha)|^2 |g(\sigma \setminus \alpha)|^2\, d\sigma$$

$$= \int_{\mathcal{P}} \int_{\mathcal{P}} \mathbb{1}_{\alpha \subset [0,t]} \mathbb{1}_{\beta \subset [t,+\infty[} |f(\alpha)|^2 |g(\beta)|^2\, d\alpha\, d\beta$$

$$\text{(by the } \mathcal{I}\text{-Lemma)}$$

$$= \int_{\mathcal{P}_{t)}} |f(\alpha)|^2\, d\alpha \int_{\mathcal{P}_{(t}} |g(\beta)|^2\, d\beta$$

$$= \|f \otimes g\|^2. \qquad \square$$

3.3 Abstract Itô calculus on Fock space

We are now ready to define the main ingredients of our structure: several differential and integral operators on the Fock space.

Projectors

For all $t > 0$ define the operator P_t from Φ to Φ by

$$[P_t f](\sigma) = f(\sigma) \mathbb{1}_{\sigma \subset [0,t]}.$$

It is clear that P_t is the orthogonal projector from Φ onto $\Phi_{t]}$.
  For $t = 0$ we define P_0 by

$$[P_0 f](\sigma) = f(\emptyset) \mathbb{1}_{\sigma = \emptyset}$$

which is the orthogonal projection onto $L^2(P_0) = \mathbb{C}\mathbb{1}$ where $\mathbb{1}$ is the vacuum of Φ:
$(\mathbb{1}(\sigma) = \mathbb{1}_{\sigma = \emptyset})$.

Gradients

For all $t \in \mathbb{R}^+$ and all f in Φ define the following function on $\mathcal{P}$:

$$[D_t f](\sigma) = f(\sigma \cup t) \mathbb{1}_{\sigma \subset [0,t]}.$$

The first natural question is: for which f does $D_t f$ lie in $\Phi = L^2(\mathcal{P})$?

Proposition 3.3. *For all $f \in \Phi$, we have*

$$\int_0^\infty \int_{\mathcal{P}} |[D_t f](\sigma)|^2 \, d\sigma \, dt = \|f\|^2 - |f(\emptyset)|^2.$$

Proof. This is again an easy application of the $\mathfrak{Z}$-Lemma:

$$\int_0^\infty \int_{\mathcal{P}} |f(\sigma \cup t)|^2 \mathbb{1}_{\sigma \subset [0,t]} \, d\sigma \, dt$$

$$= \int_{\mathcal{P}} \int_{\mathcal{P}} |f(\alpha \cup \beta)|^2 \mathbb{1}_{\#\beta = 1} \mathbb{1}_{\alpha \subset [0, \vee \beta]} \, d\alpha \, d\beta$$

$$= \int_{\mathcal{P}} \sum_{\alpha \subset \sigma} |f(\alpha \cup \sigma \setminus \alpha)|^2 \mathbb{1}_{\#(\sigma \setminus \alpha) = 1} \mathbb{1}_{\alpha \subset [0, \vee(\sigma \setminus \alpha)]} \, d\sigma$$

$$= \int_{\mathcal{P} \setminus \mathcal{P}_0} \sum_{t \in \sigma} |f(\sigma)|^2 \mathbb{1}_{\sigma \setminus t \subset [0,t]} \, d\sigma \quad \text{(this forces } t \text{ to be } \vee \sigma \text{)}$$

$$= \int_{\mathcal{P} \setminus \mathcal{P}_0} |f(\sigma)|^2 \, d\sigma = \|f\|^2 - |f(\emptyset)|^2. \qquad \square$$

This proposition implies the following: for all f in Φ, for almost all $t \in \mathbb{R}^+$ (the negligible set depends on f), the function $D_t f$ belongs to $L^2(\mathcal{P})$. Hence for *all* f in Φ, *almost all* t, $D_t f$ is an element of Φ. Nevertheless, D_t is not a well-defined operator from Φ to Φ. The only operators which can be well defined are either

$$D : L^2(\mathcal{P}) \longrightarrow L^2(\mathcal{P} \times \mathbb{R}^+)$$
$$f \longmapsto ((\sigma, t) \mapsto D_t f(\sigma))$$

which is a partial isometry; or the regularised operators D_h, for $h \in L^2(\mathbb{R}^+)$:

$$[D_h f](\sigma) = \int_0^\infty h(t)[D_t f](\sigma)\, dt.$$

But, anyway, in this course we will treat the D_t's as linear operators defined on the whole of $\varPhi$. This, in general, poses no problem; one just has to be careful in some particular situations.

Integrals

A family $(g_t)_{t \geq 0}$ of elements of $\varPhi$ is said to be an *Itô integrable process* if the following holds:

  i) $f \mapsto \|g_t\|$ is measurable

  ii) $g_t \in \varPhi_{t]}$ for all t

  iii) $\int_0^\infty \|g_t\|^2\, dt < \infty$.

If $g. = (g_t)_{t \geq 0}$ is an Itô integrable process, define

$$[\mathcal{I}(g.)](\sigma) = \begin{cases} 0 & \text{if } \sigma = \emptyset \\ g_{\vee \sigma}(\sigma-) & \text{if } \sigma \neq \emptyset. \end{cases}$$

Proposition 3.4. *For all Itô integrable process* $g. = (g_t)_{t \geq 0}$ *one has*

$$\int_\mathcal{P} |[\mathcal{I}(g.)](\sigma)|^2\, d\sigma = \int_0^\infty \|g_t\|^2\, dt < \infty.$$

Proof. Another application of the $\maltese$-Lemma (Exercise). □

Hence, for all Itô integrable process $g. = (g_t)_{t \geq 0}$, the function $\mathcal{I}(g.)$ defines an element of $\varPhi$, the *Itô integral* of the process $g..$

Recall the operator $D : L^2(\mathcal{P}) \to L^2(\mathcal{P} \times \mathbb{R}^+)$ from last subsection.

Proposition 3.5.
$$\mathcal{I} = D^*.$$

Proof.

$$\langle f, \mathcal{I}(g.) \rangle = \int_{\mathcal{P} \backslash \mathcal{P}_0} \overline{f}(\sigma) g_{\vee \sigma}(\sigma-)\, d\sigma$$

$$= \int_0^\infty \int_\mathcal{P} \overline{f}(\sigma \cup t) g_t(\sigma) \mathbb{1}_{\sigma \subset [0,t]}\, d\sigma\, dt \; (\maltese\text{-Lemma})$$

$$= \int_0^\infty \int_\mathcal{P} \overline{[D_t f]}(\sigma) g_t(\sigma)\, d\sigma\, dt$$

$$= \int_0^\infty \langle D_t f, g_t \rangle\, dt. \quad \square$$

The abstract Itô integral is a true integral

We are going to see that the Itô integral defined above can be interpreted as a true integral $\int_0^\infty g_t \, d\chi_t$ with respect to some particular family $(\chi_t)_{t\geq 0}$ in Φ.

For all $t \in \mathbb{R}^+$, define the element χ_t of Φ by

$$\begin{cases} \chi_t(\sigma) = 0 & \text{if } \#\sigma \neq 1 \\ \chi_t(s) = \mathbb{1}_{[0,t]}(s). \end{cases}$$

This family of elements of Φ has some very particular properties. The main one is the following: not only does χ_t belong to $\Phi_{t]}$ for all $t \in \mathbb{R}^+$, but also $\chi_t - \chi_s$ belongs to $\Phi_{[s,t]}$ for all $s \leq t$ (this is very easy to check from the definition). We will see later that, in some sense, $(\chi_t)_{t\geq 0}$ is the only process to satisfy this property.

Let us take an Itô integrable process $(g_t)_{t\geq 0}$ which is *simple*, that is, constant on intervals:

$$g_t = \sum_i g_{t_i} \mathbb{1}_{[t_i,t_{i+1}[}(t).$$

Define $\int_0^\infty g_t d\chi_t$ to be $\sum_i g_{t_i} \otimes (\chi_{t_{i+1}} - \chi_{t_i})$ (recall that $g_{t_i} \in \Phi_{t_i]}$ and $\chi_{t_{i+1}} - \chi_{t_i} \in \Phi_{[t_i,t_{i+1}]} \subset \Phi_{(t_i)}$). We have

$$\left[\int_0^\infty g_t \, d\chi_t \right](\sigma) = \sum_i [g_{t_i} \otimes (\chi_{t_{i+1}} - \chi_{t_i})](\sigma)$$

$$= \sum_i g_{t_i}(\sigma_{t_i})(\chi_{t_{i+1}} - \chi_{t_i})(\sigma_{(t_i)})$$

$$= \sum_i g_{t_i}(\sigma_{t_i}) \mathbb{1}_{\#\sigma_{(t_i)}=1} \mathbb{1}_{\vee \sigma_{(t_i)} \in]t_i,t_{i+1}]}$$

$$= \sum_i g_{t_i}(\sigma_{t_i}) \mathbb{1}_{\sigma - \subset [0,t_i]} \mathbb{1}_{\vee \sigma \in]t_i,t_{i+1}]}$$

$$= \sum_i g_{t_i}(\sigma-) \mathbb{1}_{\vee \sigma \in]t_i,t_{i+1}]}$$

$$= \sum_i g_{\vee\sigma}(\sigma-) \mathbb{1}_{\vee \sigma \in]t_i,t_{i+1}]}$$

$$= g_{\vee\sigma}(\sigma-).$$

Thus for simple Itô-integrable processes we have proved that

$$\mathcal{I}(g.) = \int_0^\infty g_t \, d\chi_t. \tag{8}$$

But because of the isometry formula of Proposition 3.4 we have

$$\|\mathcal{I}(g.)\|^2 = \left\| \int_0^\infty g_t \, d\chi_t \right\|^2 = \int_0^\infty \|g_t\|^2 \, dt.$$

So one can pass to the limit from simple Itô integrable processes to Itô integrable processes in general and extend the definition of this integral $\int_0^\infty g_t \, d\chi_t$. As a result, (8) holds for every Itô integrable process $(g_t)_{t \geq 0}$. So from now on we will denote the Itô integral by

$$\int_0^\infty g_t \, d\chi_t.$$

Fock space predictable representation property

If f belongs to Φ, Proposition 3.3 shows that $(D_t f)_{t \geq 0}$ is an Itô integrable process. Let us compute $\int_0^\infty D_t f \, d\chi_t$:

$$\left[\int_0^\infty D_t f \, d\chi_t \right](\sigma) = \begin{cases} 0 & \text{if } \sigma = \emptyset \\ [D_{\vee\sigma} f](\sigma-) & \text{otherwise} \end{cases}$$

$$= \begin{cases} 0 & \text{if } \sigma = \emptyset \\ f(\sigma - \cup\vee\sigma)\mathbb{1}_{\sigma - \subset [0,\vee\sigma]} & \text{otherwise} \end{cases}$$

$$= \begin{cases} 0 & \text{if } \sigma = \emptyset \\ f(\sigma) & \text{otherwise} \end{cases}$$

$$= f(\sigma) - [P_0 f](\sigma).$$

This computation together with Propositions 3.3 and 3.4 give the following fundamental result.

Theorem 3.6 (Fock space predictable representation property). *For every $f \in \Phi$ we have the representation*

$$f = P_0 f + \int_0^\infty D_t f \, d\chi_t \tag{9}$$

and

$$\|f\|^2 = |P_0 f|^2 + \int_0^\infty \|D_t f\|^2 \, dt. \tag{10}$$

The representation (9) of f as a sum of a constant and an Itô integral is unique.
The norm identity (10) polarizes as follows

$$\langle f, g \rangle = \overline{P_0 f} P_0 g + \int_0^\infty \langle D_t f, D_t g \rangle \, dt$$

for all $f, g \in \Phi$.

Proof. The only point remaining to be proved is the uniqueness property. If $f = c + \int_0^\infty g_t \, d\chi_t$ then $P_0 f = P_0 c + P_0 \int_0^\infty g_t \, d\chi_t = c$. Hence $\int_0^\infty g_t \, d\chi_t = \int_0^\infty D_t f \, d\chi_t$ that is, $\int_0^\infty (g_t - D_t f) \, d\chi_t = 0$. This implies $\int_0^\infty \|g_t - D_t f\|^2 \, dt = 0$ thus the result. $\square$

Fock space chaotic expansion property

Let h_1 be an element of $L^2(\mathbb{R}^+) = L^2(\mathcal{P}_1)$, we can define

$$\int_0^\infty h_1(t)\Omega\, d\chi_t$$

which we shall simply denote by $\int_0^\infty h_1(t)\, d\chi_t$. Note that the element f of Φ that we obtain this way is given by

$$f(\sigma) = \begin{cases} 0 & \text{if } \#\sigma \neq 1 \\ h_1(s) & \text{if } \sigma = \{s\}. \end{cases}$$

That is, we construct this way all the elements of the first particle space of Φ.

For $h_2 \in L^2(\mathcal{P}_2)$ we want to define

$$\int_{0 \leq s_1 \leq s_2} h_2(s_1, s_2)\, d\chi_{s_1}\, d\chi_{s_2}$$

where we again omit to Ω-symbol. This can be done in two ways:

• either by starting with simple h_2's and defining the iterated integral above as being

$$\sum_{s_j} \sum_{t_i \leq s_j} h_2(t_i, s_j)(\chi_{t_{i+1}} - \chi_{t_i})(\chi_{s_{j+1}} - \chi_{s_j}).$$

One proves easily (exercise) that the norm^2 of the expression above is exactly

$$\int_{0 \leq s_1 \leq s_2} |h_2(s_1, s_2)|^2\, ds_1\, ds_2;$$

so one can pass to the limit in order to define $\int_{0 \leq s_1 \leq s_2} h_2(s_1, s_2)\, d\chi_{s_1}\, d\chi_{s_2}$ for any $h_2 \in L^2(\mathcal{P}_2)$.

• or one says that $g = \int_{0 \leq s_1 \leq s_2} h_2(s_1, s_2)\, d\chi_{s_1}\, d\chi_{s_2}$ is the only $g \in \Phi$ such that the continuous linear form

$$\lambda : \Phi \longrightarrow \mathbb{C}$$

$$f \longmapsto \int_{0 \leq s_1 \leq s_2} \overline{f}(\{s_1, s_2\})h_2(s_1, s_2)\, ds_1\, ds_2$$

is of the form $\lambda(f) = \langle f, g \rangle$.

The two definitions coincide (exercise). The element of Φ which is formed this way is just the element of the second particle space associated to the function h_2.

In the same way, for $h_n \in L^2(\mathcal{P}_n)$ one defines

$$\int_{0 \leq s_1 \leq \cdots \leq s_n} h_n(s_1 \ldots s_n)\, d\chi_{s_1} \cdots d\chi_{s_n}.$$

We get

$$\left\langle \int_{0 \le s_1 \le \cdots \le s_n} h_n(s_1 \ldots s_n)\, d\chi_{s_1} \ldots d\chi_{s_n}, \right.$$

$$\left. \int_{0 \le s_1 \le \cdots \le s_m} k_m(s_1 \ldots s_m)\, d\chi_{s_1} \cdots d\chi_{s_m} \right\rangle$$

$$= \delta_{n,m} \int_{0 \le s_1 \le \cdots \le s_n} \overline{h}_n(s_1 \ldots s_n) k_n(s_1 \ldots s_n)\, ds_1 \cdots ds_n$$

For $f \in L^2(\mathcal{P})$ we define

$$\int_{\mathcal{P}} f(\sigma)\, d\chi_\sigma = f(\emptyset)\mathbb{1} + \sum_n \int_{0 \le s_1 \le \cdots \le s_n} f(\{s_1 \ldots s_n\})\, d\chi_{s_1} \cdots d\chi_{s_n}.$$

Theorem 3.7 (Fock space chaotic representation property). *For all* $f \in \Phi$ *we have*

$$f = \int_{\mathcal{P}} f(\sigma)\, d\chi_\sigma.$$

Proof. For $g \in \Phi$ we have by definition

$$\left\langle g, \int_{\mathcal{P}} f(\sigma)\, d\chi_\sigma \right\rangle$$

$$= \overline{g(\emptyset)} f(\emptyset) + \sum_n \int_{0 \le s_1 \le \cdots \le s_n} \overline{g}(\{s_n \ldots s_n\}) f(\{s_n \ldots s_n\})\, ds_1 \cdots ds_n$$

$$= \langle g, f \rangle.$$

(Details are left to the reader). ☐

$(\chi_t)_{t \ge 0}$ is the only independent increment curve in Φ

We have seen that $(\chi_t)_{t \ge 0}$ is a family in Φ satisfying

   i) $\chi_t \in \Phi_{t]}$ for all $t \in \mathbb{R}^+$;

   ii) $\chi_t - \chi_s \in \Phi_{[s,t]}$ for all $0 \le s \le t$.

These properties where fundamental ingredients for defining our Itô integral. We can naturally wonder if there are any other families $(Y_t)_{t \ge 0}$ in Φ satisfying these two properties?

If one takes $a(\cdot)$ to be a function on $\mathbb{R}^+$, and $h \in L^2(\mathbb{R}^+)$ then $Y_t = a(t)\mathbb{1} + \int_0^t h(s)\, d\chi_s$ clearly satisfies *i)* and *ii)*. But clearly, apart from multiplying every terms by a scalar factor, these families $(Y_t)_{t \ge 0}$ do not change the notions of Itô integrals. The claim now is that the above families $(Y_t)_{t \ge 0}$ are the only possible ones.

Theorem 3.8. *If* $(Y_t)_{t \ge 0}$ *is a vector process on* Φ *satisfying i) and ii) then there exist* $a : \mathbb{R}^+ \to \mathbb{C}$ *and* $h \in L^2(\mathbb{R}^+)$ *such that*

$$Y_t = a(t)\mathbb{1} + \int_0^t h(s)\, d\chi_s.$$

Proof. Let $a(t) = P_0 Y_t$. Then $\widetilde{Y}_t = Y_t - a(t)\mathbb{1}$, $t \in \mathbb{R}^+$, satisfies *i)* and *ii)* with $\widetilde{Y}_0 = 0$ (for $Y_0 = P_0 Y_0 = P_0 (Y_t - Y_0) + P_0 Y_0 = P_0 Y_t$). We can now drop the $\sim$ symbol and assume $Y_0 = 0$. Now note that $P_s Y_t = P_s Y_s + P_s (Y_t - Y_s) = P_s Y_s = Y_s$. This implies easily (exercise) that the chaotic expansion of Y_t is of the form:

$$Y_t = \int_{\mathcal{P}} \mathbb{1}_{\mathcal{P}^{t)}}(\sigma) y(\sigma) \, d\chi_\sigma \, .$$

If $\#\sigma \geq 2$, for example $\sigma = \{t_1 < t_2 < \cdots < t_n\}$, let $s < t$ be such that $t_1 < s < t_n < t$. Then

$$(Y_t - Y_s)(\sigma) = 0 \text{ for } Y_t - Y_s \in \Phi_{[s,t]} \text{ and } \sigma \not\subset [s,t] \, .$$

Furthermore

$$Y_s(\sigma) = P_s Y_s(\sigma) = \mathbb{1}_{\sigma \subset [0,s]} Y_s(\sigma) = 0 \, .$$

Thus $Y_t(\sigma) = 0$, for any $\sigma \in \mathcal{P}$ with $\#\sigma \geq 2$, any $t \in \mathbb{R}^+$. This means that $Y_t = \int_0^t y(s) \, d\chi_s$. $\square$

Higher multiplicities

When considering the Fock space $\Gamma_s(L^2(\mathbb{R}^+; \mathbb{C}^n))$ or $\Gamma_s(L^2(\mathbb{R}^+; \mathcal{G}))$ for some separable Hilbert space $\mathcal{G}$, we speak of Fock space with multiplicity n or infinite multiplicity.

The Guichardet space is then associated to the the set $\mathcal{P}_n$ of finite subsets of $\mathcal{P}$ but whose elements are given a label, a color, in $\{1, \dots, n\}$. This is also equivalent to giving oneself a family of n disjoint subsets of $\mathbb{R}^+$: $\sigma = (\sigma_1, \dots, \sigma_n)$. The norm on that Fock space is then

$$\|f\|^2 = \int_{\mathcal{P}_n} |f(\sigma)|^2 \, d\sigma$$

with obvious notations.

The universal curve $(\chi_t)_{t \geq 0}$ is replaced by a family $(\chi_t^i)_{t \geq 0}$ defined by

$$\chi_t^i(\sigma) = \begin{cases} \mathbb{1}_{[0,t]}(s) & \text{if } \sigma_j = \emptyset \text{ for all } j \neq i \text{ and } \sigma_i = \{s\} \\ 0 & \text{otherwise.} \end{cases}$$

The Fock space predictable representation is now of the form

$$f = P_0 f + \sum_i \int_0^\infty D_s^i f \, d\chi_s^i$$

with

$$\|f\|^2 = |P_0 f|^2 + \sum_i \int_0^\infty \|D_s^i f\|^2 \, ds$$

and

$$[D_s^i f](\sigma) = f(\sigma \cup \{s\}_i) 1\!\!1_{\sigma \subset [0,s]}$$

with the notation $\{s\}_i = (\emptyset, \ldots, \emptyset, \{s\}, \emptyset, \ldots, \emptyset) \in \mathcal{P}_n$.

The quantum noises are $a_j^i(t)$, labelled by $i, j = 0, 1, \ldots, n$, with the formal table

$$da_0^0(t) 1\!\!1 = dt 1\!\!1$$
$$da_0^0(t) d\chi_t^k = 0$$

$$da_i^0(t) 1\!\!1 = d\chi_t^i$$
$$da_i^0(t) d\chi_t^k = 0$$

$$da_0^i(t) 1\!\!1 = 0$$
$$da_0^i(t) d\chi_t^k = \delta_{ki} dt 1\!\!1$$

$$da_j^i(t) 1\!\!1 = 0$$
$$da_j^i(t) d\chi_t^k = \delta_{ki} d\chi_t^j.$$

3.4 Probabilistic interpretations of Fock space

In this section we present the general theory of probabilistic interpretations of Fock space. This section is not really necessary to understand the rest of the course, but the ideas coming from these notions underly the whole work.

This section needs some knowledge in the basic elements of stochastic processes, martingales and stochastic integrals. Some of that material can be found in L. Rey-Bellet's first course in this volume.

Chaotic expansions

We consider a martingale $(x_t)_{t \geq 0}$ on a probability space $(\Omega, \mathcal{F}, P)$. We take $(\mathcal{F}_t)_{t \geq 0}$ to be the natural filtration of $(x_t)_{t \geq 0}$ (the filtration is made complete and right continuous) and we suppose that $\mathcal{F} = \mathcal{F}_\infty = \vee_{t \geq 0} \mathcal{F}_t$. Such a martingale is called *normal* if $(x_t^2 - t)_{t \geq 0}$ is still a martingale for $(\mathcal{F}_t)_{t \geq 0}$. This is equivalent to saying that $\langle x, x \rangle_t = t$ for all $t \geq 0$, where $\langle \cdot, \cdot \rangle$ denotes the probabilistic angle bracket.

A normal martingale is said to satisfy the *predictable representation property* if all $f \in L^2(\Omega, \mathcal{F}, P)$ can be written as a stochastic integral

$$f = I\!\!E[f] + \int_0^\infty h_s \, dx_s$$

for a $(\mathcal{F}_t)_{t \geq 0}$-predictable process $(h_t)_t \geq 0$. Recall that

$$E[|f|^2] = |E[f]|^2 + \int_0^\infty E[|h_s|^2]\, ds$$

that is, in the $L^2(\Omega)$-norm notation:

$$\|f\|^2 = |E[f]|^2 + \int_0^\infty \|h_s\|^2\, ds\ .$$

Recall that if f_n is a function in $L^2(\Sigma_n)$, where $\Sigma_n = \{0 \le t_1 < t_2 < \cdots < t_n \in \mathbb{R}^n\} \subset (\mathbb{R}^+)^n$ is equipped with the restriction of the n-dimensional Lebesgue measure, one can define an element $I_n(f_n) \in L^2(\Omega)$ by

$$I_n(f_n) = \int_{\Sigma_n} f_n(t_1 \ldots t_n)\, dx_{t_1} \cdots dx_{t_n}$$

which is defined, with the help of the Itô isometry formula, as an iterated stochastic integral satisfying

$$\|I_n(f_n)\|^2 = \int_{\Sigma_n} |f_n(t_1 \ldots t_n)|^2\, dt_1 \cdots dt_n\ .$$

It is also important to recall that

$$\langle I_n(f_n), I_m(f_m) \rangle = 0 \ \text{ if } \ n \ne m\ .$$

The *chaotic space* of $(x_t)_{t \ge 0}$, denoted $CS(x)$, is the sub-Hilbert space of $L^2(\Omega)$ made of the random variables $f \in L^2(\Omega)$ which can be written as

$$f = E[f] + \sum_{n=1}^\infty \int_{\Sigma_n} f_n(t_1 \ldots t_n)\, dx_{t_1} \cdots dx_{t_n} \tag{11}$$

for some $f_n \in L^2(\Sigma_n)$, $n \in \mathbb{N}^*$, such that

$$\|f\|^2 = |E[f]|^2 + \sum_{n=1}^\infty \int_{\Sigma_n} |f_n(t_1 \ldots t_n)|^2\, dt_1 \cdots dt_n < \infty\ .$$

When $CS(x)$ is the whole of $L^2(\Omega)$ one says that x satisfies the *chaotic representation property*. The decomposition of f as in (11) is called the *chaotic expansion* of f.

Note that the chaotic representation property implies the predictable representation property for if f can be written as in (11) then, by putting h_t to be

$$h_t = f_1(t) + \sum_{n=1}^\infty \int_{\Sigma_n} \mathbb{1}[0,t](t_n) f_{n+1}(t_1 \ldots t_n, t)\, dx_{t_1} \cdots dx_{t_n}$$

we have

$$f = E[f] + \int_0^\infty h_t\, dx_t\ .$$

The cases where $(x_t)_{t \ge 0}$ is the Brownian motion, the compensated Poisson process or the Azéma martingale with coefficient $\beta \in [-2, 0]$, are examples of normal martingales which possess the chaotic representation property.

Isomorphism with Φ

Let us consider a normal martingale $(x_t)_{t \geq 0}$ with the predictable representation property and its chaotic space $CS(x) \subset L^2(\Omega, \mathcal{F}, P)$.

By identifying a function $f_n \in L^2(\Sigma_n)$ with a symmetric function $\tilde{f}_n$ on $(\mathbb{R}^+)^n$, one can identify $L^2(\Sigma_n)$ with $L^2_{\mathrm{sym}}((\mathbb{R}^+)^n) = L^2(\mathbb{R}^+)^{\odot n}$ (with the correct symmetric norm: $\|\tilde{f}_n\|^2_{L^2(\mathbb{R}^+)^{\odot n}} = n! \|\tilde{f}_n\|^2_{L^2(\mathbb{R}^+)^{\otimes n}}$ if one puts $\tilde{f}_n$ to be $\frac{1}{n!}$ times the symmetric expansion of f_n). It is now clear that $CS(x)$ is naturally isomorphic to the symmetric Fock space

$$\Phi = \Gamma(L^2(\mathbb{R}^+)) = \bigoplus_{n=0}^{\infty} L^2(\mathbb{R}^+)^{\odot n} .$$

The isomorphism can be explicitly written as follows:

$$U_x : \Phi \longrightarrow CS(x)$$
$$f \longmapsto Uf$$

where $f = \sum_n f_n$ with $f_n \in L^2(\mathbb{R}^+)^{\odot n}$, $n \in \mathbb{N}$, and

$$U_x f = f_0 + \sum_{n=1}^{\infty} n! \int_0^t f_n(t_1 \ldots t_n) \, dx_{t_1} \cdots dx_{t_n} .$$

If $f = \mathbb{E}[f] + \sum_{n=1}^{\infty} \int_0^t f_n(t_1 \ldots t_n) \, dx_{t_1} \cdots dx_{t_n}$ is an element of $CS(x)$, then $U_x^{-1} f = \sum_n g_n$ with $g_0 = \mathbb{E}[f]$ and $g_n = \frac{1}{n!} f_n$ symmetrized.

These isomorphisms are called the *probabilistic interpretations* of Φ. One may speak of *Brownian interpretation*, or *Poisson interpretation* ...

Structure equations

If $(x_t)_{t \geq 0}$ is a normal martingale, with the predictable representation property and if x_t belongs to $L^4(\Omega)$, for all t, then $([x,x]_t - \langle x,x \rangle_t)_{t \geq 0}$ is a $L^2(\Omega)$-martingale; so by the predictable representation property there exists a predictable process $(\psi_t)_{t \geq 0}$ such that

$$[x,x]_t - \langle x,x \rangle_t = \int_0^t \psi_s \, dx_s$$

that is,

$$[x,x]_t = t + \int_0^t \psi_s \, dx_s$$

or else

$$d[x,x]_t = dt + \psi_t \, dx_t . \tag{12}$$

This equation is called a *structure equation* for $(x_t)_{t \geq 0}$. One has to be careful that, in general, there can be many structure equations describing the same solution $(x_t)_{t \geq 0}$; there also can be several solutions (in law) to some structure equations.

What can be proved is the following:

- when $\psi_t \equiv 0$ for all t then the only solution (in law) of (12) is the Brownian motion;

- when $\psi_t \equiv c$ for all t then the only solution (in law) of (12) is the compensated Poisson process with intensity $1/c^2$;

- when $\psi_t = \beta x_{t-}$ for all t, then the only solution (in law) of (12) is the Azéma martingale with parameter β.

The importance of structure equations appears when one considers products within two different probabilistic interpretations. For exemple, let f, g be two elements of Φ and let $U_w f$ and $U_w g$ be their image in the Brownian motion interpretation $(w_t)_t \geq 0$. That is, $U_w f$ and $U_w g$ are random variables in the canonical space $L^2(\Omega)$ of the Brownian motion. They admit a natural product, as random variables: $U_w f \cdot U_w g$. If the resulting random variable is still an element of $L^2(\Omega)$ (for example if f and g are coherent vectors) then we can pull back the resulting random variable to the space Φ:

$$U_w^{-1}(U_w f \cdot U_w g).$$

This operation defines an associative product on Φ:

$$f *_w g = U_w^{-1}(U_w f \cdot U_w g)$$

called the *Wiener product*.

We could have done the same operations with the Poisson interpretation:

$$f *_p g = U_p^{-1}(U_p f \cdot U_p g),$$

this gives the *Poisson product* on Φ. One can also define an Azéma product.

The point is that one always obtains different products on Φ when considering different probabilistic interpetations. This comes frome the fact that all probabilistic interpretations of Φ have the same angle bracket $\langle x, x \rangle_t = t$ but not the same square bracket: $[x, x]_t = t + \int_0^t \psi_s\, dx_s$. The product of two random variables makes use of the square bracket: if $f = \mathbb{E}[f] + \int_0^\infty h_s\, dx_s$ and $g = \mathbb{E}[g] + \int_0^\infty k_s\, dx_s$, if $f_s = \mathbb{E}[f|\mathcal{F}_s]$ and $g_s = \mathbb{E}[g|\mathcal{F}_s]$ for all $s \geq 0$ then

$$fg = \mathbb{E}[f]\mathbb{E}[g] + \int_0^\infty f_s k_s\, dx_s + \int_0^\infty g_s h_s\, dx_s + \int_0^\infty h_s k_s d[x, x]_s$$

$$= \mathbb{E}[f]\mathbb{E}[g] + \int_0^\infty f_s k_s\, dx_s + \int_0^\infty g_s h_s\, dx_s + \int_0^\infty h_s k_s ds +$$

$$+ \int_0^\infty h_s k_s \psi_s\, dx_s .$$

For example if one takes the element χ_t of Φ, we have

$$U_w \chi_t = \int_0^\infty \mathbb{1}_{[0,t]}(s)\, dw_s = w_t \text{ the Brownian motion itself} \qquad (13)$$

and $\qquad\qquad\qquad\qquad\qquad\qquad\qquad\qquad\qquad\qquad\qquad\quad$ (14)

$$U_p \chi_t = \int_0^\infty \mathbb{1}_{[0,t]}(s)\, dx_s = x_t \text{ the compensated Poisson process itself.} \quad (15)$$

So, as $w_t^2 = 2\int_0^t w_s\, dw_s + t$ and $x_t^2 = 2\int_0^t x_s\, dx_s + t + x_t$, we have

$$\chi_t *_w \chi_t = t + 2\int_0^t \chi_s\, d\chi_s \qquad (16)$$

and

$$\chi_t *_p \chi_t = t + 2\int_0^t \chi_s\, d\chi_s + \chi_t. \qquad (17)$$

We get two different elements of Φ.

Probabilistic interpretations of the abstract Itô calculus

Let $(\Omega, \mathcal{F}, (\mathcal{F}_t)_t \geq 0, P, (x_t)_{t \geq 0})$ be a probabilistic interpretation of the Fock space Φ. Via the isomorphism described above, the space $\Phi_{t]}$ interprets as the space of $f \in CS(x.)$ whose chaotic expansion contains only functions with support included in $[0, t]$; that is, the space $CS(x) \cap L^2(\mathcal{F}_t)$. So when the chaotic expansion property holds we have $\Phi_{t]} \simeq L^2(\mathcal{F}_t)$ and thus P_t is nothing but $\mathbb{E}[\cdot\,|\mathcal{F}_t]$ (the conditional expectation) when interpreted in $L^2(\Omega)$.

The process $(\chi_t)_{t \geq 0}$ interprets as a process of random variables whose chaotic expansion is given by

$$\chi_t = \int_0^\infty \mathbb{1}_{[0,t]}(s)\, dx_s = x_t.$$

So, in any probabilistic interpretation $(\chi_t)_{t \geq 0}$ becomes the noise $(x_t)_{t \geq 0}$ itself (Brownian motion, compensated Poisson process, Azéma martingale,...). $(\chi_t)_{t \geq 0}$ is the "universal" noise, seen in the Fock space Φ.

As we have proved that the Itô integral $\mathcal{I}(g.)$ on Φ is the L^2-limit of the Riemann sums $\sum_i g_{t_i}(\chi_{t_{i+1}} - \chi_{t_i})$, it is clear that in $L^2(\Omega)$, the Itô integral interprets as the usual Itô integral with respect to $(x_t)_{t \geq 0}$.

One remark is necessary here. When one writes the approximation of the Itô integral $\int_0^\infty g_s\, dx_s$ as $\sum_i g_{t_i}(x_{t_{i+1}} - x_{t_i})$ there appear products $(g_{t_i} \cdot (x_{t_{i+1}} - x_{t_i}))$, so this notion seems to depend on the probabilistic interpretation of Φ. The point in that the product $g_{t_i} \cdot (x_{t_{i+1}} - s_{t_i})$ is not really a product. By this we mean that the Itô formula for this product does not involve any bracket term:

$$g_{t_i}(x_{t_{i+1}} - x_{t_i}) = \int_{t_i}^{t_{i+1}} g_{t_i}\, dx_s$$

so it gives rise to the same formula whatever is the probabilistic interpretation $(x_t)_{t\geq 0}$. Only is involved the tensor product structure: $\Phi \simeq \Phi_{t_i]} \otimes \Phi_{[t_i}$; the product $g_{t_i}(x_{t_{i+1}} - x_{t_i})$ is only a tensor product $g_{t_i} \otimes (x_{t_{i+1}} - x_{t_i})$ in this structure. This tensor product structure is common to all the probabilistic interpretations.

We have seen that $\int_0^\infty g_t \, d\chi_t$ interprets as the usual Itô integral $\int_0^\infty g_t \, d\chi_t$ in any probabilistic interpretation $(x_t)_{t\geq 0}$. Thus the representation

$$f = P_0 f + \int_0^\infty D_s f \, d\chi_x$$

of Theorem 3.6 is just a Fock space expression of the predictable representation property. The process $(D_t f)_{t\geq 0}$ is then interpreted as the predictable process that represents f in his predictable representation.

4 Quantum stochastic calculus

We now leave the probabilistic intepretations of the Fock space and we enter into the theory of quantum noises itself, with its associated theory of integration.

4.1 An heuristic approach to quantum noise

Adaptedness

When trying to define "quantum stochastic integrals" of operators on Φ, mimicking integral representations such as in the Toy Fock space, we have to consider integrals of the form

$$\int_0^t H_s \, dM_s$$

where $(H_t)_{t\geq 0}$ and $(M_t)_{t\geq 0}$ are families of operators on Φ.

The first natural idea is to consider approximations of the above by Riemann sums:

$$\sum_i H_{t_i} \left(M_{t_{i+1}} - M_{t_i} \right),$$

but, immediatly, this kind of definition faces two difficulties:

i) The operators we are going to consider are not in general bounded and therefore the above sum may lead us to serious domain problems.

ii) The operators we consider H_s, M_s need not commute in general and we can naturally wonder why we should not give the preference to sums like

$$\sum_i \left(M_{t_{i+1}} - M_{t_i} \right) H_{t_i},$$

or even more complicated forms.

This means that, at this stage of the theory, we have to make concessions: we cannot integrate any operator process with respect to any operator process. But this should not be a surprise, already in the classical theory of stochastic calculus one can only integrate predictable processes against semimartingales.

The first step in this integration theory consists is obtained by applying, at the operator level, a construction similar to the one of the Itô integral on Φ with respect to $(\chi_t)_{t \geq 0}$. Indeed, recall the decomposition of the Fock space Φ:

$$\Phi = \Phi_{s]} \otimes \Phi_{[s,t]} \otimes \Phi_{[t}$$

for all $s \leq t$.

If there exist operator families $(X_t)_{t \geq 0}$ on Φ with the property that $X_t - X_s$ acts on $\Phi_{s]} \otimes \Phi_{[s,t]} \otimes \Phi_{[t}$ as $I \otimes K_{s,t} \otimes I$ and if we consider operator families $(H_t)_{t \geq 0}$ such that H_t is of the form $H_t \otimes I$ on $\Phi_{t]} \otimes \Phi_{[t}$ then the Riemann sums

$$\sum_i H_{t_i} \left(X_{t_{i+1}} - X_{t_i} \right)$$

are well-defined and unambiguous for the products

$$H_{t_i} \left(X_{t_{i+1}} - X_{t_i} \right)$$

are not true compositions of operators anymore but just tensor products (just like for vectors in the Itô integral):

$$\left(H_{t_i} \otimes I \right) \left(I \otimes \left(X_{t_{i+1}} - X_{t_i} \right) \otimes I \right) = H_{t_i} \otimes \left(X_{t_{i+1}} - X_{t_i} \right).$$

In particular, there are no more domain problem added by the composition of operators, no more commutation problem.

Families of operators of the form $H_t \otimes I$ on $\Phi_{t]} \otimes \Phi_{[t}$ are obvious to construct. They are called *adapted processes of operators*. The true definition of adapted processes of operators, in the case of unbounded operators, are actually not that simple. They are exactly what is stated above, in the spirit, but this requires a more careful definition. We do not develop these refinements in this course (see [9]).

The existence of non-trivial operator families $(X_t)_{t \geq 0}$ on Φ with the property that $X_t - X_s$ acts on $\Phi_{s]} \otimes \Phi_{[s,t]} \otimes \Phi_{[t}$ as $I \otimes K_{s,t} \otimes I$ for all $s \leq t$ is not so clear.

The three quantum noises: heuristics

We call "*quantum noise*" any processes of operators on Φ, say $(X_t)_{t \geq 0}$, such that, for all $t_i \leq t_{i+1}$, the operator $X_{t_{i+1}} - X_{t_i}$ acts as $I \otimes k \otimes I$ on $\Phi_{t_i]} \otimes \Phi_{[t_i,t_{i+1}]} \otimes \Phi_{[t_{i+1}}$.

Let us consider the operator $dX_t = X_{t+dt} - X_t$. It acts only on $\Phi_{[t,t+dt]}$. The chaotic representation property of Fock space (Theorem 3.7) shows that this part of the Fock space is generated by the vacuum $\mathbb{1}$ and by $d\chi_t = \chi_{t+dt} - \chi_t$. Hence dX_t is determined by its value on $\mathbb{1}$ and on $d\chi_t$. These values have to remain in $\Phi_{[t,t+dt]}$

and to be integrators also, that is $d\chi_t$ or $dt\mathbb{1}$ (denoted dt). As a consequence the only *irreducible* noises are given by

	$d\chi_t$	$\mathbb{1}$
da_t°	$d\chi_t$	0
da_t^-	dt	0
da_t^+	0	$d\chi_t$
$da_t^\times$	0	dt

These are four noises and not three as announced, but we shall see later that $da_t^\times$ is just dtI.

The three quantum noises: serious business

Recall the definitions of creation, annihilation and differential second quantization operators on Φ. For any $h \in L^2(\mathbb{R}^+)$, any operator H on $L^2(\mathbb{R}^+)$ and any symmetric tensor product $u_1 \circ \ldots \circ u_n$ in Φ we put

$$a^+(h)u_1 \circ \ldots \circ u_n = h \circ u_1 \circ \ldots \circ u_n$$

$$a^-(h)u_1 \circ \ldots \circ u_n = \sum_{i=1}^{n} <h,\, u_i> u_1 \circ \ldots \circ \widehat{u}_i \circ \ldots \circ u_n$$

$$\Lambda(H)u_1 \circ \ldots \circ u_n = \sum_{i=1}^{n} u_1 \circ \ldots \circ Hu_i \circ \ldots \circ u_n.$$

In the case where $H = \mathcal{M}_h$ is the multiplication operator by h we write $a^\circ(h)$ for $\Lambda(H)$.

An easy computation shows that $a^+(h), a^-(h), a^\circ(h)$ are closable operators whose domain contains $\mathcal{E}$, the space of coherent vectors, and which satisfy

$$<\varepsilon(u),\, a^+(h)\varepsilon(v)> = \int_0^\infty \overline{u}(s)\, h(s)\, ds <\varepsilon(u),\, \varepsilon(v)>$$

$$<\varepsilon(u),\, a^-(h)\varepsilon(v)> = \int_0^\infty \overline{v}(s)\, h(s)\, ds <\varepsilon(u),\, \varepsilon(v)>$$

$$<\varepsilon(u),\, a^\circ(h)\varepsilon(v)> = \int_0^\infty \overline{u}(s)v(s)\, h(s)\, ds <\varepsilon(u),\, \varepsilon(v)>.$$

One can also easily obtain the above explicit formulas:

$$[a^+(h)f] = \sum_{s \in \sigma} h(s)\, f(\sigma \setminus s)$$

$$[a^-(h)f] = \int_0^\infty h(s)\, f(\sigma \cup s)\, ds$$

$$[a^\circ(h)f] = \sum_{s \in \sigma} h(s)\, f(\sigma).$$

For any $t \in {I\!\!R}^+$ and any $\varepsilon = +, -, \circ$ we put $a^\varepsilon(h_{t]}) = a^\varepsilon(h{1\!\!1}_{[0,t]})$. It is then easy to check from the definitions that any operator process $(X_t)_{t \geq 0}$ of the form

$$X_t = a(t)I + a^+(f_{t]}) + a^-(g_{t]}) + a^\circ(k_{t]})$$

is a quantum noise (in order to avoid domain problems we have to ask that h, k belong to $L^2({I\!\!R}^+)$ and k belongs to $L^\infty({I\!\!R}^+)$).

The previous heuristical discussion seems to says that they should be the only ones. This result is intuitively simple, but its proof is not so simple (see [17]), we do not develop it here.

Theorem 4.1. *A family of closable operators* $(X_t)_{t \geq 0}$ *defined on* $\mathcal{E}$ *is a quantum noise if and only if there exist a function a on* ${I\!\!R}^+$, *functions* $f, g \in L^2({I\!\!R}^+)$ *and a function* $k \in L^\infty({I\!\!R}^+)$ *such that*

$$X_t = a(t)I + a^+(f_{t]}) + a^-(g_{t]}) + a^\circ(k_{t]})$$

for all t. □

Putting $a_t^\varepsilon = a^\varepsilon({1\!\!1}_{[0,t]})$, we can see that all the quantum noises are determined by the four processes $(a_t^\varepsilon)_{t \geq 0}$, for $\varepsilon = +, -, \circ, \times$ where we have put $a_t^\times = tI$.

It is with respect to these four operator processes that the quantum stochastic integrals are defined.

4.2 Quantum stochastic integrals

Heuristic approach

Let us now formally consider a quantum stochastic integral

$$T_t = \int_0^t H_s da_s^\varepsilon$$

with respect to one of the four above noises. Let it act on a vector process

$$f_t = P_t f = \int_0^t D_s f \, d\chi_s \text{ (we omit the expectation } P_0 f \text{ for the moment).}$$

The result is a process of vectors $(T_t f_t)_{t \geq 0}$ in Φ. We claim that one can expect the family $(T_t f_t)_{t \geq 0}$ to satisfy an Itô-like integration by part formula:

$$d(T_t f_t) = T_t df_t + (dT_t)f_t + (dT_t)(df_t)$$
$$= T_t(D_t f \, d\chi_t) + (H_t da_t^\varepsilon)f_t + (H_t da_t^\varepsilon)(D_t f \, d\chi_t).$$

There are three reasons for that claim:

i) This is the continuous version of the quantum Itô formula obtained in discrete time (Theorem 2.2).

ii) Quantum stochastic calculus contains in particular the classical one, it should then satisfy the same kind of Itô integration by part formula.

ii) More convincing: if one considers an operator process $(H_t)_{t \geq 0}$ which is simple (i.e. constant by intervals) and a vector process $(D_t f)_{t \geq 0}$ which is simple too, then the integrated form of the above identity is exactly true (Exercise for very motivated readers!).

In the tensor product structure $\Phi = \Phi_{t]} \otimes \Phi_{[t}$ this formula writes

$$d(T_t f_t) = (T_t \otimes I)(D_t f \otimes d\chi_t) + (H_t \otimes da_t^\varepsilon)(f_t \otimes \mathbb{1}) + (H_t \otimes da_t^\varepsilon)(D_t f \otimes d\chi_t),$$

that is,

$$d(T_t f_t) = T_t D_t f_t \otimes d\chi_t + H_t f_t \otimes da_t^\varepsilon \mathbb{1} + H_t D_t f \otimes da_t^\varepsilon \, d\chi_t. \tag{18}$$

In the right hand side one sees three terms; the first one always remains and is always the same. The other two depend on the heuristic table satisfied by the quantum noises. Integrating (18) and using the quantum noise table one gets

$$T_t f_t = \int_0^t T_s D_s f \, d\chi_s + \begin{cases} \int_0^t H_s D_s f \, d\chi_s & \text{if } \varepsilon = 0 \\ \int_0^t H_s P_s f \, d\chi_s & \text{if } \varepsilon = + \\ \int_0^t H_s D_s f \, ds & \text{if } \varepsilon = - \\ \int_0^t H_s P_s f \, ds & \text{if } \varepsilon = \times. \end{cases} \tag{19}$$

A correct definition

We want to exploit formula (19) as a definition of the quantum stochastic integrals $T_t = \int_0^t H_s da_s^\varepsilon$.

Let $(H_t)_{t \geq 0}$ be an adapted process of operators on Φ, let $(T_t)_{t \geq 0}$ be another one. One says that (19) is *meaningful* for a given $f \in \Phi$ if

- $P_t f \in \mathrm{Dom}\, T_t$;
- $D_s f \in \mathrm{Dom}\, T_s, s \leq t$ and $\int_0^t \|T_s D_s f\|^2 \, ds < \infty$;

$$\bullet \begin{cases} \text{if } \varepsilon = \circ, D_s f \in \mathrm{Dom}\, H_s, s \leq t \text{ and } \int_0^t \|H_s D_s f\|^2 \, ds < \infty \\ \text{if } \varepsilon = +, P_s f \in \mathrm{Dom}\, H_s, s \leq t \text{ and } \int_0^t \|H_s P_s f\|^2 \, ds < \infty \\ \text{if } \varepsilon = -, D_s f \in \mathrm{Dom}\, H_s, s \leq t \text{ and } \int_0^t \|H_s D_s f\| \, ds < \infty \\ \text{if } \varepsilon = \times, P_s f \in \mathrm{Dom}\, H_s, s \leq t \text{ and } \int_0^t \|H_s P_s f\| \, ds < \infty \end{cases}$$

One says that (19) is *true* if the equality holds.

A subspace $\mathcal{D} \subset \Phi$ is called an *adapted domain* if for all $f \in \mathcal{D}$ and all (almost all) $t \in \mathbb{R}^+$, one has

$$P_t f \text{ and } D_t f \in \mathcal{D}.$$

There are many examples of adapted domains. All the domains we shall meet during this course are adapted:

- $\mathcal{D} = \Phi$ itself is adapted;
- $\mathcal{D} = \mathcal{E}$ is adapted; even more $\mathcal{D} = \mathcal{E}(\mathcal{M})$ is adapted once $1\!\!1_{[0,t]}\mathcal{M} \subset \mathcal{M}$ for all t.
- The space of *finite particles* $\Phi_f = \{f \in L^2(\mathcal{P});\ f(\sigma) = 0 \text{ for } \#\sigma > N, \text{ for some } N \in I\!\!N\}$ is adapted.
- All the Fock scales $\Phi^{(a)} = \{f \in L^2(\mathcal{P});\ \int_{\mathcal{P}} a^{\#\sigma}|f(\sigma)|^2 \, d\sigma < \infty\}$, for $a \geq 1$, are adapted.
- Maassen's space of test vectors: $\{f \in L^2(\mathcal{P});\ f(\sigma) = 0 \text{ for } \#\sigma \not\subset [0,T], \text{for some } T \in I\!\!R^+, \text{ and } |f(\sigma)| \leq CM^{\#\sigma} \text{ for some } C, M\}$ is adapted.

The above equation (19) is the definition of the quantum stochastic integrals that we shall follow and apply along this course. The definition is exactly formulated as follows.

Let $(H_t)_{t \geq 0}$ be an adapted process of operators defined on an adapted domain $\mathcal{D}$. One says that a process $(T_t)_{t \geq 0}$ is the *quantum stochastic integral*

$$T_t = \int_0^t H_s \, da_s^\varepsilon$$

on the domain $\mathcal{D}$, if (19) is meaningfull and true for all $f \in \mathcal{D}$.

We now have to give at least one criterion for the existence of a solution to equation (19). When considering the domain $\mathcal{E}$ there is a simple characterization.

Theorem 4.2. *Let* $(H_t)_{t \geq 0}$ *be an adapted process of operators defined on* $\mathcal{E}$. *If for every* $u \in L^2(I\!\!R^+)$ *and every* $t \in I\!\!R^+$ *we have*

$$
\begin{cases}
\displaystyle\int_0^t |u(s)|^2 \, \|H_s\varepsilon(u)\|^2 \, ds < \infty & \text{if } \varepsilon = \circ \\[2ex]
\displaystyle\int_0^t \|H_s\varepsilon(u)\|^2 \, ds < \infty & \text{if } \varepsilon = + \\[2ex]
\displaystyle\int_0^t |u(s)| \, \|H_s\varepsilon(u)\| \, ds < \infty & \text{if } \varepsilon = \\[2ex]
\displaystyle\int_0^t \|H_s\varepsilon(u)\| \, ds < \infty & \text{if } \varepsilon = \times
\end{cases}
$$

is satisfied. Then the corresponding equation (19) for

$$\int_0^t H_s \, da_s^\varepsilon$$

admits a unique solution on $\mathcal{E}$ *which satisfies*

$$<\varepsilon(u),\ \int_0^t H_s\, da_s^\varepsilon\, \varepsilon(v)> = \begin{cases} \int_0^t \overline{u}(s) v(s) <\varepsilon(u),\ H_s\varepsilon(v)> ds & \text{if } \varepsilon = \circ \\[2mm] \int_0^t \overline{u}(s) <\varepsilon(u),\ H_s\varepsilon(v)> ds & \text{if } \varepsilon = + \\[2mm] \int_0^t v(s) <\varepsilon(u),\ H_s\varepsilon(v)> ds & \text{if } \varepsilon = - \\[2mm] \int_0^t <\varepsilon(u),\ H_s\varepsilon(v)> ds & \text{if } \varepsilon = \times. \end{cases}$$

$$\tag{20}$$

Furthermore, any operator T_t which satisfies (20) for some $(H_t)_{t\geq 0}$ is of the form $T_t = \int_0^t H_s\, da_s^\varepsilon$ in the sense of the definition (19).

Proof. Let $(H_t)_{t\geq 0}$ be an adapted process satisfying the above condition for some ε. We shall prove that (19) admits a unique solution by using a usual Picard method. Let us write it for the case $\varepsilon = \circ$ and leave the three other cases to the reader.

For $u \in L^2(\mathbb{R}^+)$, one can easily check that $D_t\varepsilon(u) = u(t)\varepsilon(u_{t]})$ for almost all t, where $u_{t]}$ means $u\mathbb{1}_{[0,t]}$. This means that, in order to construct the desired quantum stochastic integral on $\mathcal{E}$, we have to solve the equation

$$T_t\varepsilon(u_{t]}) = \int_0^t u(s) T_s\varepsilon(u_{s]})\, d\chi_s + \int_0^t u(s) H_s\varepsilon(u_{s]})\, d\chi_s. \tag{21}$$

Let $x_t = T_t\varepsilon(u_{t]})$, $t \geq 0$. We have to solve

$$x_t = \int_0^t u(s) x_s\, d\chi_s + \int_0^t u(s) H_s\varepsilon(u_{s]})\, d\chi_s.$$

Put $x_t^0 = \int_0^t u(s) H_s\varepsilon(u_{s]})\, d\chi_s$ and

$$x_t^{n+1} = \int_0^t u(s) x_s^n\, d\chi_s + \int_0^t u(s) H_s\varepsilon(u_{s]})\, d\chi_s.$$

Let $y_t^0 = x_t^0$ and $y_t^{n+1} = x_t^{n+1} - x_t^n = \int_0^t u(s) y_s^n\, d\chi_s$. We have

$$\|y_t^{n+1}\|^2 = \int_0^t |u(s)|^2 \|y_s^n\|^2\, ds$$

$$= \int_0^t \int_0^{t_2} |u(t_1)|^2 |u(t_2)|^2 \|y_{t_1}^{n-1}\|^2\, dt_1\, dt_2$$

$$\vdots$$

$$= \int_{0\leq t_1\leq\cdots\leq t_n\leq t} |u(t_1)|^2 \cdots |u(t_n)|^2 \|y_{t_1}^0\|^2\, dt_1\cdots dt_n$$

$$= \int_{0\leq t_1\leq\cdots\leq t_n\leq t} |u(t_1)|^2 \cdots |u(t_n)|^2 \int_0^{t_1} |u(s)|^2 \|H_s\varepsilon(u_{s]})\|^2\, ds\, dt_1\cdots dt_n$$

$$\leq \int_0^t |u(s)|^2 \|H_s\varepsilon(u_{s]})\|^2\, ds\ \frac{\left(\int_0^t |u(s)|^2\, ds\right)^n}{n!}.$$

From this estimate one easily sees that the sequences

$$x_t^n = \sum_{k=0}^{n} y_t^k, \quad n \in \mathbb{N}, \ t \in \mathbb{R}^+$$

are Cauchy sequences in Φ. Let us call $x_t = \lim_{n \to +\infty} x_t^n$. One also easily sees, from the same estimate, that

$$\int_0^t |u(s)|^2 \|x_s\|^2 \, ds < \infty \quad \text{for all } t \in \mathbb{R}^+.$$

Passing to the limit in equality (21), one gets

$$x_t = \int_0^t u(s) x_s \, d\chi_s + \int_0^t u(s) H_s \varepsilon(u_{s]}) \, d\chi_s.$$

Define operators T_t on $\Phi_{t]}$ (more precisely on $\mathcal{E} \cap \Phi_{t]}$) by putting $T_t \varepsilon(u_{t]}) = x_t$. We leave to the reader to check that this defines (by linear extension) an operator on $\mathcal{E} \cap \Phi_{t]}$ (use the fact that any finite family of coherent vectors is free). Extend the operator T_t to $\mathcal{E}$ by adaptedness:

$$T_t \varepsilon(u) = T_t \varepsilon(u_{t]}) \otimes \varepsilon(u_{[t}).$$

We thus get a solution to (19). Uniqueness is easily obtained by Gronwall's lemma.

Let us now prove that this solution satisfies the announced identity. We have

$$\langle \varepsilon(v_{t]}), T_t \varepsilon(u_{t]}) \rangle = \int_0^t \overline{v}(s) u(s) \langle \varepsilon(v_{t]}), T_s \varepsilon(u_{t]}) \rangle \, ds$$

$$+ \int_0^t \overline{v}(s) u(s) \langle \varepsilon(v_{s]}), H_s \varepsilon(u_{s]}) \rangle \, ds.$$

Put $\alpha_t = \langle \varepsilon(v_{t]}), T_t \varepsilon(u_{t]}) \rangle$, $t \in \mathbb{R}^+$. We have

$$\alpha_t = \int_0^t \overline{v}(s) u(s) \alpha_s \, ds + \int_0^t \overline{v}(s) u(s) \langle \varepsilon(v_{s]}), H_s \varepsilon(u_{s]}) \rangle \, ds$$

that is,

$$\frac{d}{dt} \alpha_t = \overline{v}(t) u(t) \alpha_t + \overline{v}(t) u(t) \langle \varepsilon(v_{t]}), H_t \varepsilon(u_{t]}) \rangle.$$

Or else

$$\alpha_t = e^{\int_0^t \overline{v}(s) u(s) \, ds} \int_0^t \overline{v}(s) u(s) \langle \varepsilon(v_{s]}), H_s \varepsilon(u_{s]}) \rangle e^{-\int_0^s \overline{v}(k) u(k) \, dk} \, ds$$

$$= \int_0^t \overline{v}(s) u(s) \langle \varepsilon(v_{s]}), H_s \varepsilon(u_{s]}) \rangle e^{-\int_s^t \overline{v}(k) u(k) \, dk} \, ds$$

$$= \int_0^t \overline{v}(s) u(s) \langle \varepsilon(v_{s]}), H_s \varepsilon(u_{s]}) \rangle \langle \varepsilon(v_{[s,t]}), \varepsilon(u_{[s,t]}) \rangle \, ds$$

$$= \int_0^t \overline{v}(s) u(s) \langle \varepsilon(v_{t]}), H_s \varepsilon(u_{t]}) \rangle \, ds \quad \text{(by adaptedness)}.$$

The converse direction is easy to obtain by reversing the above arguments. $\quad\square$

Let us now see how equation (19) can provide a solution on Φ_f, the space of finite particles. We still only take the example $T_t = \int_0^t H_s \, da_s^\circ$ (the reader may easily check the other three cases). The following computation are only made algebraically, without taking much care about integrability or domain problems. We have the equation

$$T_t f_t = \int_0^t T_s D_s f \, d\chi_s + \int_0^t H_s D_s f \, d\chi_s.$$

Let $f = \mathbb{1}$. This implies (as $D_t \mathbb{1} = 0$ for all t)

$$T_t \mathbb{1} = 0 .$$

Let $f = \int_0^\infty f_1(s) \, d\chi_s$ for $f_1 \in L^2(\Sigma_1)$. We have

$$T_t f_t = \int_0^t T_s f_1(s) \mathbb{1} \, d\chi_s + \int_0^t H_s f_1(s) \mathbb{1} \, d\chi_s$$

$$= 0 + \int_0^t f_1(s) H_s \mathbb{1} \, d\chi_s.$$

Let $f = \int_{0 \le t_1 \le t_2} f_2(t_1, t_2) \, d\chi_{t_1} \, d\chi_{t_2}$ for $f_2 \in L^2(\Sigma_2)$. We have

$$T_t f_t = \int_0^t T_s \int_0^s f_2(t_1, s) \, d\chi_{t_1} \, d\chi_s + \int_0^t H_s \int_0^s f_2(t_1, s) \, d\chi_{t_1} \, d\chi_s$$

$$= \int_0^t \int_0^s f_2(u, s) H_u \mathbb{1} \, d\chi_u \, d\chi_s + \int_0^t H_s \int_0^s f_2(u, s) \, d\chi_u \, d\chi_s.$$

This way, one sees that, by induction on the chaoses, one can derive the action of T_t on Φ_f.

Let us now give the formulas for the formal adjoint of a quantum stochastic integral. We do not discuss here the very difficult problem of the domain of the adjoint of a quantum stochastic integral and the fact that it is a quantum stochastic integral or not. In the case of the domain $\mathcal{E}$ the reader may easily derive conditions for this adjoint to exist on $\mathcal{E}$.

$$\left(\int_0^\infty H_s \, da_s^\circ \right)^* = \int_0^\infty H_s^* \, da_s^\circ$$

$$\left(\int_0^\infty H_s \, da_s^+ \right)^* = \int_0^\infty H_s^* \, da^-$$

$$\left(\int_0^\infty H_s \, da_s^- \right)^* = \int_0^\infty H_s^* \, da_s^+$$

$$\left(\int_0^\infty H_s \, da_s^\times \right)^* = \int_0^\infty H_s^* \, da_s^\times$$

We now have a useful theorem which often helps to extend the domain of a quantum stochastic integral when it is already defined on $\mathcal{E}$.

Theorem 4.3 (Extension theorem). *If* $(T_t)_{t\geq 0}$ *is an adapted process of operators on* Φ *which admits an integral representation on* $\mathcal{E}$ *and such that the adjoint process* $(T_t^*)_{t\geq 0}$ *admits an integral representation on* $\mathcal{E}$. *Then the integral representations of* $(T_t)_{t\geq 0}$ *and* $(T_t^*)_{t\geq 0}$ *can be extended everywhere equation (19) is meaningful.*

Before proving this theorem, we shall maybe be clear about what it exactly means. The hypotheses are that:

- $T_f = \int_0^t H_s^\circ \, da_s^\circ + \int_0^t H_s^+ \, da_s^+ + \int_0^t H_s^- \, da_s^- + \int_0^t H_s^\times \, da_s^\times$ on $\mathcal{E}$. This in particular means that

$$\int_0^t |u(s)|^2 \|H_s^\circ \varepsilon(u_{s]})\|^2 + \|H_s^+ \varepsilon(u_{s]})\|^2 + |u(s)|\, \|H_s^- \varepsilon(u_{s]})\| + \|H_s^\times \varepsilon(u_{s]})\|\, ds$$

is finite for all $t \in \mathbb{R}^+$, all $u \in L^2(\mathbb{R}^+)$.

- The assumption on the adjoint simply means that

$$\int_0^t |u(s)|^2 \|H_s^{0*} \varepsilon(u_{s]})\|^2 + \|H_s^{-*} \varepsilon(u_{s]})\|^2 + |u(s)|\, \|H_s^{+*} \varepsilon(u_{s]})\|$$

$$+ \|H_s^{\times *} \varepsilon(u_{s]})\|\, ds < \infty$$

for all $t \in \mathbb{R}^+$ and all u in $L^2(\mathbb{R}^+)$.

The conclusion is that for all $f \in \Phi$, such that equation (19) is meaningful (for $(T_t)_{t\geq 0}$ or for $(T_t^*)_{t\geq 0}$), then the equality (19) will be valid.

Let us take an example. Let $J_t \varepsilon(u) = \varepsilon(-u_{t]}) \otimes \varepsilon(u_{[t})$. It is an adapted process of operators on Φ which is made of unitary operators, and $J_t^2 = I$. We leave as an exercise to check the following points.

- The quantum stochastic integral $B_t = \int_0^t J_s \, da_s^-$ is well defined on $\mathcal{E}$, the quantum stochastic integral $B_t^* = \int_0^t J_s da_s^+$ is well defined on $\mathcal{E}$ and is the adjoint of B_t (on $\mathcal{E}$);
- We have $J_t = I - 2\int_0^t J_s \, da_s^\circ$;
- If $X_t = -2\int_0^t X_s \, da_s^\circ$ then $X_t \equiv 0$ for all t.
- Altogether this gives
$$B_t J_t + J_t B_t = 0 \; ;$$

- We conclude that $B_t B_t^* + B_t^* B_t = tI$.

The last identity shows that B_t is a bounded operator with norm smaller that $\sqrt{t}$. Now, we know that, for all $f \in \mathcal{E}$ we have

$$B_t f_t = \int_0^t B_s D_s f \, d\chi_s + \int_0^t J_s D_s f \, ds. \tag{22}$$

We know that the adjoint of B_t can be represented as a Quantum stochastic integral on $\mathcal{E}$. Hence the hypotheses of the Extension Theorem hold.

For which $f \in \Phi$ do we have all the terms of equation (19) being well defined? The results above easily show that for *all* $f \in \Phi$ the quantities $B_t f_t$, $\int_0^t B_s D_s f \, d\chi_s$, $\int_0^t J_s D_s f \, ds$ are well defined. Hence the extension theorem says that equation (19) is valid for all $f \in \Phi$. The same holds for B_t^*. The integral representation of $(B_t)_{t \geq 0}$ (and $(B_t^*)_{t \geq 0}$) is valid on *all* Φ.

Let us now prove the extension theorem.

Proof. Let $f \in \Phi$ be such that all the terms of equation (19) are meaningful. Let $(f_n)_n$ be a sequence in $\mathcal{E}$ which converges to f. Let $g \in \mathcal{E}$. We have

$$\left| \langle g, T_t f_t - \int_0^t T_s D_s f \, d\chi_s - \int_0^t H_s^\circ D_s f \, d\chi_s \right.$$
$$\left. - \int_0^t H_s^+ P_s f \, d\chi_s - \int_0^t H_s^- D_s f \, ds - \int_0^t H_s^\times P_s f \, ds \rangle \right|$$

$$\leq |\langle g, T_t P_t (f - f_n) \rangle| + \left| \langle g, \int_0^t T_s D_s (f - f_n) \, d\chi_s \rangle \right|$$

$$+ \left| \langle g, \int_0^t H_s^\circ D_s (f - f_n) \, d\chi_s \rangle \right| + \left| \langle g, \int_0^t H_s^\times P_s (f - f_n) \, ds \rangle \right|$$

$$+ |\langle g, \int_0^t H_s^- D_s (f - f_n) \, ds \rangle| + |\langle g, \int_0^t H_s^\times P_s (f - f_n) \, ds \rangle|$$

$$\leq \|T_t^* g\| \, \|f - f_n\| + \int_0^t |\langle T_s^* D_s g, D_s (f - f_n) \rangle| \, ds$$

$$+ \int_0^t |\langle H_s^{\circ *} D_s g, D_s (f - f_n) \rangle| \, ds + \int_0^t |\langle H_s^{+ *} D_s g, P_s (f - f_n) \rangle| \, ds$$

$$+ \int_0^t |\langle H_s^{- *} g, D_s (f - f_n) \rangle| \, ds + \int_0^t |\langle H_s^{\times *} g, P_s (f - f_n) \rangle| \, ds$$

$$\leq \left[\|T_t^* g\| + \int_0^t \|T_s^* D_s g\|^2 \, ds + \int_0^t \|H_s^{\circ *} D_s g\|^2 \, ds + \int_0^t \|H_s^{+ *} D_s g\| \, ds \right.$$

$$\left. + \int_0^t \|H_s^{- *} g\|^2 \, ds + \int_0^t \|H_s^{\times *} g\| \, ds \right] \|f - f_n\| \, .$$

The theorem is proved. $\square$

Quantum stochastic integrals satisfy a quantum Itô formula, that is, they are stable under composition and the integral representation of the composition is given by a Itô-like integration by part formula.

The complete quantum Itô formula with correct domain assumptions is a rather heavy theorem. We shall give a complete statement of it later on, but for the moment we state under a form which is sufficient for many applications.

Let

$$T_t = \int_0^t H_s \, da_s^\varepsilon$$

and

$$S_t = \int_0^t K_s \, da_s^\nu$$

be two quantum stochastic integral processes. Then, on any domain where each term is well defined we have

$$T_t S_t = \int_0^t T_s \, dS_s + \int_0^t dT_s \, S_s + \int_0^t dT_s \, dS_s$$

in the sense

$$T_t S_t = \int_0^t T_s K_s \, da_s^\nu + \int_0^t H_s S_s \, da_s^\varepsilon + \int_0^t H_s K_s \, da_s^\varepsilon da_s^\nu$$

where the quadratic terms $da_s^\varepsilon da_s^\nu$ are given by the following Itô table:

	da^+	da^-	da°	$da^\times$
da^+	0	0	0	0
da^-	$da^\times$	0	da^-	0
da°	da^+	0	da°	0
$da^\times$	0	0	0	0

This quantum Itô formula will be proved in section 5.1 in the case of quantum stochastic integrals having the whole of Φ as a domain.

Maximal solution

This section is not necessary for the understanding of the rest of the course, it is addressed to readers motivated by fine domain problems on quantum stochastic integrals.

We have not yet discussed here the existence of solution to equation (19) in full generality. That is, for a given adapted process of operators $(H_t)_{t \geq 0}$ and a given $\varepsilon \in \{+, -, \circ, \times\}$ we consider the associated equation (19). We then wonder

i) if there always exists a solution $(T_t)_{t \geq 0}$;

ii) if the solution is always unique;

iii) on which maximal domain that solution is defined.

The complete answer to these three questions has been given in [9]. It is a long and difficult result for which we need to completely revisit the whole theory of quantum stochastic calculus and the notion of adaptedness. Here we shall just give the main result.

For $\sigma = \{t_1 < \ldots < t_n\} \in \mathcal{P}$ we put

$$D_\sigma = D_{t_1} \ldots D_{t_n}$$

with $D_\emptyset = I$.

Consider the following operators on Φ

$$[\Lambda_t^\circ(H.)f](\sigma) = \sum_{s \in \sigma_{t]}} [H_s D_s D_{\sigma_{(s}} f](\sigma_s)$$

$$[\Lambda_t^+(H.)f](\sigma) = \sum_{s \in \sigma_{t]}} [H_s P_s D_{\sigma_{(s}} f](\sigma_s)$$

$$[\Lambda_t^-(H.)f](\sigma) = \int_0^t [H_s D_s D_{\sigma_{(s}} f](\sigma_s)\, ds$$

$$[\Lambda_t^\times(H.)f](\sigma) = \int_0^t [H_s P_s D_{\sigma_{(s}} f](\sigma_s)\, ds$$

together with their maximal domain Dom $\Lambda_t^\varepsilon(H.)$, that is, the space of $f \in \Phi$ such that the above expression is well-defined and square integrable as a function of σ.

We then have the following complete characterization (see [2]).

Theorem 4.4. *For every adapted process* $(H_t)_{t \geq 0}$ *of operators on* Φ *and every* $\varepsilon \in \{\circ, +, -, \times\}$, *the following assertions are equivalent.*

i) $(T_t)_{t \geq 0}$ *is a solution of the equation (19).*

ii) $(T_t)_{t \geq 0}$ *is the restriction of* $\Lambda_t^\varepsilon(H.)$ *to a stable subspace of* Dom $\Lambda_t^\varepsilon(H.)$.

This result means that with the above formulas and above domains we have

i) the explicit action of any quantum stochastic integral on any vector of its domain

ii) the maximal domain of that operator

iii) the right to use equation (19) on that domain without restriction (every term is well-defined).

4.3 Back to probabilistic interpretations

Multiplication operators

Consider a probabilistic interpretation $(\Omega, \mathcal{F}, P, (x_t)_{t \geq 0})$ of the Fock space, which is described by a structure equation

$$d[x, x]_t = dt + \psi_t\, dx_t.$$

The operator M_{x_t} on Φ of multiplication by x_t (for this interpretation) is a particular operator on Φ. It is adapted at time t. The process $(M_{x_t})_{t \geq 0}$ is an adapted process of operators on Φ. Can we represent this process as a sum of quantum stochastic integrals?

If one denotes by M_{ψ_t} the operator of multiplication by ψ_t (for the $(x_t)_{t \geq 0}$-product again) we have the following:

Theorem 4.5.

$$M_{x_t} = a_t^+ + a_t^- + \int_0^t M_{\psi_t}\, da_t^\circ.$$

Proof. Let us be clear about domains: the domain of M_{x_t} is exactly the space of $f \in \Phi$ such that $x_t \cdot U_x f$ belongs to $L^2(\Omega)$ (recall that U_x is the isomorphism $U_x : \Phi \to L^2(\Omega)$).

Let us now go to the proof of the theorem. We have

$$x_t f = \int_0^\infty x_{s \wedge t} D_s f\, dx_s + \int_0^t P_s f\, dx_s + \int_0^t D_s f\, ds + \int_0^t \psi_s D_s f\, dx_s$$

by the usual Itô formula. That is, on Φ

$$M_{x_t} f = \int_0^\infty M_{x_{s \wedge t}} D_s f\, d\chi_s + \int_0^t P_s f\, d\chi_s + \int_0^t D_s f\, ds + \int_0^t M_{\psi_s} D_s f\, d\chi_s$$

which is exactly equation (19) for the quantum stochastic process $X_t = a_t^+ + a_t^- + \int_0^t M_{\psi_t}\, da_t^\circ$. $\square$

In particular we have obtained the following very important results.

- The multiplication operator by the Brownian motion is $a_t^+ + a_t^-$.
- The multiplication operator by compensated Poisson process is $a_t^+ + a_t^- + a_t^\circ$.
- The multiplication operator by the β-Azéma martingale is the unique solution

of

$$X_t = a_t^+ + a_t^- + \int_0^t \beta X_s\, da_s^\circ.$$

Once again, as in the discrete time setup, we have obtained in a single structure, the Fock space Φ, a very simple way to represent many different classical noises that have nothing to do together. Furthermore their representation is obtained by very simple combinations of the three quantum noises. The three quantum noises appear as very natural (their form, together with the process $a_t^\times$, a kind of basis for local operator processes on the continuous tensor product structure of Φ), and they constitute basic bricks from which one can recover the main classical noises.

5 The algebra of regular quantum semimartingales

In this section we present several developments of the definitions of quantum stochastic integrals. These developments make great use of the versatility of our definitions, in particular the fact that quantum stochasitc integrals can a priori be defined on any kind of domain.

5.1 Everywhere defined quantum stochastic integrals

A true quantum Itô formula

With our definition of quantum stochastic integrals defined on any stable domain, we may meet quantum stochastic integrals that are defined on the whole of Φ. Let us recall a few facts. An adapted process of bounded operators $(T_t)_{t\geq 0}$ on Φ is said to have the integral representation

$$T_t = \sum_{\varepsilon=\{0,+,-,\times\}} \int_0^t H_s^\varepsilon \, da_s^\varepsilon$$

on the *whole of* Φ if, for all $f \in \Phi$ one has

$$\int_0^t \|T_s D_s f\|^2 + \|H_s^\circ D_s f\|^2 + \|H_s^+ P_s f\|^2 + \|H_s^- D_s f\| + \|H_s^\times P_s f\| \, ds < \infty$$

for all $t \in I\!\!R^+$ (the H_t^ε are bounded operators) and

$$T_t P_t f = \int_0^t T_s D_s f \, d\chi_s + \int_0^t H_s^\circ D_s f \, d\chi_s + \int_0^t H_s^+ P_s f \, d\chi_s + \int_0^t H_s^- D_s f \, ds$$
$$+ \int_0^t H_s^\times P_s f \, ds.$$

If we have two such processes $(S_t)_{t\geq 0}$ and $(T_t)_{t\geq 0}$ one can compose them. As announced previously with the quantum Itô formula, the resulting process $(S_t T_t)_{t\geq 0}$ is also representable as a sum of quantum stochastic integrals on the whole of Φ.

Theorem 5.1. *If $T_t = \sum_\varepsilon \int_0^t H_s^\varepsilon \, da_s^\varepsilon$ and $S_t = \sum_\varepsilon \int_0^t K_s^\varepsilon \, da_s^\varepsilon$ are everywhere defined quantum stochastic integrals, then $(S_t T_t)_{t\geq 0}$ is everywhere representable as a sum of quantum stochastic integrals:*

$$S_t T_t = \int_0^t (S_s H_s^\circ + K_s^\circ T_s + K_s^\circ H_s^\circ) \, da_s^\circ + \int_0^t (S_s H_s^+ + K_s^+ T_s + K_s^\circ H_s^+) \, da_s^+$$
$$+ \int_0^t (S_s H_s^- + K_s^- T_s + K_s^- H_s^\circ) \, da_s^- + \int_0^t (S_s H_s^\times + K_s^\times T_s + K_s^- H_s^+) \, da_s^\times.$$

Before proving this theorem we will need the following preliminary result.

Lemma 5.2. *Let $g_t = \int_0^t v_s \, ds$ be an adapted process of vectors of Φ, with $\int_0^t \|v_s\| \, ds < \infty$ for all t. Let $(S_t)_{t\geq 0}$ be as in Theorem 5.1. Then*

$$S_t g_t = \int_0^t S_s v_s \, ds + \int_0^t K_s^+ g_s \, d\chi_s + \int_0^t K_s^\times g_s \, ds.$$

Proof. As S_t is bounded we have (details are left to the reader)

$$S_t g_t = S_t \int_0^t v_s \, ds = \int_0^t S_t v_s \, ds$$

$$= \int_0^t S_t \left(P_0 v_s + \int_0^s D_u v_s \, d\chi_u \right) ds$$

$$= \int_0^t S_t P_0 v_s \, ds + \int_0^t \left[\int_0^s S_u D_u v_s \, d\chi_u + \int_0^s K_u^\circ D_u v_s \, d\chi_u \right.$$

$$+ \int_0^s K_u^- D_u v_s \, ds + \int_0^t K_u^+ P_u \int_0^s D_v v_s \, d\chi_v \, d\chi_u$$

$$\left. + \int_0^t K_u^\times P_u \int_0^s D_v v_s \, d\chi_v \, du \right] ds$$

$$= \int_0^t S_t P_0 v_s \, ds + \int_0^t \left[S_s \int_0^s D_u v_s \, d\chi_u + \int_s^t K_u^+ \int_0^s D_v v_s d\chi_v \, d\chi_u \right.$$

$$\left. + \int_s^t K_u^\times \int_0^s D_v v_s \, d\chi_v \, du \right] ds$$

$$= \int_0^t S_s v_s \, ds + \int_0^t \int_s^t K_u^+ v_s \, d\chi_u \, ds + \int_0^t \int_s^t K_u^\times v_s \, du \, ds$$

$$= \int_0^t S_s v_s \, ds + \int_0^t \int_0^u K_u^+ v_s \, ds \, d\chi_u + \int_0^t \int_0^u K_u^\times v_s \, ds \, du$$

$$= \int_0^t S_s v_s \, ds + \int_0^t K_u^+ \int_0^u v_s \, ds \, d\chi_u + \int_0^t K_u^\times \int_0^u v_s \, ds \, du$$

$$= \int_0^t S_s v_s \, ds + \int_0^t K_u^+ g_u \, d\chi_u + \int_0^t K_u^\times g_u \, du.$$

This proves the Lemma. $\quad\square$

We now prove the theorem.

Proof. We just compute the composition, using Lemma 5.2

$$T_t f_t = \int_0^t T_s D_s f \, d\chi_s + \int_0^t H_s^\circ D_s f \, d\chi_s + \int_0^t H_s^+ P_s f \, d\chi_s$$

$$+ \int_0^t H_s^- D_s f \, ds + \int_0^t H_s^\times P_s f \, ds.$$

Hence

$$S_t T_t f_t = \int_0^t S_s \left[T_s D_s f + H_s^\circ D_s f + H_s^+ P_s f \right] d\chi_s$$

$$+ \int_0^t K_s^\circ \left[T_s D_s f + H_s^\circ D_s f + H_s^+ P_s f \right] d\chi_s + \int_0^t K_s^- \left[T_s D_s f + H_s^\circ D_s f + H_s^+ P_s f \right] ds$$

$$+ \int_0^t K_s^+ \left[\int_0^s T_u D_u f \, d\chi_u + \int_0^s H_u^\circ D_u f \, d\chi_u + \int_0^s H_u^+ P_u f \, d\chi_u \right] d\chi_s$$

$$+ \int_0^t K_s^\times \left[\int_0^s T_u D_u f \, d\chi_u + \int_0^s H_u^\circ D_u f \, d\chi_u + \int_0^s H_u^+ P_u f \, d\chi_u \right] du$$

$$+ \int_0^t S_s \left[H_s^- D_s f + H_s^\times P_s f \right] ds + \int_0^t K_s^+ \left[\int_0^s H_u^- D_u f + \int_0^s H_u^\times P_u f \, du \right] d\chi_s$$

$$+ \int_0^t K_s^\times \left[\int_0^s H_u^- D_u f + \int_0^s H_u^\times P_u f \, du \right] ds$$

$$= \int_0^t S_s T_s D_s f \, d\chi_s + \int_0^t \left[S_s H_s^\circ + K_s^\circ T_s + K_s^\circ H_s^\circ \right] D_s f \, d\chi_s$$

$$+ \int_0^t \left[S_s H_s^+ + K_s^+ T_s + K_s^\circ H_s^+ \right] P_s f \, d\chi_s + \int_0^t \left[S_s H_s^- + K_s^- T_s + K_s^- H_s^\circ \right] D_s f \, ds$$

$$+ \int_0^t \left[S_s H_s^\times + K_s^\times T_s + K_s^- H_s^+ \right] P_s f \, ds \, .$$

This proves the theorem. $\square$

A family of examples

We have seen $B_t = \int_0^t J_s \, da_s^-$ as an example of everywhere defined quantum stochastic integrals. This example belongs to a larger family of examples which we shall present here.

Let $\mathcal{S}$ be the set of *bounded* adapted processes of operators $(T_t)_{t \geq 0}$ on Φ such that

$$T_t = \sum_\varepsilon \int_0^t H_s^\varepsilon \, da_s^\varepsilon \quad \text{on } \mathcal{E},$$

all the operators H_s^ε being bounded and

$$\begin{cases} t \mapsto \|H_t^\circ\| \in L^\infty_{\text{loc}}(I\!R^+) \\ t \mapsto \|H_t^+\| \in L^2_{\text{loc}}(I\!R^+) \\ t \mapsto \|H_t^-\| \in L^2_{\text{loc}}(I\!R^+) \\ t \mapsto \|H_t^\times\| \in L^1_{\text{loc}}(I\!R^+) \, . \end{cases}$$

With these conditions, we claim that $t \mapsto \|T_t\|$ has to be in $L^\infty_{\text{loc}}(I\!R^+)$. Indeed, the operator $\int_0^t H_s^\times \, da_s^\times$ satisfies

$$\int_0^t H_s^\times \, da_s^\times f = \int_0^t H_s^\times f \, ds$$

hence it is a bounded operator, with norm dominated by $\int_0^t \|H_s^\times\| \, ds$, which is a locally bounded function of t. The difference $M_t = T_t - \int_0^t H_s^\times \, da_s^\times$ is thus a martingale of bounded operators, that is $P_s M_t P_s = M_s P_s$ for all $s \leq t$. But as M is a martingale we have $\|M_s f_s\| = \|P_s M_t f_s\| \leq \|M_t f_s\|$ for $s \leq t$. Hence $t \mapsto \|M_t\|$ is locally bounded. Thus, so is $t \mapsto \|T_t\|$.

With all these informations, it is easy to check that the integral representation of $(T_t)_{t \geq 0}$, as well as the one of $(T_t^*)_{t \geq 0}$, can be extended on the whole of Φ by the extension Theorem (Theorem 4.3).

5.2 The algebra of regular quantum semimartingales

It is an algebra

As all elements of $\mathcal{S}$ are everywhere defined quantum stochastic integrals, one can compose them and use the quantum Itô formula (Theorem 5.1.

Theorem 5.3. $\mathcal{S}$ *is a $*$-algebra for the adjoint and composition operations.*

Proof. Let

$$T_t = \int_0^t H_s^\circ \, da_s^\circ + \int_0^t H_s^+ \, da_s^+ + \int_0^t H_s^- \, da_s^- + \int_0^t H_s^\times \, da_s^\times$$

and

$$S_t = \int_0^t K_s^\circ \, da_s^\circ + \int_0^t K_s^+ \, da_s^+ + \int_0^t K_s^- \, da_s^- + \int_0^t K_s^\times \, da_s^\times$$

be two elements of $\mathcal{S}$. The adjoint process $(T_t^*)_{t \geq 0}$ is given by

$$T_t^* = \int_0^t H_s^{\circ *} \, da_s^\circ + \int_0^t H_s^{-*} \, da_s^+ + \int_0^t H_s^{+*} \, da_s^- + \int_0^t H_s^{\times *} \, da_s^\times \, .$$

It is straightforward to check that it belongs to $\mathcal{S}$. The Itô formula for the composition of two elements of $\mathcal{S}$ gives

$$S_t T_t = \int_0^t \left[S_s H_s^\circ + K_s^\circ T_s + K_s^\circ H_s^\circ \right] da_s^\circ$$
$$+ \int_0^t \left[S_s H_s^+ + K_s^+ T_s + K_s^\circ H_s^+ \right] da_s^+$$
$$+ \int_0^t \left[S_s H_s^- + K_s^- T_s + K_s^- H_s^\circ \right] da_s^-$$
$$+ \int_0^t \left[S_s H_s^\times + K_s^\times T_s + K_s^- H_s^+ \right] da_s^\times \, .$$

From the conditions on the maps $t \mapsto \|S_t\|$, $t \mapsto \|T_t\|$, $t \mapsto \|K_t^\varepsilon\|$ and $t \mapsto \|H_t^\varepsilon\|$, it is easy to check that the coefficients in the representation of $(S_t T_t)_t \geq 0$ are bounded operators that satisfy the norm conditions for being in $\mathcal{S}$. For example, the coefficient of $da_t^\times$ satisfies

$$\int_0^t \|S_s H_s^\times + K_s^\times T_s + K_s^- H_s^+\| \, ds$$

$$\leq \sup_{s \leq t} \|S_s\| \int_0^t \|H_s^\times\| \, ds + \sup_{s \leq t} \|T_s\| \int_0^t \|K_s^\times\| \, ds$$

$$+ \left(\int_0^t \|K_s^-\|^2 \, ds \right)^{1/2} \left(\int_0^t \|H_s^+\|^2 \, ds \right)^{1/2}$$

hence it is locally integrable.   □

Thus $\mathcal{S}$ is a nice space of quantum semimartingales that one can compose without bothering about any domain problem, one can pass to the adjoint, one can use formula (19) on the whole of Φ.

A characterization

A problem comes from the definition of $\mathcal{S}$. Indeed, it is in general difficult to know if a process of operators is representable as quantum stochastic integrals; it is even more difficult to know the regularity of its coefficients. We know that $\mathcal{S}$ is not empty, as it contains $B_t = \int_0^t J_s \, da_s^-$ that we have met above. It is natural to wonder how large that space is. It is natural to seek for a characterization of $\mathcal{S}$ that depends only on the process $(T_t)_t \geq 0$.

One says that a process $(T_t)_t \geq 0$ of *bounded* adapted operators is a *regular quantum semimartingale* is there exists a locally integrable function h on $\mathbb{R}$ such that for all $r \leq s \leq t$, all $f \in \mathcal{E}$ one has (where $f_r = P_r f$)

i) $\|T_t f_r - T_s f_r\|^2 \leq \|f_r\|^2 \int_s^t h(u) \, du$;

ii) $\|T_t^* f_r - T_s^* f_r\|^2 \leq \|f_r\|^2 \int_s^t h(u) \, du$;

iii) $\|P_s T_t f_r - T_s f_r\| \leq \|f_r\| \int_s^t h(u) \, du$.

Theorem 5.4. *A process $(T_t)_t \geq 0$ of bounded adapted operators is a regular quantum semimartingale if and only if it belongs to $\mathcal{S}$.*

Proof. Showing that elements of $\mathcal{S}$ satisfy the three estimates that define regular quantum semimartingales is straightforward. We leave it as an exercise.

The interesting part is to show that a regular quantum semimartingale is representable as quantum stochastic integrals and belongs to $\mathcal{S}$. We will only sketch that proof, as the details are rather long and difficult to develop.

Let $x_t = T_t f_r$ for $t \geq r$ (r is fixed, t varies). It is an adapted process of vectors on Φ. It satisfies

$$\|P_s x_t - x_s\| \le \|f_r\| \int_s^t h(u) \, du.$$

This condition is a Hilbert space analogue of a condition in classical probability that defines particular semimartingales: the quasimartingales. O. Enchev [18] has provided a Hilbert space extension of this result and we can deduce from his result that $(x_t)_{t \ge r}$ can be written

$$x_t = m_t + \int_0^t k_s \, ds$$

where m is a martingale in Φ $(P_s m_t = m_s)$ and h is an adapted process in Φ such that $\int_0^t \|k_s\| \, ds < \infty$.

Thus $P_s x_t - x_s = \int_s^t P_s k_u \, du$ and we have

$$\left\| \int_0^t P_s k_u \, du \right\| \le \|f_r\| \int_0^t h(u) \, du, \quad \text{for all } r \le s \le t.$$

Actually k_u depends linearly on f_r. The inequality above implies (difficult exercise) that

$$\|k_u(f_r)\| \le \|f_r\| h(u) .$$

Hence k_u is a bounded operator on $\Phi_{u]}$, we extend it as a bounded adapted operator $H_u^\times$.

Let $M_t = T_t - \int_0^t H_u^\times \, da_u^\times$, $t \in \mathbb{R}^+$. It is easy to check, from what we have already done, that $(M_t)_t \ge 0$ is a martingale of bounded operator (Hint: compute $P_s M_t f_r - M_s f_r$). It is easy to check that $(M_t)_t \ge 0$ also satisfies the conditions $i)$ and $ii)$ of the definition of regular quantum semimartingales, with another function h, say h'.

Now, let $(y_t)_{t \ge r}$ be $(M_t f_r)_{t \ge r}$. It is a martingale of vectors in Φ. Thus it can be represented as

$$y_t - y_s = \int_s^t \xi_u \, d\chi_u.$$

The vector ξ_u depends linearly on f_r and we have

$$\int_0^t \|\xi_u(f_r)\|^2 \, du \le \|f_r\|^2 \int_s^t h'(u) \, du \ \text{(by } i)).$$

Thus ξ_u extends to a unique adapted bounded operator H_u^+ on Φ. Doing the same with $(M_t^* f_r)_{t \ge r}$ gives an adapted process of operators (bounded): $(H_u^-)_{u \ge 0}$.

Let $f \in \Phi$, let $f_t = P_t f$ and define

$$X_t f_t = T_t f_t - \int_0^t T_s D_s f \, d\chi_s - \int_0^t H_s^+ P_s f \, d\chi_s - \int_0^t H_s^- D_s f \, ds - \int_0^t H_s^\times P_s f \, ds.$$

One easily checks that each X_t commutes with all the P_u's, $u \in \mathbb{R}^+$. Let us consider a bounded operator H on Φ such that $P_u H = H P_u$ for all $u \in \mathbb{R}^+$. Notice that for almost all t, all $a \le t \le b$, all f one has

$$D_t H(P_b f - P_a f) = D_t P_b H f - D_t P_a H f = D_t H f$$

for $D_t P_s = \begin{cases} D_t & \text{if } t \le s \\ 0 & \text{if } t > s. \end{cases}$

Define $\widetilde{H}_t^\circ$ by

$$\widetilde{H}_t^\circ f_t = D_t \int_a^b P_u f \, d\chi_u - H f_t \quad \text{for any } a \le t \le b.$$

By computing $\int_a^b \|\widetilde{H}_t^\circ f_t\|^2 \, dt$ one easily checks that $\widetilde{H}_t^\circ$ is bounded with locally bounded norm. Moreover we have

$$H g = \int_0^\infty H D_s f \, d\chi_s + \int_0^\infty \widetilde{H}_s^\circ D_s f \, d\chi_s .$$

That is exactly $H = \int_0^\infty \widetilde{H}_s^\circ \, da_s^\circ$.

Actually we have (almost) proved the following nice characterization.

Theorem 5.5. *Let T be a bounded operator on Φ. The following are equivalent.*

i) $T P_t = P_t T$ for all $t \in \mathbb{R}^+$;

ii) $T = \lambda I + \int_0^\infty H_s \, da_s^\circ$ on the whole of Φ.

Applying this to X_t, we finally get, putting $H_s^\circ = \widetilde{H}_s^\circ + X_s$

$$T_t f_t = \int_0^t T_s D_s f \, d\chi_s + \int_0^t H_s^\circ D_s f \, d\chi_s + \int_0^t H_s^+ P_s f \, d\chi_s$$
$$+ \int_0^t H_s^- D_s f \, ds + \int_0^t H_s^\times P_s f \, ds.$$

This is equation (19) for the announced integral representation.   □

6 Approximation by the toy Fock space

In this section, we are back to the spin chain setting. As announced in the first section of this course, we will show that the toy Fock space $\mathrm{T}\Phi$ can be embedded into the Fock space Φ in such a way that it constitutes an approximation of it and of its basic operators.

6.1 Embedding the toy Fock space into the Fock space

Let $\mathcal{S} = \{0 = t_0 < t_1 < \cdots < t_n < \cdots\}$ be a partition of $\mathbb{R}^+$ and $\delta(\mathcal{S}) = \sup_i |t_{i+1} - t_i|$ be the diameter of $\mathcal{S}$. For $\mathcal{S}$ being fixed, define $\Phi_i = \Phi_{[t_i, t_{i+1}]}$, $i \in \mathbb{N}$. We then have $\Phi \simeq \bigotimes_{i \in \mathbb{N}} \Phi_i$. For all $i \in \mathbb{N}$, define

$$X_i = \frac{\chi_{t_{i+1}} - \chi_{t_i}}{\sqrt{t_{i+1} - t_i}} \in \Phi_i \,,$$

$$a_i^- = \frac{a_{t_{i+1}}^- - a_{t_i}^-}{\sqrt{t_{i+1} - t_i}} \circ P_{1]} \,,$$

$$a_i^\circ = a_{t_{i+1}}^\circ - a_{t_i}^\circ \,,$$

$$a_i^+ = P_{1]} \circ \frac{a_{t_{i+1}}^+ - a_{t_i}^+}{\sqrt{t_{i+1} - t_i}} \,,$$

where $P_{1]}$ is the orthogonal projection onto $L^2(\mathcal{P}_1)$ and where the above definition of a_i^+ is understood to be valid on Φ_i only, with a_i^+ being the identity operator I on the other Φ_j's (the same is automatically true for a_i^-, a_i°).

Proposition 6.1. *With the above notations we have*

$$\begin{cases} a_i^- X_i = 1\!\!1 \\ a_i^- 1\!\!1 = 0 \end{cases}$$

$$\begin{cases} a_i^\circ X_i = X_i \\ a_i^\circ 1\!\!1 = 0 \end{cases}$$

$$\begin{cases} a_i^+ X_i = 0 \\ a_i^+ 1\!\!1 = X_i \end{cases}.$$

Proof. As $a_t^- 1\!\!1 = a_t^\circ 1\!\!1 = 0$ it is clear that $a_i^- 1\!\!1 = a_i^\circ 1\!\!1 = 0$. Furthermore, $a_t^+ 1\!\!1 = \chi_t$ thus

$$a_i^+ 1\!\!1 = P_{1]} \frac{\chi_{t_{i+1}} - \chi_{t_i}}{\sqrt{t_{i+1} - t_i}} = X_i \,.$$

Furthermore, by (19) we have

$$a_i^- X_i = \frac{1}{t_{i+1} - t_i} \left(a_{t_{i+1}}^- - a_{t_i}^- \right) \int_{t_i}^{t_{i+1}} \mathbb{1} \, d\chi_t$$

$$= \frac{1}{t_{i+1} - t_i} \left[\int_{t_i}^{t_{i+1}} \left(a_t^- - a_{t_i}^- \right) \mathbb{1} \, d\chi_t + \int_{t_i}^{t_{i+1}} \mathbb{1} \, dt \right]$$

$$= \frac{1}{t_{i+1} - t_i} \left(0 + t_{i+1} - t_i \right) = \mathbb{1} \, ;$$

$$a_i^\circ X_i = \frac{1}{t_{i+1} - t_i} \left(a_{t_{i+1}}^\circ - a_{t_i}^\circ \right) \int_{t_i}^{t_{i+1}} \mathbb{1} \, d\chi_t$$

$$= \frac{1}{t_{i+1} - t_i} \left[\int_{t_i}^{t_{i+1}} \left(a_t^\circ - a_{t_i}^\circ \right) \mathbb{1} \, d\chi_t + \int_{t_i}^{t_{i+1}} \mathbb{1} \, d\chi_t \right]$$

$$= \frac{1}{t_{i+1} - t_i} \left(\chi_{t_{i+1}} - \chi_{t_i} \right) = X_i \, ;$$

$$a_i^+ X_i = \frac{1}{t_{i+1} - t_i} P_{1]} \left(a_{t_{i+1}}^+ - a_{t_i}^+ \right) \int_{t_i}^{t_{i+1}} \mathbb{1} \, d\chi_t$$

$$= \frac{1}{t_{i+1} - t_i} P_{1]} \left[\int_{t_i}^{t_{i+1}} \left(a_t^+ - a_{t_i}^+ \right) \mathbb{1} \, d\chi_t + \int_{t_i}^{t_{i+1}} \int_{t_i}^{t} \mathbb{1} \, d\chi_s \, d\chi_t \right]$$

$$= \frac{2}{t_{i+1} - t_i} P_{1]} \int_{t_i}^{t_{i+1}} \int_{t_i}^{t} \mathbb{1} \, d\chi_s \, d\chi_t$$

$$= 0 \, .$$

These are the announced relations. □

Thus the action of the operators a_i^ε on the X_i and on $\mathbb{1}$ is similar to the action of the corresponding operators on the toy Fock spaces. We are now going to construct the toy Fock space inside Φ. We are still given a fixed partition $\mathcal{S}$. Define $T\Phi(\mathcal{S})$ to be the space of vectors $f \in \Phi$ which are of the form

$$f = \sum_{A \in \mathcal{P}_{I\!N}} f(A) X_A$$

(with $\|f\|^2 = \sum_{A \in \mathcal{P}_{I\!N}} |f(A)|^2 < \infty$). The space $T\Phi(\mathcal{S})$ can be clearly identified to the toy Fock space $T\!\Phi$; the operators a_i^ε, $\varepsilon \in \{+, -, 0\}$, act on $T\Phi(\mathcal{S})$ exactly in the same way as the corresponding operators on $T\!\Phi$. We have completely embedded the toy Fock space into the Fock space.

6.2 Projections on the toy Fock space

Let $\mathcal{S} = \{0 = t_0 < t_1 < \cdots < t_n < \cdots\}$ be a fixed partition of $I\!\!R^+$. The space $T\Phi(\mathcal{S})$ is a closed subspace of Φ. We denote by $I\!\!E[\cdot / \mathcal{F}(\mathcal{S})]$ the operator of orthogonal projection from Φ onto $T\Phi(\mathcal{S})$.

Proposition 6.2. *If $\mathcal{S} = \{0 = t_0 < t_1 < \cdots < t_n < \cdots\}$ and if $f \in \Phi$ is of the form*

$$f = \int_{0<s_1<\cdots<s_m} f(s_1,\ldots,s_m)\,d\chi_{s_1}\cdots d\chi_{s_m}$$

then

$$\mathbb{E}\left[f/\mathcal{F}(\mathcal{S})\right] = \sum_{i_1<\cdots<i_m\in\mathbb{N}} \frac{1}{\sqrt{t_{i_1+1}-t_{i_1}}\cdots\sqrt{t_{i_m+1}-t_{i_m}}}$$

$$\int_{t_{i_1}}^{t_{i_1+1}}\cdots\int_{t_{i_m}}^{t_{i_m+1}} f(s_1,\ldots,s_m)\,ds_1\cdots ds_m\; X_{i_1}\cdots X_{i_m}\,.$$

Proof. The quantity f_n on the right handside of the above identity is clearly an element of $T\Phi(\mathcal{S})$. We have, for $A=\{i_1\ldots i_k\}$

$$\langle f, X_A\rangle =$$

$$= \frac{\delta_{k,m}}{\sqrt{t_{i_1+1}-t_{i_1}}\cdots\sqrt{t_{i_m+1}-t_{i_m}}}\Big\langle \int_{0<s_1<\cdots<s_m} f(s_1,\ldots,s_m)\,d\chi_{s_1}\cdots d\,\chi_{s_m},$$

$$\int_{t_{i_1}}^{t_{i_1+1}}\cdots\int_{t_{i_m}}^{t_{i_m+1}} \mathbb{1}\,d\chi_{s_1}\cdots d\chi_{s_m}\Big\rangle$$

$$= \frac{\delta_{k,m}}{\sqrt{t_{i_1+1}-t_{i_1}}\cdots\sqrt{t_{i_m+1}-t_{i_m}}}\int_{t_{i_1}}^{t_{i_1+1}}\cdots\int_{t_{i_m}}^{t_{i_m+1}} \overline{f}(s_1,\ldots,s_m)\,ds_1\cdots ds_m\,.$$

On the other hand we have

$$\langle f_n, X_A\rangle = \delta_{k,m}\frac{1}{(t_{i_1+1}-t_{i_1})^{3/2}\cdots(t_{i_m+1}-t_{i_m})^{3/2}}$$

$$\times\int_{t_{i_1}}^{t_{i_1+1}}\cdots\int_{t_{i_m}}^{t_{i_m+1}}\overline{f}(s_1,\ldots,s_m)\,ds_1\cdots ds_m\left\|\left(\chi_{t_{i_1+1}}-\chi_{t_{i_1}}\right)-\left(\chi_{t_{i_m+1}}-\chi_{t_{i_m}}\right)\right\|^2$$

$$= \delta_{k,m}\frac{1}{\sqrt{t_{i_1+1}-t_{i_1}}\cdots\sqrt{t_{i_m+1}-t_{i_m}}}$$

$$\times\int_{t_{i_1}}^{t_{i_1+1}}\cdots\int_{t_{i_m}}^{t_{i_m+1}}\overline{f}(s_1,\ldots,s_m)\,ds_1\cdots ds_m\,.$$

This proves our proposition. □

The following identities could also have been used as natural definitions of the operators a_i^ε on $T\Phi(\mathcal{S})$.

Proposition 6.3. *For any partition $\mathcal{S}$ and any $f\in\mathcal{D}$ we have*

$$a_i^\circ\,\mathbb{E}\left[f/\mathcal{F}(\mathcal{S})\right] = \mathbb{E}\left[\left(a_{t_{i+1}}^\circ - a_{t_i}^\circ\right)f/\mathcal{F}(\mathcal{S})\right]$$

$$\sqrt{t_{i+1}-t_i}\,a_i^\pm\,\mathbb{E}\left[f/\mathcal{F}(\mathcal{S})\right] = \mathbb{E}\left[\left(a_{t_{i+1}}^\pm - a_{t_i}^\pm\right)f/\mathcal{F}(\mathcal{S})\right]\,.$$

Proof. Let us take f of the form

$$f = \int_{0<s_1<\cdots<s_m} f(s_1,\ldots,s_m)\, d\chi_{s_1}\cdots d\chi_{s_m} .$$

Then

$$\left(a^\circ_{t_{i+1}} - a^\circ_{t_i}\right) f = \int_{0<s_1<\cdots<s_m} \left|\{s_1,\ldots,s_m\}\cap[t_i,t_{i+1}]\right| f(s_1,\ldots,s_m)\, d\chi_{s_1}\cdots d\chi_{s_m}$$

$$\mathbb{E}\left[(a^\circ_{t_{i+1}} - a^\circ_{t_i})f/\mathcal{F}(\mathcal{S})\right]$$

$$= \sum_{j_1<\cdots<j_m\in\mathbb{N}} \frac{1}{\sqrt{t_{j_1+1}-t_{j_1}}\cdots\sqrt{t_{j_m+1}-t_{j_m}}} \int_{t_{j_1}}^{t_{j_1+1}}\cdots\int_{t_{j_m}}^{t_{j_m+1}}$$

$$\times \left|\{s_1,\ldots,s_m\}\cap[t_i,t_{i+1}]\right| f(s_1,\ldots,s_m)\, ds_1\cdots ds_m\, X_{j_1}\cdots X_{j_m}$$

$$= \sum_{j_1<\cdots<j_m\in\mathbb{N}} \frac{1}{\sqrt{t_{j_1+1}-t_{j_1}}\cdots\sqrt{t_{j_m+1}-t_{j_m}}}\mathbb{1}_{i\in\{j_1\cdots j_m\}}$$

$$\int_{t_{j_1}}^{t_{j_1+1}}\cdots\int_{t_{j_m}}^{t_{j_m+1}} f(s_1,\ldots,s_m)\, ds_1\cdots ds_m\, X_{j_1}\cdots X_{j_m}$$

$$= a^\circ_i \sum_{j_1<\cdots<j_m\in\mathbb{N}} \frac{1}{\sqrt{t_{j_1+1}-t_{j_1}}\cdots\sqrt{t_{j_m+1}-t_{j_m}}}$$

$$\int_{t_{j_1}}^{t_{j_1+1}}\cdots\int_{t_{j_m}}^{t_{j_m+1}} f(s_1,\ldots,s_m)\, ds_1\cdots ds_m\, X_{j_1}\cdots X_{j_m}$$

$$= a^\circ_i\, \mathbb{E}\left[f/\mathcal{F}(\mathcal{S})\right] .$$

In the same way

$$\left(a^-_{t_{i+1}} - a^-_{t_i}\right) f = \int_{0<s_1<\cdots<s_{m-1}} \int_{t_i}^{t_{i+1}} f(\{s_1,\ldots,s_{m-1}\}\cup s)\, ds\, d\chi_{s_1}\cdots d\chi_{s_{m-1}}$$

$$\mathbb{E}\big[(a_{t_{i+1}}^- - a_{t_i}^-)f/\mathcal{F}(\mathcal{S})\big]$$

$$= \sum_{j_1 < \cdots < j_{m-1} \in \mathbb{N}} \frac{1}{\sqrt{t_{j_1+1} - t_{j_1}} \cdots \sqrt{t_{j_{m-1}+1} - t_{j_{m-1}}}} \int_{t_{j_1}}^{t_{j_1+1}} \cdots \int_{t_{j_{m-1}}}^{t_{j_{m-1}+1}} \int_{t_i}^{t_{i+1}}$$

$$\times f(\{s_1, \ldots, s_{m-1}\} \cup s)\, ds\, ds_1 \cdots ds_{m-1}\, X_{j_1} \cdots X_{j_{m-1}}$$

$$= \sum_{j_1 < \cdots < j_{m-1} \in \mathbb{N}} \sum_{k=0}^{m-1} \mathbb{1}_{0 < j_1 < \cdots < j_k < i < j_{k+1} < \cdots < j_{m-1}}$$

$$\times \frac{1}{\sqrt{t_{j_1+1} - t_{j_1}} \cdots \sqrt{t_{j_{m-1}+1} - t_{j_{m-1}}}}$$

$$\int_{t_{j_1}}^{t_{j_1+1}} \cdots \int_{t_{j_k}}^{t_{j_k+1}} \int_{t_i}^{t_{i+1}} \int_{t_{j_{k+1}}}^{t_{j_{k+1}+1}} \cdots \int_{t_{j_{m-1}}}^{t_{j_{m-1}+1}} f(s_1, \ldots, s_k, s, s_{k+1} \cdots s_{m-1})$$

$$\times ds_1 \cdots ds_k\, ds\, ds_{k+1} \cdots ds_{m-1}\, X_{j_1} \cdots X_{j_{m-1}}$$

$$= \sqrt{t_{i+1} - t_i} \sum_{j_1 < \cdots < j_m \in \mathbb{N}} \frac{1}{\sqrt{t_{j_1+1} - t_{j_1}} \cdots \sqrt{t_{j_m+1} - t_{j_m}}}$$

$$\int_{t_{j_1}}^{t_{j_1+1}} \cdots \int_{t_{j_m}}^{t_{j_m+1}} f(s_1, \ldots, s_m)\, ds_1 \cdots ds_m\, \mathbb{1}_{i \in \{j \ldots j_m\}} X_{j_1} \cdots \widehat{X}_i \cdots X_{j_m}$$

$$= \sqrt{t_{i+1} - t_i}\, a_i^-\, \mathbb{E}\big[f/\mathcal{F}(\mathcal{S})\big].$$

Finally,

$$(a_{t_{i+1}}^+ - a_{t_i}^+)f = \sum_{k=0}^{n} \int_{0 < s_1 < \cdots < s_k < s < s_{k+1} < \cdots < s_m} \mathbb{1}_{[t_i, t_{i+1}]}(s)$$

$$\times f(s_1, \ldots, s_m)\, d\chi_{s_1} \cdots d\chi_{s_k}\, d\chi_s\, d\chi_{s_{k+1}} \cdots d\chi_{s_m}.$$

$$\mathbb{E}\big[(a_{t_{i+1}}^+ - a_{t_i}^+)f/\mathcal{F}(\mathcal{S})\big]$$

$$= \sum_{j_1 < \cdots < j_{m+1} \in \mathbb{N}} \frac{1}{\sqrt{t_{j_1+1} - t_{j_1}} \cdots \sqrt{t_{j_{m+1}+1} - t_{j_{m+1}}}} \sum_{k=0}^{n} \int_{t_{j_1}}^{t_{j_1+1}} \cdots \int_{t_{j_{m+1}}}^{t_{j_{m+1}+1}}$$

$$\times \mathbb{1}_{[t_i, t_{i+1}]}(t_{j_{k+1}}) f(s_1, \ldots, \widehat{s_{k+1}} \cdots s_{m+1})\, ds_1 \cdots ds_{m+1}\, X_{j_1} \cdots X_{j_{m+1}}$$

$$= \sum_{j_1 < \cdots < j_{m+1} \in \mathbb{N}} \frac{1}{\sqrt{t_{j_1+1} - t_{j_1}} \cdots \sqrt{t_{j_{m+1}+1} - t_{j_{m+1}}}}\, \mathbb{1}_{i \in \{j_1 \ldots j_{m+1}\}}$$

$$\int_{t_{j_1}}^{t_{j_1+1}} \cdots \int_{t_{j_{m+1}}}^{t_{j_{m+1}+1}} f(s_1, \ldots, \widehat{s_i} \ldots s_{m+1})\, ds_1 \cdots ds_{m+1}\, X_{j_1} \cdots X_{j_{m+1}}$$

$$= \sum_{j_1 < \cdots < j_m \in \mathbb{N}} \sqrt{t_{i+1} - t_i}\, \frac{1}{\sqrt{t_{j_1+1} - t_{j_1}} \cdots \sqrt{t_{j_m+1} - t_{j_m}}}$$

$$\int_{t_{j_1}}^{t_{j_1+1}} \cdots \int_{t_{j_m}}^{t_{j_m+1}} f(s_1, \ldots, s_m)\, ds_1 \cdots ds_m\, \mathbb{1}_{i \notin \{j_1 \ldots j_m\}} X_{j_1} \cdots X_{j_m} X_i$$

$$= \sqrt{t_{i+1} - t_i}\, a_i^+\, \mathbb{E}\big[f/\mathcal{F}(\mathcal{S})\big].$$

We have proved all the announced relations. □

6.3 Approximations

We are now going to prove that the Fock space Φ and its basic operators a_t^+, a_t^-, a_t° can be approximated by the toy Fock spaces $T\Phi(\mathcal{S})$ and their basic operators a_i^+, a_i^-, a_i°.

We are given a refining sequence $(\mathcal{S}_n)_{n\in\mathbb{N}}$ of partitions whose diameter $\delta(\mathcal{S}_n)$ tends to 0 when n tends to $+\infty$. Let $T\Phi(n) = T\Phi(\mathcal{S}_n)$ and $P_n = \mathbb{E}[\cdot/\mathcal{F}(\mathcal{S}_n)]$, for all $n\in\mathbb{N}$.

Theorem 6.4.

i) For every $f\in\Phi$ there exists a sequence $(f_n)_{n\in\mathbb{N}}$ such that $f_n\in T\Phi(n)$, for all $n\in\mathbb{N}$, and $(f_n)_{n\in\mathbb{N}}$ converges to f in Φ.

ii) If $\mathcal{S}_n = \{0 = t_0^n < t_1^n < \cdots < t_k^n < \cdots\}$, then for all $t\in\mathbb{R}^+$, the operators $\sum_{i;t_i^n\leq t} a_i^\circ$, $\sum_{i;t_i^n\leq t}\sqrt{t_{i+1}^n - t_i^n}\,a_i^-$ and $\sum_{i;t_i^n\leq t}\sqrt{t_{i+1}^n - t_i^n}\,a_i^+$ converge strongly on $\mathcal{D}$ to a_t°, a_t^- and a_t^+ respectively.

iii) With the same notations as in ii), for all $t\in\mathbb{R}^+$, the operators $\sum_{i;t_i^n\leq t} a_i^\circ\, P_n$, $\sum_{i;t_i^n\leq t}\sqrt{t_{i+1}^n - t_i^n}\,a_i^-\, P_n$ and $\sum_{i;t_i^n\leq t}\sqrt{t_{i+1}^n - t_i^n}\,a_i^+\, P_n$ converge strongly on $\mathcal{D}$ to a_t°, a_t^- and a_t^+ respectively.

Proof. i) As the $\mathcal{S}_n$ are refining then the $(P_n)_n$ forms an increasing family of orthogonal projection in Φ. Let $P_\infty = \bigvee_n P_n$. Clearly, for all $s \leq t$, we have that $\chi_t - \chi_s$ belongs to $\mathrm{Ran}P_\infty$. But by the construction of the Itô integral and by Theorem 5, we have that the $\chi_t - \chi_s$ generate Φ. Thus $P_\infty = I$. Consequently if $f\in\Phi$, the sequence $f_n = P_n f$ satisfies the statements.

ii) The convergence of $\sum_{i,t_i^n\leq t} a_i^\circ$ and $\sum_{i,t_i^n\leq t}\sqrt{t_{i+1}^n - t_i^n}\,a_i^-$ to a_t° and a_t^- respectively is clear from the definitions. Let us check the case of a^+. We have, for $f\in\mathcal{D}$

$$\left[\sum_{i;t_i^n\leq t}\sqrt{t_{i+1}^n - t_i^n}\,a_i^+ f\right](\sigma) = \sum_{i;t_i^n\leq t}\mathbb{1}_{|\sigma\cap[t_i^n,t_{i+1}^n]|=1}\sum_{s\in\sigma\cap[t_i^n,t_{i+1}^n]} f(\sigma\setminus\{s\})\,.$$

Put $t^n = \inf\{t_i^n\in\mathcal{S}_n\,; t_i^n \geq t\}$. We have

$$\left\| \sum_{i; t_i^n \le t} \sqrt{t_{i+1}^n - t_i^n} \, a_i^+ - a_t^+ f \right\|^2$$

$$= \int_{\mathcal{P}} \left| \sum_{i; t_i^n \le t} \mathbb{1}_{|\sigma \cap [t_i^n, t_{i+1}^n]| = 1} \sum_{s \in \sigma \cap [t_i^n, t_{i+1}^n]} f(\sigma \setminus \{s\}) - \sum_{s \in \sigma \cap [0, t]} f(\sigma \setminus \{s\}) \right|^2 d\sigma$$

$$\le 2 \int_{\mathcal{P}} \left| \sum_{s \in \sigma \cap [t, t]} f(\sigma \setminus \{s\}) \right|^2 d\sigma +$$

$$+ 2 \int_{\mathcal{P}} \left| \sum_{i; t_i^n \le t} \mathbb{1}_{|\sigma \cap [t_i^n, t_{i+1}^n]| \ge 2} \sum_{s \in \sigma \cap [t_i^n, t_{i+1}^n]} f(\sigma \setminus \{s\}) \right|^2 d\sigma.$$

For any fixed σ, the terms inside each of the integrals above converge to 0 when n tends to $+\infty$. Furthermore we have, for n large enough,

$$\int_{\mathcal{P}} \left| \sum_{s \in \sigma \cap [t, t^n]} f(\sigma \setminus \{s\}) \right|^2 d\sigma \le \int_{\mathcal{P}} |\sigma| \sum_{\substack{s \in \sigma \\ s \le t+1}} |f(\sigma \setminus \{s\})|^2 \, d\sigma$$

$$= \int_0^{t+1} \int_{\mathcal{P}} (|\sigma| + 1) |f(\sigma)|^2 \, d\sigma \, ds$$

$$\le (t + 1) \int_{\mathcal{P}} (|\sigma| + 1) |f(\sigma)|^2 \, d\sigma$$

which is finite for $f \in \mathcal{D}$;

$$\int_{\mathcal{P}} \left| \sum_{i; t_i^n \le t} \mathbb{1}_{|\sigma \cap [t_i^n, t_{i+1}^n]| \ge 2} \sum_{s \in \sigma \cap [t_i^n, t_{i+1}^n]} f(\sigma \setminus \{s\}) \right|^2 d\sigma$$

$$\le \int_{\mathcal{P}} \left(\sum_{i; t_i^n \le t} \mathbb{1}_{|\sigma \cap [t_i^n, t_{i+1}^n]| \ge 2} \left| \sum_{s \in \sigma \cap [t_i^n, t_{i+1}^n]} f(\sigma \setminus \{s\}) \right| \right)^2 d\sigma$$

$$\le \int_{\mathcal{P}} \left(\sum_{i; t_i^n \le t} \sum_{s \in \sigma \cap [t_i^n, t_{i+1}^n]} |f(\sigma \setminus \{s\})| \right)^2 d\sigma$$

$$= \int_{\mathcal{P}} \left(\sum_{\substack{s \in \sigma \\ s \le t^n}} |f(\sigma \setminus \{s\})| \right)^2 d\sigma$$

$$= \int_{\mathcal{P}} |\sigma| \sum_{\substack{s \in \sigma \\ s \le t^n}} |f(\sigma \setminus \{s\})|^2 \, d\sigma$$

$$\le (t + 1) \int_{\mathcal{P}} (|\sigma| + 1) |f(\sigma)|^2 \, d\sigma$$

in the same way as above. So we can apply Lebesgue's theorem. This proves *ii)*.

iii) By Proposition 6.3, we have for all $f \in \mathcal{D}$

$$\sum_{i; t_i^n \le t} \sqrt{t_{i+1}^n - t_i^n} \, a_i^+ \, P_n f = P_n a_{t^n}^+ f \, .$$

Consequently

$$\left\| \sum_{i;t_i^n \leq t} \sqrt{t_{i+1}^n - t_i^n} \, a_i^+ \, P_n f - a_t^+ f \right\|^2$$

$$\leq 2 \|a_t^+ f - P_n a_t^+ f\|^2 + 2 \|P_n(a_t^+ f - a_{t^n}^+ f)]\|^2$$

$$\leq 2 \|a_t^+ f - P_n a_t^+ f\|^2 + 2 \|a_t^+ f - a_{t^n}^+ f\|^2$$

which tends to 0 as n tends to $+\infty$.

The cases of a° and a^- are obtained in the same way. $\square$

6.4 Probabilistic interpretations

Recall that the operator of multiplication by the Brownain motion in the Fock space Φ is

$$W_t = a_t^+ + a_t^-$$

and the operator of Poisson multiplication by the Poisson process is

$$N_t = a_t^+ + a_t^- + a_t^\circ + tI \ .$$

Let us consider an approximation of the Fock space Φ by toy Fock spaces $T\Phi(n)$, $n \in \mathbb{N}$.

Theorem 6.5. *On $T\Phi(n)$, let $X_i = a_i^+ + a_i^-$, $i \in \mathbb{N}$. Then, for all $t \in \mathbb{R}^+$; we have that*

$$\sum_{i;t_i \leq t} \sqrt{t_{i+1} - t_i} \, X_i$$

converges strongly to W_t.

Proof. The proof is immediate from Theorem 6.4. $\square$

Let $S_n = \{i/n \; ; \; i \in \mathbb{N}\}$.

Theorem 6.6. *On $T\Phi(n)$, let $X_i = a_i^+ + a_i^- + c_n a_i^\circ$, $i \in \mathbb{N}$ be associated to the coefficient $p_n = 1/n$. Then, for all $t \in \mathbb{R}^+$, we have that*

$$\frac{1}{\sqrt{n}} \sum_{i;t_i \leq t} X_i$$

converges strongly to $X_t = N_t - tI$, the operator of multiplication by the compensated Poisson process.

Proof. If $p_n = 1/n$, then $q_n = 1 - 1/n$ and $c_n = \frac{1 - 2/n}{\sqrt{1/n - 1/n^2}} = \frac{n-2}{\sqrt{n-1}}$. Thus $c_n/\sqrt{n}$ converges to 1. Now,

$$\frac{1}{\sqrt{n}} \sum_{i;t_i \le t} X_i = \sum_{i;t_i \le t} \frac{1}{\sqrt{n}} a_i^+ + \frac{1}{\sqrt{n}} a_i^- + \frac{c_n}{\sqrt{n}} a_i^\circ$$

$$= \sum_{i;t_i \le t} \sqrt{t_{i+1} - t_i}(a_i^+ + a_i^-) + \frac{c_n}{\sqrt{n}} \sum_{i;t_i \le t} a_i^\circ$$

which clearly converges to $a_t^+ + a_t^- + a_t^\circ$ by Theorem 6.4 □

6.5 The Itô tables

This section is heuristic, but it gives a good idea of why the discrete quantum Itô table is a discrete approximation of the usual one, though they seem different. Let $S_n = \{i/n \; ; \; i \in \mathbb{N}\}$. Let $\tilde{a}_i^+ = 1/\sqrt{n} \, a_i^+$, $\tilde{a}_i^- = 1/\sqrt{n} \, a_i^-$ and $\tilde{a}_i^\circ = a_i^\circ$. The Theorem 6.4 shows that $\tilde{a}_i^\varepsilon$ is a good approximation of da_t^ε, where $t = t_i$. Now the discrete Itô table becomes

$\longrightarrow$	$\tilde{a}_i^+$	$\tilde{a}_i^-$	$\tilde{a}_i^\circ$
$\tilde{a}_i^+$	0	$\frac{1}{n}\tilde{a}_i^\circ$	0
$\tilde{a}_i^-$	$\frac{1}{n}I - \frac{1}{n}\tilde{a}_i^\circ$	0	$\tilde{a}_i^-$
$\tilde{a}_i^\circ$	$\tilde{a}_i^+$	0	$\tilde{a}_i^\circ$.

But

1) $\frac{1}{n}\tilde{a}_i^\circ$ is not an infinitesimal for $\sum_{i;t_i \le t} \frac{1}{n}\tilde{a}_i^\circ$ is almost $\frac{1}{n}a_t^\circ$ which converges to 0. Thus $\frac{1}{n}\tilde{a}_i^\circ$ can be considered to be 0 in this table;

2) $\frac{1}{n}I$ is simply $dt\,I$, that is $(t_{i+1} - t_i)I$. Thus at the limit this table becomes

$\longrightarrow$	da_t^+	da_t^-	da_t°
da_t^+	0	0	0
da_t^-	$dt\,I$	0	da_t^-
da_t°	da_t^+	0	da_t° .

That is, the usual Itô table.

These heuristic arguments have been made rigourous in [33].

7 Back to repeated interactions

We are now ready to come back to repeated quantum interactions and to give an idea of what happens in the limit $h \to 0$.

Recall our evolution equation on $\mathcal{H}_0 \otimes \bigotimes_{\mathbb{N}} \mathbb{C}^{N+1}$:

$$V_{n+1} = U_{n+1}V_n \tag{23}$$

of section I.

7.1 Unitary dilations of completely positive semigroups

In this section, we will show that equations such as (23) appear naturally in a general setup and allow one to obtain natural unitary dilations of completely positive semigroups in discrete time.

Consider a discrete semigroup $(P_n)_{n \in \mathbb{N}}$ of completely positive maps on $\mathcal{B}(\mathcal{H}_0)$, that is,

$$P_n(X) = \ell^n(X)$$

where ℓ is a completely positive, weakly continuous map on $\mathcal{B}(\mathcal{H}_0)$.

In the sequel we always assume that $\ell(I) = I$. By Kraus' theorem (see [30], Proposition 29.8) this means that ℓ is of the form

$$\ell(X) = \sum_{i=0}^{N} V_i^* X V_i$$

for some N and some family (V_i) of bounded operators on $\mathcal{H}_0$ such that $\sum_i V_i^* V_i = I$. Of course the indexation is *a priori* indifferent to the specificity of the value $i = 0$. The special role played by one of the values will appear later on.

Let $\mathbb{E}_0$ be the partial trace on $\mathcal{H}_0$ defined by

$$< \phi, \mathbb{E}_0(H) \psi > = < \phi \otimes \Omega, H \psi \otimes \Omega >$$

for all $\phi, \psi \in \mathcal{H}_0$ and every operator H on $\mathcal{H}_0 \otimes \mathbb{TP}$.

Theorem 7.1. *For any completely positive map*

$$\ell(X) = \sum_{i=0}^{N} V_i^* X V_i$$

on $\mathcal{B}(\mathcal{H}_0)$ there exists a unitary operator $\mathbb{L}$ on $\mathcal{H}_0 \otimes \mathbb{C}^{N+1}$ such that the associated unitary family of automorphisms

$$j_n(H) = u_n^* H u_n$$

(where u_n is associated to $\mathbb{L}$ by (23)) satisfies

$$\mathbb{E}_0(j_n(X \otimes I)) = P_n(X),$$

for all $n \in \mathbb{N}$.

Proof. Consider a decomposition of $\mathcal{L}$ of the form

$$\ell(X) = \sum_{i=0}^{N} V_i^* X V_i$$

for a family (V_i) of bounded operators on $\mathcal{H}_0$ such that $\sum_{i=0}^{N} V_i^* V_i = I$.

We claim that there exists a unitary operator $\mathbb{L}$ on $\mathcal{H}_0 \otimes \mathbb{C}^{N+1}$ of the form

$$
\mathbb{L} = \begin{pmatrix} V_1 & \cdots\cdots \\ V_2 & \cdots\cdots \\ \vdots & \vdots & \vdots \\ V_N & \cdots\cdots \end{pmatrix}.
$$

Indeed, the condition $\sum_{i=0}^{N} V_i^* V_i = I$ guarantees that the m first columns of $\mathbb{L}$ (where $m = \dim \mathcal{H}_0$) constitute an orthonormal family of $\mathcal{H}_0 \otimes \mathbb{C}^{N+1}$. We can thus fill the matrix by completing it into an orthonormal basis of $\mathcal{H}_0 \otimes \mathbb{C}^{N+1}$; this makes out a unitary, $(N+1) \times (N+1)$ matrix $\mathbb{L}$ on $\mathcal{H}_0$, of which we denote the coefficients by $(A_j^i)_{i,j=0,\ldots,N}$; with this notation we have for all i, $A_0^i = V_{i+1}$. To this matrix $\mathbb{L}$ we associate a family $(\mathbb{L}_i)_{i \geq 0}$ of ampliations as explained in section

Now, for every operator H on $\mathcal{H}_0 \otimes \mathbb{C}^{N+1}$, put

$$
j_n(H) = u_n^* H u_n.
$$

It satisfies

$$
j_{n+1}(H) = u_n^* \mathbb{L}_{n+1}^* H \, \mathbb{L}_{n+1} u_n.
$$

We consider this relation for an operator H of the form $H = X \otimes I$, where X is an operator on $\mathcal{H}_0$. Write $\mathbb{L}_{n+1}^*(X \otimes I)\mathbb{L}_{n+1}$ in $(\mathcal{H}_0 \otimes \mathrm{T\!\Phi}_{n]}) \otimes \mathbb{C}_{n+1}^{N+1}$; it is simply

$$
\mathbb{L}_{n+1}^*(X \otimes I)\mathbb{L}_{n+1} =
$$

$$
= \begin{pmatrix} (A_0^0)^* & (A_0^1)^* & \cdots \\ (A_1^0)^* & (A_1^1)^* & \cdots \\ \vdots & \vdots & \vdots \\ (A_N^0)^* & (A_N^1)^* & \cdots \end{pmatrix} \begin{pmatrix} X & 0 & \cdots & 0 \\ 0 & X & \cdots & 0 \\ \vdots & \vdots & & \vdots \\ 0 & 0 & \cdots & X \end{pmatrix} \begin{pmatrix} A_0^0 & A_1^0 & \cdots \\ A_0^1 & A_1^1 & \cdots \\ \vdots & \vdots & \vdots \\ A_0^N & A_1^N & \cdots \end{pmatrix}
$$

which is easily seen to be the matrix $(B_j^i(X))_{i,j=0,\ldots,N}$ with

$$
B_j^i(X) = \sum_{k=0}^{N} (A_j^k)^* X A_i^k.
$$

Note that, more precisely, the operator $\mathbb{L}_{n+1}^*(X \otimes I)\mathbb{L}_{n+1}$ is written in $(\mathcal{H}_0 \otimes \mathrm{T\!\Phi}_{n]}) \otimes \mathbb{C}_{n+1}^{N+1}$ as the matrix $(B_j^i(X) \otimes I)_{i,j=0,\ldots,N}$. The operator u_n, in turn, acts only on $\mathcal{H}_0 \otimes \mathbb{C}_n^{N+1}$, so that $u_n^* \mathbb{L}_{n+1}^*(X \otimes I)\mathbb{L}_{n+1} u_n$ can be written in $(\mathcal{H}_0 \otimes \mathrm{T\!\Phi}_{n]}) \otimes \mathbb{C}_{n+1}^{N+1}$ as $(u_n^*(B_j^i(X) \otimes I)u_n)_{i,j=0,\ldots,N}$; simply put, we have proved that

$$
(j_{n+1}(X \otimes I))_j^i = j_n(B_j^i(X) \otimes I)
$$

because both terms act as the identity beyond $(\mathcal{H}_0 \otimes \mathrm{T\!\Phi}_{n]}) \otimes \mathbb{C}_{n+1}^{N+1}$.

Consider now $T_n(X) = \mathbb{E}_0(j_n(X \otimes I))$. We have

$$\begin{aligned}
<\phi, T_{n+1}(X)\psi> &= <\phi \otimes \Omega, j_{n+1}(X \otimes I)\psi \otimes \Omega> \\
&= <\phi, (j_{n+1}(X \otimes I))_0^0 \psi> \\
&= <\phi, j_n(B_0^0(X) \otimes I)\psi> \\
&= <\phi, T_n(B_0^0(X))\psi>;
\end{aligned}$$

now remember that for all i, $A_0^i = V_{i+1}$. This implies that $B_0^0(X) = \ell(X)$. The above proves that $T_{n+1}(X) = T_n(\ell(X))$ for any n and the theorem follows. $\quad\square$

7.2 Convergence to Quantum Stochastic Differential Equations

We now describe the convergence of these discrete time evolutions to continuous time ones.

Quantum stochastic differential equations

We do not develop here the whole theory of Q.S.D.E., this will be done in F. Fagnola's course much more precisely, but we just give an idea of what they are.

Quantum stochastic differential equations are equations of the form

$$dU_t = \sum_{i,j} L_j^i U_t \, da_j^i(t), \tag{22}$$

with initial condition $U_0 = I$. The above equation has to be understood as an integral equation

$$U_t = I + \int_0^t \sum_{i,j} L_j^i U_t \, da_j^i(t),$$

for operators on $\mathcal{H}_0 \otimes \Phi$, the operators L_j^i being bounded operators on $\mathcal{H}_0$ alone which are ampliated to $\mathcal{H}_0 \otimes \Phi$.

The main motivation and application of that kind of equation is that it gives an account of the interaction of the small system $\mathcal{H}_0$ with the bath Φ in terms of quantum noise perturbation of a Schödinger-like equation. Indeed, the first term of the equation

$$dU_t = L_0^0 U_t \, dt + \dots$$

describes the induced dynamics on the small system, all the other terms are quantum noises terms.

One of the main application of equations such as (22) is that they give explicit constructions of unitary dilations of semigroups of completely positive maps of $\mathcal{B}(\mathcal{H}_0)$ (see [H-P]). Let us only recall one of the main existence, uniqueness and boundedness theorems connected to equations of the form (22). The literature is huge about those equations; we refer to [Par] for the result we mention here.

Theorem 7.2. *If $\mathcal{H}_0$ is finite dimensional then the quantum stochastic differential equation*

$$dU_t = \sum_{i,j} L_j^i U_t \, da_j^i(t),$$

with $U_0 = I$, admits a unique solution defined on the space of coherent vectors.

The solution $(U_t)_{t \geq 0}$ is made of unitary operators if and only if there exist, on $\mathcal{H}_0$, a bounded self-adjoint H, bounded operators S_j^i, $i, j = 1, \ldots, N$, such that the matrix $(S_j^i)_{i,j}$ is unitary, and bounded operators L_i, $i = 1, \ldots, N$ such that, for all $i, j = 1, \ldots, N$

$$L_0^0 = -(iH + \frac{1}{2} \sum_k L_k^* L_k)$$

$$L_i^0 = L_i$$

$$L_0^i = -\sum_k L_k^* S_j^k$$

$$L_j^i = S_j^i - \delta_{ij} I.$$

If the operators L_j^i are of this form then the unitary solution $(U_t)_{t \geq 0}$ of the above equation exists even if $\mathcal{H}_0$ is only assumed to be separable.

Convergence theorems

In this section we study the asymptotic behaviour of the solutions of an equation

$$u_{n+1} = \mathbb{L}_{n+1} u_n;$$

if the matrix $\mathbb{L}(h)$ converges (with a particular normalization) as h tends to zero and prove that, in the limit, the solutions of such equations converge to solutions of quantum stochastic differential equations of the form (22). Notice that we no longer assume that $\mathbb{L}(h)$ has been conveniently constructed for our needs; in particular $\mathbb{L}$ is not assumed to be unitary.

Let h be a parameter in $\mathbb{R}^+$, which is thought of as representing a small time interval. Let $\mathbb{L}(h)$ be an operator on $\mathcal{H}_0 \otimes \mathbb{C}^{N+1}$, with coefficients $\mathbb{L}_j^i(h)$ as a $(N+1) \times (N+1)$ matrix of operators on $\mathcal{H}_0$. Let $u_n(h)$ be the associated solution of

$$u_{n+1}(h) = \mathbb{L}_{n+1}(h) u_n(h).$$

In the following we will drop dependency in h and write simply $\mathbb{L}$ or u_n. Besides, we denote

$$\varepsilon_{ij} = \frac{1}{2}(\delta_{0i} + \delta_{0j})$$

for all $i, j = 0, \ldots, N$.

Theorem 7.3. *Assume that there exist operators L_j^i on $\mathcal{H}_0$ such that*

$$\lim_{h \to 0} \frac{\mathbb{L}_j^i(h) - \delta_{ij} I}{h^{\varepsilon_{ij}}} = L_j^i$$

for all $i, j = 0, \ldots, N$, where convergence is in operator norm. Assume that the quantum stochastic differential equation

$$dU_t = \sum_{i,j} L_j^i U_t \, da_j^i(t)$$

with initial condition $U_0 = I$ has a solution $(U_t)_{t \geq 0}$ which is a process of bounded operators with a locally uniform norm bound.

Then, for all t, for every ϕ, ψ in $L^\infty([0, t])$, the quantity

$$< a \otimes \varepsilon(\phi), \, \mathbb{E}_\mathcal{S} u_{[t/h]} \mathbb{E}_\mathcal{S} \, b \otimes \varepsilon(\psi) >$$

converges to

$$< a \otimes \varepsilon(\phi), \, U_t \, b \otimes \varepsilon(\psi) >$$

when h goes to 0.

Moreover, the convergence is uniform for a, b in any bounded ball of $\mathcal{K}$, uniform for t in a bounded interval of $\mathbb{R}_+$.

If furthermore $\|u_k\|$ is locally uniformly bounded in the sense that, for any t in $\mathbb{R}_+$, $\{\|u_k(h)\|, \ k \leq t/h\}$ is bounded for any h, then $u_{[t/h]}$ converges weakly to U_t on all $\mathcal{H}_0 \otimes \Phi$.

Remarks

– This is where we particularize the index zero : the above hypotheses of convergence simply mean that, among the coefficients of $\mathbb{L}$,

$$(\mathbb{L}_0^0(h) - I)/h \text{ converges,}$$

$$\mathbb{L}_j^i(h)/\sqrt{h} \text{ converges if either } i \text{ or } j \text{ is zero,}$$

$$\mathbb{L}_j^i(h) - \delta_{i,j} \text{ converges if neither } i \text{ nor } j \text{ is zero}$$

and we recover the fact that the 0 index must relate to the small system, on which the considered time scale is different from the time scale of the reservoir.

– The assumption that $\mathcal{H}_0$ is finite dimensional is only needed in order to ensure that the quantum stochastic differential equation has a solution; if for example the L_j^i's are of the form described in Theorem 8 then the separability of $\mathcal{H}_0$ is enough.

For our example where $\mathbb{L}$ is given by

$$\mathbb{L} = \begin{pmatrix} \cos \alpha & 0 & 0 & -\sin \alpha \\ 0 & 1 & 0 & 0 \\ 0 & 0 & 1 & 0 \\ \sin \alpha & 0 & 0 & \cos \alpha \end{pmatrix}$$

with $\alpha = \sqrt{h}$, since for all h the matrix $\mathbb{L}(h)$ is unitary, we get that for all t, $u_{[t/h]}$ converges strongly to U_t where $(U_t)_{t \in \mathbb{R}_+}$ is the solution of

$$dU_t = -\frac{1}{2} V^* V \, U_t \, dt + V U_t \, da_1^0(t) - V^* U_t \, da_0^1(t)$$

with $V = \begin{pmatrix} 0 & 0 \\ 1 & 0 \end{pmatrix}$; this is the evolution associated to the spontaneous decay into the ground state in the Wigner-Weisskopf model for the two-level atom.

8 Bibliographical comments

The mathematical theory of quantum stochastic calculus was first developed by Hudson and Parthasarathy [25]. They defined quantum stochastic integrals on the space of coherent vectors. They also defined and solved the first class of quantum Langevin equations. They finaly proved that quantum Langevin equations allow to construct unitary dilations of any completely positive semigroups. No need to say that this article is a fundamental one, which started a whole theory.

An extension of their quantum stochastic calculus, trying to go further than the domain of coherent vectors was proposed by Belavkin and later by Lindsay ([15], [26]). Their definitions were making use of the Malliavin gradient (and was constrained by its domain) and the Skorohod integral.

The definition of quantum stochastic integrals as in subsection 4.2 is due to Attal and Meyer ([10] and later developed in [5]). The main point with that approach was the absence of arbitrary domain constraints. The discovery of the quantum semi-martingales by Attal in [4] was a direct consequence of that approach and of and anterior work of Parthasarathy and Sinha on regular quantum martingales ([31]).

The maximal definition of quantum stochastic integrals and unification of the different approaches, as in subsection 4.2 was given by Attal and Lindsay ([9]).

The theorem showing rigorously that there are only 3 quantum noises is due to Coquio ([17]).

The notion of Toy Fock space with its probabilistic interpretations in terms of random walks was developed by Meyer ([29]). The concrete realization of the Toy Fock as a subspace and an approximation of Φ is due to Attal ([2]) and has been developed much further by Pautrat ([32], [33]). These different works led to the proof of the convergence of repeated interactions to quantum stochastic differential equation ([11]).

References

1. L. Accardi, A. Frigerio, Y.G. Lu, *Quantum Langevin equation in the weak coupling limit*, Quantum probability and applications V (Heidelberg, 1988), p. 1–16, Lecture Notes in Math., 1442, Springer, Berlin, 1990.

2. S. Attal, *Quantum Noise Theory*, book to appear, Springer Verlag.

3. S. Attal, *Approximating the Fock space with the toy Fock space* Séminaire de Probabilités, XXXVI, p. 477–491, Lecture Notes in Math., 1801, Springer Verlag, Berlin, 2003.

4. S. Attal, *An algebra of non-commutative bounded semimartingales. Square and angle quantum brackets*, Journal of Functional Analysis 124 (1994), no. 2, p. 292–332.

5. S. Attal, P.-A. Meyer, *Interprétation probabiliste et extension des intégrales stochastiques non commutatives*, Séminaire de Probabilités, XXVII, p. 312–327, Lecture Notes in Math., 1557, Springer Verlag, Berlin, 1993.

6. S. Attal, M. Emery, *Equations de structure pour des martingales vectorielles* Séminaire de Probabilités, XXVIII, p. 256–278, Lecture Notes in Math., 1583, Springer, Berlin, 1994.

7. S. Attal, A. Joye, *Weak coupling limit for repeated quantum interactions*, preprint.

8. S. Attal, A. Joye, *The Langevin equation for a quantum heat bath*, preprint.

9. S. Attal, J.M. Lindsay, *Quantum stochastic calculus with maximal operator domains*, The Annals of Probability 32 (2004), p. 488–529.

10. S. Attal, P.A. Meyer, *Interprétation probabiliste et extension des intégrales stochastiques non commutatives*, Séminaire de Probabilités XXVII, Lect. Notes in Math. 1557, p. 312–327.

11. S. Attal, Y. Pautrat, *From repeated to continuous quantum interactions*, Annales Henri Poincaré (Physique Théorique), to appear (2005).

12. A. Barchielli, *Measurement theory and stochastic differential equations in quantum mechanics*, Physical Review A (3) 34 (1986), no. 3,P. 1642–1649.

13. A. Barchielli, *Some stochastic differential equations in quantum optics and measurement theory: the case of counting processes*, Stochastic evolution of quantum states in open systems and in measurement processes (Budapest, 1993), p. 1–14, World Sci. Publishing, River Edge, NJ, 1994.

14. A. Barchielli, V.P. Belavkin, *Measurements continuous in time and a posteriori states in quantum mechanics*, Journal of Physics A 24 (1991), no. 7,P. 1495–1514.

15. V.P. Belavkin, *A quantum non-adapted Itô formula and non stationnary evolutions in Fock scale*, Quantum Probability and Related Topics VI, World Scientific (1991), p. 137-180.

16. J. Bellissard, R. Rebolledo, D. Spehner, W. von Waldenfels, *The quantum flow of electronic transport*, preprint.

17. A. Coquio, *Why are there only three quantum noises?*, Probability Theory and Related Fields 118 (2000), p. 349–364.

18. O. Enchev, *Hilbert-space-valued quasimartingales*, Boll. Un. Mat. Ital. B (7) 2 (1988), p. 19–39.

19. F. Fagnola, R. Rebolledo, *A view on stochastic differential equations derived from quantum optics*, Stochastic models (Guanajuato, 1998),P. 193–214, Aportaciones Mat. Investig., 14, Soc. Mat. Mexicana, Mxico, 1998.

20. F. Fagnola, R. Rebolledo, C. Saavedra, *Quantum flows associated to master equations in quantum optics*, Journal of Mathematical Physics 35 (1994), no. 1, P. 1–12.

21. G. W. Ford, M. Kac, P. Mazur, *Statistical mechanics of assemblies of coupled oscillators*, Journal of Mathematical Physisics 6 (1965),p. 504–515.

22. G. W. Ford, J. T. Lewis, R. F. O'Connell, *Quantum Langevin equation*, Physical Review A (3), vol 37 (1988), no. 11, p. 4419–4428.

23. C. W. Gardiner, P. Zoller, *Quantum noise. A handbook of Markovian and non-Markovian quantum stochastic methods with applications to quantum optics.*, Second edition. Springer Series in Synergetics. Springer-Verlag, Berlin, 2000.

24. H. Grabert, P. Schramm, G-L. Ingold, *Quantum Brownian motion: the functional integral approach*, Physical Reports 168 (1988), no. 3, p. 115–207.

25. R.L. Hudson, K.R. Parthasarathy, *Quantum Itô's formula and stochastic evolutions* Communications in Mathematical Physics, 93 (1984), P. 301–323.

26. J.M. Lindsay, *Quantum and non-causal stochastic calculus*, Probability Theory and Related Fields 97 (1993), p. 65-80.

27. J.M. Lindsay, H. Maassen, *Stochastic calculus for quantum Brownian motion of non-minimal variance—an approach using integral-sum kernel operators*, Mark Kac Seminar on Probability and Physics Syllabus 1987–1992 (Amsterdam, 1987–1992), 97–167, CWI Syllabi, 32, Math. Centrum, Centrum Wisk. Inform., Amsterdam, 1992.

28. H. Maassen, P. Robinson, *Quantum stochastic calculus and the dynamical Stark effect*, Reports in Math. Phys. 30 (1991), 185–203.

29. P.-A. Meyer, *Quantum Probability for Probabilists*, 2nd edition, Lecture Notes in mathematics 1538, Springer-Verlag, Berlin, 1993.

30. K.R. Parthasarathy, *An Introduction to Quantum Stochastic Calculus*, Monographs in Mathematics 85, Birkhäuser (1992).

31. K.R. Parthasarathy, K.B. Sinha *Stochastic integral representation of bounded quantum martingales on Fock space*, Journal of Functional Analysis, 67 (1986), p. 126-151.

32. Y. Pautrat, *Integral and kernel representation of operators on the infinite dimensional Toy Fock space*, Seminaire de Probabilités, Springer Lecture Notes in Mathematics (to appear).

33. Y. Pautrat, *Pauli matrices and quantum Itô formula*, Mathematical Physics, Analysis and Geometry (to appear).

Complete Positivity and the Markov structure of Open Quantum Systems*

Rolando Rebolledo

Facultad de Matemáticas, Universidad Católica de Chile,
Casilla 306 Santiago 22, Chile
e-mail: rrebolle@puc.cl

1 Introduction: a preview of open systems in Classical Mechanics 149
    1.1   Introducing probabilities 152
    1.2   An algebraic view on Probability 154

2   Completely positive maps 157

3   Completely bounded maps 162

4   Dilations of CP and CB maps 163

5   Quantum Dynamical Semigroups and Markov Flows 168

6   Dilations of quantum Markov semigroups 173
    6.1   A view on classical dilations of QMS 174
    6.2   Towards quantum dilations of QMS 180

References 181

1 Introduction: a preview of open systems in Classical Mechanics

We denote by $\Sigma \subseteq \mathbb{R}^3 \times \mathbb{R}^3$ the state space of a single particle mechanical system, that is each element $x = (q, p) = (q_1, q_2, q_3, p_1, p_2, p_3) \in \Sigma$ corresponds to the pair of *position* and *momentum* of a particle, which is supposed to have mass m.

In Newtonian Mechanics, the homogeneous dynamics is entirely characterised by the *Hamilton* operator

$$H(x) = \frac{1}{2m} |p|^2 + V(q), \ (x \in \Sigma), \tag{1}$$

---
* Partially supported by FONDECYT grant 1030552 and CONICYT/ECOS exchange program

where $|\cdot|$ denotes here the euclidian norm in $\mathbb{R}^3$. This allows to write the initial value problem which characterises the evolution of states as

$$
\begin{cases}
q_i' = \dfrac{\partial H}{\partial p_i}(x), \\[2mm]
p_i' = -\dfrac{\partial H}{\partial q_i}(x), \ 1 \le i \le 3, \\[2mm]
x(0) = x_0.
\end{cases}
\tag{2}
$$

If I denotes the 3×3 identity matrix call

$$
J = \begin{pmatrix} 0 & I \\ -I & 0 \end{pmatrix}.
\tag{3}
$$

Then (2) becomes

$$
x' = J\nabla H(x), \ x(0) = x_0,
\tag{4}
$$

where ∇ is the customary notation for the gradient of a function.

We will construct a different representation of our system, which will prepare notations for the sequel. Call $\Omega = D([0, \infty[, \Sigma)$ the space of functions $\omega = (\omega(t); \ t \ge 0)$ defined in $[0, \infty[$ with values in Σ, which have left-hand limits $(\omega(t-) = \lim_{s \to t, \ s<t} \omega(s))$ and are right-continuous $(\omega(t+) = \lim_{s \to t, \ s>t} \omega(s) = \omega(t))$ on each $t \ge 0$ (with the convention $\omega(0-) = \omega(0)$).

Define $X_t(\omega) = \omega(t)$ for all $t \ge 0$. So that for each trajectory $\omega \in \Omega$, and any time $t \ge 0$, $X_t(\omega)$ is the state of the system at time t when it follows the trajectory ω. We can write $X_t(\omega) = (Q_t(\omega), P_t(\omega))$, where $Q_t, P_t : \Omega \to \mathbb{R}^3$ represent respectively, the position and momentum applications.

Thus, (2) may be written

$$
dX_t(\omega) = J\nabla H(X_t(\omega))dt, \ X_0(\omega) = x_0.
$$

There is no great change in this writing of the equations of motion, however let us agree that such expression is a short way of writing an integral equation that is

$$
X_t(\omega) = x_0 + \int_0^t J\nabla H(X_s(\omega))ds.
$$

Solutions of the above equation are obviously continuous and differentiable. Moreover they preserve the total energy of the system:

$$
H(X_t(\omega)) = H(x_0),
\tag{5}
$$

for all $t \ge 0$.

But our basic space of trajectories Ω allows discontinuities. Thus, we may modify our simple model by introducing kicks and suppose for simplicity that $\Sigma \subset \mathbb{R}^2$. Assume for instance that at a given time t_0 the particle collides with another object which introduces an instantaneous modification (force) on the momentum. Mathematically that variation on the momentum is given by a *jump* at time t_0, that is

$\Delta P_{t_0}(\omega) = P_{t_0}(\omega) - P_{t_0-}(\omega)$. From the physical point of view, we have changed the system: we no more have a single particle but a two-particle system. In the new system the jump in the momentum of the first particle is $(-1)\times$ the jump in the momentum of the second particle via the law of *conservation of the momentum*. Suppose that the magnitude of the jump in the momentum of the colliding particle (the instantaneous force) is $c > 0$, and call $\xi(\omega)$ its sign, that is $\xi(\omega) = 1$ if the main particle is pushed forward, $\xi(\omega) = -1$ if it is pushed backwards. We then have

$$\Delta P_{t_0}(\omega) = \xi(\omega)c = c\Delta V_{t_0}(\omega),$$

where $V_t(\omega) = \xi(\omega)\mathbf{1}_{[t_0,\infty[}(t)$ and $\mathbf{1}_{[t_0,\infty[}$ is the characteristic function of $[t_0,\infty[$ (or the Heaviside function at t_0). The function $t \mapsto V_t(\omega)$ has finite variations on bounded intervals of the real line. Integration with respect to V corresponds to the customary Lebesgue-Stieltjes theory which turns out to be rather elementary in this case: if f is a right-continuous function,

$$F(t) = \int_{]0,t]} f(s)dV_s(\omega) = \sum_{0 < s \le t} f(s)\Delta V_s(\omega),$$

which allows to use the short-hand writing $dF = f(t)dV(t)$. Thus the equation of motion is written simply

$$dX_t(\omega) = J\nabla H(X_t(\omega))dt + \sigma(X_t)dV_t, \ X_0(\omega) = x_0,$$

where

$$\sigma(x) = \begin{pmatrix} 0 \\ c \end{pmatrix}.$$

$$dP_t(\omega) = cdV_t(\omega).$$

More generally, we let assume that σ is a function of the state of the system, and denote by $K(x)$ a function such that $\sigma(x) = J\nabla K(x)$, for instance $K(x) = \begin{pmatrix} 1 \\ -cq \end{pmatrix}$. This yields to

$$dX_t(\omega) = J\nabla H(X_t(\omega))dt + J\nabla K(X_t)dV_t, \ X_0(\omega) = x_0. \tag{6}$$

Take $h > 0$ and consider times $T_n^h = nh$. We suppose that a sequence of impulses take place at times $T_1^h(\omega) < T_2^h(\omega) < \ldots < T_n^h(\omega) < \ldots$. Then, the process V becomes

$$V_t^h(\omega) = \sum_{n=0}^{\infty} \xi_n(\omega)\mathbf{1}_{[T_n^h(\omega), T_{n+1}^h(\omega)[}(t) = \sum_{n=0}^{[t/h]} \xi_n,$$

where the sequence $(\xi_n(\omega))_{n \in \mathbb{N}}$ takes values in $\{-1, 1\}$.

Assume that the masses of the colliding particles are all identical to c_h. The energy dissipated during the collisions will be proportional to

$$c_h^2 \sum_{n=1}^{[t/h]} |\xi_n(\omega)|^2 = c_h^2[t/h],$$

since $|\xi_n(\omega)|^2 = 1$. To keep the dissipated energy finite as $h \to 0$, we need to choose c_h proportional to $\sqrt{h}$. Let us examine what happens to X_t, which we denote X_t^h to underline the dependence on t. Notice that

$$X_t^h(\omega) = X_0^h(\omega) + \int_0^t J\nabla H(X_s^h(\omega))ds + \begin{pmatrix} 0 \\ un \end{pmatrix} \sqrt{h}V_t^h(\omega)$$

$$= X_0^h(\omega) + \int_0^t J\nabla H(X_s^h(\omega))ds + \begin{pmatrix} 0 \\ un \end{pmatrix} \sqrt{h} \sum_{n=0}^{[t/h]} \xi_n(\omega).$$

Now we are faced to the following problem: from one hand, the dissipated energy is $h[t/h]$ which tends to t if $h \to 0$; but we currently have no tools to prove that X_t^h converges. To cope with this problem we need to modify the mathematical framework of our study by introducing probabilities.

1.1 Introducing probabilities

Consider the space Ω introduced before, endowed with the sigma-algebra $\mathcal{F}$ generated by its open subsets, the Borel sigma-algebra.

To solve the limit problem stated in the previous section, we consider a probability measure $\mathbb{P}$ for which the sequence $(\xi_n)_{n \in \mathbb{N}}$ satisfies:

- ξ_n is $\mathbb{P}$-independent of ξ_m for all n, m;
- $\mathbb{P}(\xi_n = \pm 1) = \dfrac{1}{2}$ for all n.

Under these hypothesis we obtain that the characteristic function, or Fourier transform of $M_t^h = \sqrt{h}V_t^h$ is

$$\mathbb{E}\left(e^{iuM_t^h}\right) = \prod_{i=1}^{[t/h]} \mathbb{E}\left(e^{iu\sqrt{h}\xi}\right) = \left(\cos(u\sqrt{h})\right)^{[t/h]}.$$

The last expression is equivalent to $(1 - u^2h/2)^{[t/h]}$ as $h \to 0$, thus

$$\lim_{h \to 0} \mathbb{E}\left(e^{iuM_t^h}\right) = e^{-\frac{u^2 t}{2}}. \tag{7}$$

In Classical Probability Theory, the above result is known as the Central Limit Theorem, for the random variables M_t^h: they converge in distribution towards a normal (or Gaussian) random variable with zero mean and variance t.

However, that result can be improved.

We concentrate on the equation satisfied by the algebraic flow. If the trajectories satisfy the equations

$$dX_t^h = J\nabla H(X_t^h)dt + J\nabla K(X_t^h)dM_t^h, \tag{8}$$

the flow $j^h_t(f) = f(X^h_t)$ satisfies

$$dj^h_t(f) = j^h_t(\mathbf{ad}_H(f))dt + j^h_t(\mathbf{ad}_K(f))dM^h_t + j^h_t\left(\frac{1}{2}\mathbf{ad}^2_K(f)\right)dM^h_t dM^h_t, \quad (9)$$

where $\mathbf{ad}_H(f) = \{H, f\}$ the Poisson bracket, and $\mathbf{ad}^2_K(\cdot) = \{K, \{K, \cdot\}\}$. This can be rewritten in the form

$$dj_t(f) = j^h_t(Lf)dt + j^h_t(\mathbf{ad}_K(f))dM^h_t + j^h_t\left(\frac{1}{2}\mathbf{ad}^2_K(f)\right)(dM^h_t dM^h_t - dt), \quad (10)$$

where,

$$Lf = \frac{1}{2}\mathbf{ad}^2_K(f) + \mathbf{ad}_H(f). \quad (11)$$

We notice that the processes M^h are *square integrable martingales* with respect to the family of σ-algebras $\mathcal{F}^h_t$ generated by the variables ξ_k, $k \leq [t/h]$. We recall that to each square integrable martingale M one can associate a unique predicatble increasing process A such that $M^2 - A$ is a martingale. The Brownian motion is characterized by its associated increasing process: it is a continuous martingale for which the associated process is $A_t = t$. Thus, as a shortcut, we denote $dM_t dM_t$ the measure dA_t. and we say that *the Itô table $dM^h_t dM^h_t$ converges to dA_t* if the process A^h_t converges to A_t. Now we can allow M^h to be a general family of square integrabe martingales.

Theorem 1.1. *Assume that for all $\epsilon > 0$, $t \geq 0$ it holds*

$$\mathbb{E}(\sum_{s \leq t} |\Delta M^h_s|^2 \mathbf{1}_{|\Delta M^h_s| > \epsilon}) \to 0, \quad (12)$$

as $h \to 0$, then the Itô table $dM^h_t dM^h_t$ converges in probability to dt if and only if the processes M^h converge in distribution towards a Brownian Motion.

This theorem is a particular version of the Central Limit Theorem for Martingales (Rebolledo 1980).

We thus obtain that the limit equation for the trajectories is of the form

$$dX_t = J\nabla H(X_t)dt + J\nabla K(X_t)dW_t, \quad X_0 = x. \quad (13)$$

While that of the algebraic flow is

$$dj_t(f) == j_t(Lf)dt + j_t(\mathbf{ad}_K(f))dW_t, \quad (14)$$

where Lf is given by (11).

The semigroup which corresponds to this dynamics is given by

$$T_t f(x) = \mathbb{E}(j_t(f)|X_0 = x). \quad (15)$$

And its generator is L.

The basic setting for Quantum Theory of a closed system is that of a complex separable Hilbert space $\mathfrak{h}$: observables are selfadjoint elements of the algebra $\mathcal{B}(\mathfrak{h})$ of all linear bounded operators on $\mathfrak{h}$, states are assimilated to density matrices, or positive trace-class operators ρ with $\operatorname{tr}(\rho) = 1$. The dynamics is given by a group of unitary operators $(U_t)_{t \in \mathbb{R}}$. Assume for simplicity that $U_t = e^{-itH}$ with $H = H^* \in \mathcal{B}(\mathfrak{h})$.

In this case, the evolution equation is simply:

$$dU_t = -iHU_t dt. \qquad (16)$$

The flow is the group of automorphisms $j_t : \mathcal{B}(\mathfrak{h}) \to \mathcal{B}(\mathfrak{h})$ given by

$$j_t(x) = U_t^* x U_t, \quad (x \in \mathcal{B}(\mathfrak{h})). \qquad (17)$$

The equation of this flow is

$$dj_t(x) = j_t(i[H, x])dt, \qquad (18)$$

and we define the semigroup as $T_t(x) = j_t(x)$.

The cahellenge is to provide a mathematical framework where one can include the formalism of quantum Mechanics and that of classical Probability Theory. This is required to properly speak about quantum open systems.

1.2 An algebraic view on Probability

An algebra $\mathfrak{A}$ on the complex field $\mathbb{C}$ is a vector space endowed with a product, $(a, b) \in \mathfrak{A} \times \mathfrak{A} \mapsto ab \in \mathfrak{A}$, such that

1. $a(b + c) = ab + ac$,
2. $a(\beta b) = \beta ab = (\beta a)b$,
3. $a(bc) = (ab)c$,

for all a, b, c in $\mathfrak{A}$, $\beta \in \mathbb{C}$.

Definition 1.2. *A $*$-algebra is an algebra $\mathfrak{A}$ on the complex field $\mathbb{C}$ endowed with and involution $* : \mathfrak{A} \to \mathfrak{A}$ such that*

1. $(\alpha a + \beta b)^* = \overline{\alpha} a^* + \overline{\beta} b^*$,
2. $(a^*)^* = a$,
3. $(ab)^* = b^* a^*$,

*for all $a, b \in \mathfrak{A}$, $\alpha, \beta \in \mathbb{C}$. Elements of the form $a = b^* b$ are called **positive**, they form the cone of positive elements denoted by $\mathfrak{A}^+$. This cone introduces a partial order on the algebra: $a \leq b$ if $b - a \in \mathfrak{A}^+$, for all $a, b \in \mathfrak{A}$.*

*A $*$-algebra $\mathfrak{D}$ satisfies Daniell's condition, equivalently we say it is a D^*-algebra, if it contains a unit 1; for any $a \in \mathfrak{D}^+$ there exists $\lambda > 0$ such that $a \leq \lambda 1$, and any increasing net $(a_\alpha)_{\alpha \in I}$ of positive elements with an upper bound in $\mathfrak{D}^+$ has a **least upper bound** $\sup_{\alpha \in I} a_\alpha$ in $\mathfrak{D}^+$.*

*An **algebraic probability space** is a couple $(\mathfrak{A}, \mathbb{E})$ where $\mathfrak{A}$ is a $*$-algebra on the complex field endowed with a unit 1 and $\mathbb{E} : \mathfrak{A} \to \mathbb{C}$ is a linear form, called a state, such that*

(S1) $\mathbb{E}\,(a^*a) \geq 0$, *for all* $a \in \mathfrak{A}$ *($\mathbb{E}\,(\cdot)$ is positive),*

(S2) $\mathbb{E}\,(1) = 1$.

We denote $\mathfrak{S}(\mathfrak{A})$ *the convex set of all states defined over the algebra* $\mathfrak{A}$.

Given another $*$-algebra $\mathfrak{B}$, *a* **random variable** *on* $\mathfrak{A}$ *with values on* $\mathfrak{B}$ *is a* $*$-*homomorphism*

$$j : \mathfrak{B} \to \mathfrak{A}.$$

Such a random variable, defines an image-state on $\mathfrak{B}$, **the law** *of* j, *by* $\mathbb{E}_j(B) = \mathbb{E}\,(j(B))$, $B \in \mathfrak{B}$.

This is the more general setting in which essential definitions for a Probability Theory can be given. As it is, one can hardly obtain interesting properties unless further conditions on both the involved $*$-algebras and states being assumed. Let us show first that the classical case is well included in this new theoretical framework.

Example 1.3. Given a measurable space $(\Omega, \mathcal{F})$, consider the algebra $\mathfrak{A} = b\mathcal{F}$ of all bounded measurable complex functions. This is a D^*-algebra. The corresponding algebraic probability space is then $(\mathfrak{A}, \mathbb{E})$ where $\mathbb{E}\,(X) = \int_\Omega X(\omega)d\mathbb{P}(\omega)$, for each $X \in \mathfrak{A}$, $\mathbb{P}$ being a probability measure on $(\Omega, \mathcal{F})$.

Moreover, to a classical complex-valued random variable $X \in \mathfrak{A}$ corresponds an algebraic random variable as follows. Consider the algebra $\mathfrak{B}$ of all bounded borelian functions f and define $j_X(f) = f(X)$. The map

$$j_X : \mathfrak{B} \to \mathfrak{A},$$

is clearly a $*$-homomorphism. In this case, the law $\mathbb{E}_{j_X}(f) = \mathbb{E}\,(f(X))$, defined on $\mathfrak{B}$, determines a measure on $\mathbb{C}$ endowed with its borelian σ-algebra which is the classical distribution of the random variable X.

Definition 1.4. *Given an algebraic probability space* $(\mathfrak{A}, \mathbb{E})$, *the state* $\mathbb{E}$ *is* **normal** *if for any increasing net* $(x_\alpha)_{\alpha \in I}$ *of* $\mathfrak{A}^+$ *with least upper bound* $\sup_\alpha x_\alpha$ *in* $\mathfrak{A}$ *it holds*

$$\mathbb{E}\left(\sup_\alpha x_\alpha \right) = \sup_\alpha \mathbb{E}\,(x_\alpha).$$

We denote $\mathfrak{S}_n(\mathfrak{A})$ *the set of all normal states on the algebra* $\mathfrak{A}$. *A* **pure state** *is an element* $\mathbb{E} \in \mathfrak{S}_n(\mathfrak{A})$ *for which the only positive linear functionals majorized by* $\mathbb{E}$ *are of the form* $\lambda\mathbb{E}$ *with* $0 \leq \lambda \leq 1$.

Any projection $p \in \mathfrak{A}$, *that is* $p^2 = p$, *is called* **an event**. *However, the set of projections in* $\mathfrak{A}$ *could be rather poor and in some cases reduced to the trivial elements* 0 *and* 1.

Proposition 1.5. *Given a* $*$-algebra $\mathfrak{A}$, *pure states are the extremal points of the convex set* $\mathfrak{S}_n(\mathfrak{A})$.

Proof. Let $\mathbb{E}$ be a pure state and suppose that $\mathbb{E} = \lambda\mathbb{E}_1 + (1-\lambda)\mathbb{E}_2$ with $\mathbb{E}_i \in \mathfrak{S}_n(\mathfrak{A})$, $(i = 1, 2)$, and $0 < \lambda < 1$. Then $\mathbb{E}_1 \leq \mathbb{E}$ contradicting that $\mathbb{E}$ is a pure state. Thus $\mathbb{E}$ is an extremal point of the convex set $\mathfrak{S}_n(\mathfrak{A})$.

Now, if $\mathbb{E}$ is extremal, let suppose that there exists a non trivial positive linear functional $\varphi \leq \mathbb{E}$. We may assume $0 < \varphi(1) < 1$, otherwise we replace φ by $\lambda\varphi$ with $0 < \lambda < 1$. Define

$$\mathbb{E}_1 = \frac{1}{\varphi(1)}\varphi, \quad \mathbb{E}_2 = \frac{1}{1 - \varphi(1)}(\mathbb{E} - \varphi).$$

Both, $\mathbb{E}_1$ and $\mathbb{E}_2$ are states and $\mathbb{E} = \varphi(1)\mathbb{E}_1 + (1 - \varphi(1))\mathbb{E}_2$. This is a contradiction since $\mathbb{E}$ is extremal. Thus $\mathbb{E}$ is a pure state.

Before going on, let us say a word about the notation of states. In Probability the notation $\mathbb{E}$ is more appealing, however in the tradition of operator algebras, states are oftenly denoted by greek letters like ω. We will use both notations depending on the kind of properties we wish to emphasize.

In the previous example, $\mathfrak{A}$ contained non trivial events: all elements $p = 1_E$, with $E \in \mathfrak{F}$. We will see later a commutative algebraic probability space which has no nontrivial projections, but we first give a prototype of a non-commutative probability space.

Example 1.6. Consider the algebra $\mathfrak{A} = \mathfrak{M}_n(\mathbb{C}^n)$ the space of $n \times n$-matrices acting on the space $\mathfrak{h} = \mathbb{C}^n$. Given a positive density matrix with unit trace ρ, one defines a state as $\mathbb{E}(A) = \text{tr}(\rho A)$. Thus, $(\mathfrak{A}, \mathbb{E})$ is an example of a non-commutative algebraic probability space.

An *observable* here is any self-adjoint operator X. Take $\mathfrak{B} = \mathfrak{A}$, and U a unitary transformation of $\mathbb{C}^n$. Then $j(B) = U^*BU$, $B \in \mathfrak{B}$ defines a random variable.

In this case, $\mathfrak{A}$ has non-trivial events: any projection defined on $\mathbb{C}^n$.

Example 1.7. Consider a compact space Ω endowed with its borelian σ-algebra $\mathfrak{B}(\Omega)$ and a Radon probability measure $\mathbb{P}$ (or Radon expectation $\mathbb{E}(\cdot)$). Now, take $\mathfrak{A} = C(\Omega, \mathbb{C})$ the algebra of continuous complex-valued functions. The space $(\mathfrak{A}, \mathbb{E})$ is a commutative probability space with no nontrivial events. This is an important space which is frequently used when studying classical dynamical systems.

The above is an example of an important class of *-algebras, the class of C^*-algebras.

Definition 1.8. *A *-algebra $\mathfrak{A}$ endowed with a norm $\|\cdot\|$ is a Banach *-algebra if it is complete with respect to the topology defined by this norm, which is referred to as the* uniform topology, *and $\|a\| = \|a^*\|$, for all $a \in \mathfrak{A}$. A Banach *-algebra is a C^*-algebra if moreover,*

$$\|a^*a\| = \|a\|^2, \tag{19}$$

for all $a \in \mathfrak{A}$.

A subspace S of a unital C^*-algebra, is called an operator system *if for any $s \in S$ one has $s^* \in S$ and $1 \in S$.*

A von Neumann algebra on a Hilbert space $\mathfrak{h}$ is a *-subalgebra $\mathfrak{M}$ of $\mathcal{B}(\mathfrak{h})$ which is weakly closed. This is equivalent to $\mathfrak{M} = \mathfrak{M}''$ where the right hand term denotes the bicommutant. Another equivalent characterization is that $\mathfrak{M}$ is the dual of a Banach space denoted $\mathfrak{M}_*$ and called its predual.

2 Completely positive maps

We start by extending the classical notion of a *transition kernel* in Probability Theory. Let be given two measurable spaces $(E_i, \mathcal{E}_i)$, $(i = a, b)$, and a kernel $P(x, dy)$ from E_b to E_a. That is, $P : E_b \times \mathcal{E}_a$ is such that

- $x \mapsto P(x, A)$ is measurable from E_b in $[0, 1]$ for any $A \in \mathcal{E}_a$;
- $A \mapsto P(x, A)$ is a probability on $(E_a, \mathcal{E}_a)$ for all $x \in E_b$.

We denote $\mathcal{A}$ (respectively $\mathcal{B}$) the algebra of all complex bounded measurable functions defined on E_a (resp. E_b). These are *-algebras (they have an involution given by the operation of complex conjugation) with unit. Moreover, they are C^*-algebras since they are complete for the topology defined by the uniform norm. The kernel P defines a linear map Φ_P from $\mathcal{A}$ to $\mathcal{B}$ given by $\Phi_P(a) = Pa$, where

$$Pa(x) = \int_{E_b} P(x, dy)a(y),$$

for all $a \in \mathcal{A}$, $x \in E_b$.

It is worth noticing that Φ_P is a positive map, moreover it satisfies a stronger property: for any finite collection of elements $a_i \in \mathcal{A}$, $b_i \in \mathcal{B}$, $(i = 1, \ldots, n)$, the function

$$\sum_{i,j=1}^n b_i \Phi_P(a_i \bar{a}_j) \bar{b}_j, \tag{20}$$

is positive. Indeed, for any fixed $x \in E_b$, $P(x, \cdot)$ is positive definite, so that for any collection $\alpha_1, \ldots, \alpha_n$ of complex numbers, the sum

$$\sum_{i,j} \alpha_i \bar{\alpha}_j P(a_i \bar{a}_j)(x),$$

is positive. It is enough to choose $\alpha_i = b_i(x)$, $(i = 1, \ldots, n)$, to obtain (20).

Now, take a probability measure μ on $(E_b, \mathcal{E}_b)$, call $\Omega = E_b \times E_a$, $\mathcal{F} = \mathcal{E}_b \otimes \mathcal{E}_a$ and define a probability $\mathbb{P}$ on $(\Omega, \mathcal{F})$ given by

$$\mathbb{E}(b \otimes a) = \int_{E_b} \mu(dx) b(x) Pa(x), \tag{21}$$

where $a \in \mathcal{A}$, $b \in \mathcal{B}$.

Under the probability $\mathbb{P}$, the random variables (X_b, X_a), given by the coordinate maps on $E_b \times E_a$, satisfy the following property: μ is the distribution of X_b and $P(x, dy)$ is the conditional probability of X_a given that $X_b = x$.

We now study the construction of $L^2(\Omega, \mathcal{F}, \mathbb{P})$. With this purpose consider the family of random variables of the form $X = \sum_{i=1}^{n} b_i \otimes a_i$ with $a_i \in \mathcal{A}, b_i \in \mathcal{B}$, $(i = 1, \ldots, n)$. The scalar product of two of such elements, is

$$\langle X^{(1)}, X^{(2)} \rangle = \int_{E_b} \mu(dx) \sum_{i,j} b_i^{(1)}(x) P(a_i^{(1)} \overline{a_j^{(2)}})(x) \overline{b_j^{(2)}}(x). \qquad (22)$$

Notice that (20) is needed if one wants to define the scalar product through (22). Within this commutative framework, the property (20) is granted by the positivity of the kernel. This fails in the non-commutative case.

Definition 2.1. *Let be given two $*$-algebras $\mathfrak{A}$, $\mathfrak{B}$ and an operator system $\mathcal{S}$ which is a subspace of $\mathfrak{A}$. A linear map $\Phi : \mathcal{S} \to \mathfrak{B}$ is* **completely positive** *if for any two finite collections $a_1, \ldots, a_n \in \mathcal{S}$ and $b_1, \ldots, b_n \in \mathfrak{B}$, the element*

$$\sum_{i,j=1}^{n} b_i^* \Phi(a_i^* a_j) b_j \in \mathfrak{B},$$

is positive.

The set of all completely positive maps from $\mathcal{S}$ to $\mathfrak{B}$ is denoted $\mathbf{CP}(\mathcal{S}, \mathfrak{B})$.

We restrict our attention to C^*-algebras and recall that a *representation of a C^*–algebra* $\mathfrak{A}$ is a couple $(\pi, \mathfrak{k})$, where $\mathfrak{k}$ is a complex Hilbert space and π is a $*$–homomorphism of $\mathfrak{A}$ and the algebra of all bounded linear operators on $\mathfrak{k}$, $\mathcal{B}(\mathfrak{k})$.

Remark 2.2. Assume that the algebra $\mathfrak{B}$ is included in $\mathcal{B}(\mathfrak{h})$ say, for a given complex separable Hilbert space $\mathfrak{h}$. Then the positivity of the element

$$\sum_{i,j=1}^{n} b_i^* \Phi(a_i^* a_j) b_j,$$

introduced before is equivalent to

$$\sum_{i,j=1}^{n} \langle u, b_i^* \Phi(a_i^* a_j) b_j u \rangle \geq 0,$$

for all $u \in \mathfrak{h}$. Equivalently,

$$\sum_{i,j=1}^{n} \langle u_i, \Phi(a_i^* a_j) u_j \rangle \geq 0, \qquad (23)$$

for all collection of vectors $u_1, \ldots, u_n \in \mathfrak{h}$.

Two complementary results, one due to Arveson and the second proved by Stinespring, show that complete positivity is always derived from positivity in the commutative case. More precisely,

Theorem 2.3. *Given a positive map* $\Phi : \mathfrak{A} \to \mathfrak{B}$, *it is completely positive if at least one of the two conditions below is satisfied*

(a) $\mathfrak{A}$ *is commutative (Stinespring [35]);*
(b) $\mathfrak{B}$ *is commutative (Arveson [3]).*

Proof. (a) Suppose $\mathfrak{B} \subseteq B(\mathfrak{h})$ for a complex and separable Hilbert space $\mathfrak{h}$. Since $\mathfrak{A}$ is a commutative C^*-algebra containing a unit $\mathbf{1}$, it is isomorphic to the espace of continuous functions defined on a compact Hausdorff space (the spectrum $\sigma(\mathfrak{A})$ of $\mathfrak{A}$). So that any element $a \in \mathfrak{A}$ is identified with a continuous function $a(x)$, $x \in \sigma(\mathfrak{A})$. Therefore, since the map Φ is positive, linear, and $\Phi(\mathbf{1}) = 1$, it follows that for all $u, v \in \mathfrak{h}$, there exists a complex-valued Baire measure with finite total variation $\mu_{u,v}$ such that

$$\langle v, \Phi(a)u \rangle = \int_{\sigma(\mathfrak{A})} d\mu_{u,v}(x)a(x).$$

Take now arbitrary vectors $u_1, \ldots, u_n \in \mathfrak{h}$. Define

$$d\mu = \sum_{i,j} |d\mu_{u_i,u_j}|,$$

where the vertical bars denote total variation of the corresponding measure. Then each measure μ_{u_i,u_j} is absolutely continuous with respect to the positive measure μ. Let h_{u_i,u_j} denote the Radon-Nykodim derivative of μ_{u_i,u_j}. Put $u = \sum_i \lambda_i u_i$, then

$$d\mu_{u,u} = \left(\sum_{i,j} \overline{\lambda_i}\lambda_j h_{u_i,u_j} \right) d\mu,$$

and since both measures $\mu_{u,u}$ and μ are positive, it follows that

$$\sum_{i,j} \overline{\lambda_i}\lambda_j h_{u_i,u_j} \geq 0,$$

μ-almost surely for all finite collection $\lambda_1, \ldots, \lambda_n$ of complex numbers. Furthermore, for any collection $a_1, \ldots, a_n \in \mathfrak{A}$,

$$\sum_{i,j} \langle u_i, \Phi(a_i^* a_j)u_j \rangle = \int_{\sigma(\mathfrak{A})} d\mu(x) \left(\sum_{i,j} \overline{a_i(x)}a_j(x)h_{u_i,u_j}(x) \right) \geq 0.$$

Thus, Φ is completely positive.
(b) If $\mathfrak{B}$ is commutative, we identify elements b of $\mathfrak{B}$ with continuous functions $b(x)$. Thus, given arbitrary collections $a_1, \ldots, a_n \in \mathfrak{A}, b_1, \ldots, b_n \in \mathfrak{B}$,

$$\sum_{i,j} \overline{b_i(x)}\Phi(a_i^* a_j)b_j(x) = \Phi\left(\left[\sum_k b_k(x)a_k \right]^* \left[\sum_k b_k(x)a_k \right] \right) \geq 0.$$

Thus, the notion of complete positivity attains its full sense in the pure noncommutative framework, that is, when both $\mathfrak{A}$ and $\mathfrak{B}$ are non abelian. For each $n \geq 1$, let denote $\mathcal{M}_n(\mathfrak{A})$ the algebra of all $n \times n$-matrices $(a_{i,j})$, where $a_{i,j} \in \mathfrak{A}$. Moreover, to any linear map $\Phi : \mathfrak{A} \to \mathfrak{B}$, we associate the map $\Phi_n : \mathcal{M}_n(\mathfrak{A}) \to \mathcal{M}_n(\mathfrak{B})$ defined by

$$\Phi_n((a_{i,j})) = (\Phi(a_{i,j})). \tag{24}$$

The following characterization follows imediately from the definition.

Proposition 2.4. *Given two C^*-algebras $\mathfrak{A}$ and $\mathfrak{B}$, a linear map $\Phi : \mathfrak{A} \to \mathfrak{B}$ is completely positive if and only if $\Phi_n : \mathcal{M}_n(\mathfrak{A}) \to \mathcal{M}_n(\mathfrak{B})$ is positive for all $n \geq 1$.*

Definition 2.5. *A linear map $\Phi : \mathfrak{A} \to \mathfrak{B}$ is n-positive if $\Phi_n : \mathcal{M}_n(\mathfrak{A}) \to \mathcal{M}_n(\mathfrak{B})$ is positive.*

As we will see in the next section, the study of a linear map $\Phi : \mathfrak{A} \to \mathfrak{B}$ through the induced sequence of maps (Φ_n) is a powerful procedure. Especially because we can use well-known features about matrix algebra to obtain results for linear maps between C^*-algebras.

For instance, consider a Hilbert space $\mathfrak{h}$, positive operators $P, Q \in \mathcal{B}(\mathfrak{h})$, and A any bounded operator. Take $\lambda \in \mathbb{C}$, and vectors $u, v \in \mathfrak{h}$. Compute

$$\left\langle \begin{pmatrix} \lambda u \\ v \end{pmatrix}, \begin{pmatrix} P & A \\ A^* & Q \end{pmatrix} \begin{pmatrix} \lambda u \\ v \end{pmatrix} \right\rangle = \lambda^2 \langle u, Pu \rangle + \overline{\lambda}\langle u, Av \rangle + \lambda \overline{\langle u, Av \rangle} + \langle v, Qv \rangle.$$

Thus the right-hand term is positive if and only if

$$|\langle u, Av \rangle|^2 \leq \langle u, Pu \rangle \langle v, Qv \rangle. \tag{25}$$

From this elementary computation we derive that the matrix

$$\begin{pmatrix} P & A \\ A^* & Q \end{pmatrix}, \tag{26}$$

is positive if and only if (25) holds. As a result we obtain:

Proposition 2.6. *Let $\mathfrak{A}$ and $\mathfrak{B}$ be two C^*-algebras. We assume that $\mathfrak{A}$ has a unit.*

1. *Suppose $a \in \mathfrak{A}$, then $\|a\| \leq 1$ if and only if the matrix*

$$\begin{pmatrix} 1 & a \\ a^* & 1 \end{pmatrix},$$

is positive in $\mathcal{M}_2(\mathfrak{A})$.
2. *Let $b \in \mathfrak{A}$ be a positive element of $\mathfrak{A}$. Then $a^*a \leq b$ if and only if the matrix*

$$\begin{pmatrix} 1 & a \\ a^* & b \end{pmatrix},$$

is positive in $\mathcal{M}_2(\mathfrak{A})$.

3. Suppose that $\mathcal{B}$ is also unital and that $\Phi : \mathfrak{A} \to \mathcal{B}$ is a 2-positive linear map which preserves the unit. Then Φ is contractive.

4. Let Φ be a unital 2-positive linear map as before. Then $\Phi(a)^*\Phi(a) \le \Phi(a^*a)$, for all $a \in \mathfrak{A}$. This is known as the **Schwartz inequality for 2-positive maps**.

Proof. 1. Taking a representation $(\pi, \mathfrak{h})$ of $\mathfrak{A}$, let $A = \pi(a)$, $P = Q = 1$ in (26), which is positive if and only if $|\langle u, Av \rangle|^2 \le \|u\|^2 \|v\|^2$ for all $u, v \in \mathfrak{h}$. This is equivalent to the condition $\|a\| \le 1$.

2. Similarly, choosing $P = 1$, $Q = \pi(b)$, $A = \pi(a)$, the positivity of the matrix (26) is equivalent to $|\langle u, Av \rangle|^2 \le \|u\|^2 \langle u, Qv \rangle$, which holds if and only if $\|a^*a\| = \|A\|^2 \le \|Q^{1/2}\|^2 = \|b^{1/2}\|^2$, that is, $a^*a \le b$.

3. Notice that for any $a \in \mathfrak{A}$ such that $\|a\| \le 1$,

$$\Phi_2 \begin{pmatrix} 1 & a \\ a^* & 1 \end{pmatrix} = \begin{pmatrix} 1 & \Phi(a) \\ \Phi(a)^* & 1, \end{pmatrix}$$

is positive, so that $\|\Phi(a)\| \le 1$.

4. For any element $a \in \mathfrak{A}$, the product

$$\begin{pmatrix} 1 & a \\ 0 & 0 \end{pmatrix}^* \begin{pmatrix} 1 & a \\ 0 & 0 \end{pmatrix} = \begin{pmatrix} 1 & a \\ a^* & a^*a \end{pmatrix},$$

is positive. Thus,

$$\begin{pmatrix} 1 & \Phi(a) \\ \Phi(a^*) & \Phi(a^*a) \end{pmatrix} \ge 0.$$

By part 2 before, we obtain that $\Phi(a)^*\Phi(a) \le \Phi(a^*a)$.

Let $\mathfrak{f}$ be a finite-dimensional space. The algebra $\mathcal{B}(\mathfrak{f})$ is isomorphic to the algebra of $n \times n$ matrices $\mathcal{M}_n$ for some n. Suppose that $\mathcal{S}$ is an operator system contained in a C^*-algebra $\mathfrak{A}$ and let $\Phi : \mathfrak{A} \to \mathcal{B}(\mathfrak{f})$ be a completely positive map. An extension theorem due to Krein shows that any positive map defined on $\mathcal{S}$ with values in $\mathbb{C}$ can be extended to all of $\mathfrak{A}$. So that, for all m, the positive map $\Phi_m : \mathcal{M}_m(\mathcal{S}) \to \mathcal{M}_n$ can be extended to $\mathcal{M}_m(\mathfrak{A})$. This means that the completely positive map $\Phi : \mathcal{S} \to \mathcal{M}_n$ can be extended to all of $\mathfrak{A}$ that is, there exists a CP map $\Psi : \mathfrak{A} \to \mathcal{M}_n$ such that $\Psi|_{\mathcal{S}} = \Phi$. The following crucial result proved by Arveson gives the main extension theorem for CP maps.

Theorem 2.7 (Arveson). *Let $\mathfrak{A}$ be a C^*-algebra, $\mathcal{S}$ an operator system contained in $\mathfrak{A}$ and $\Phi : \mathcal{S} \to \mathcal{B}(\mathfrak{h})$ a completely positive map. Then there exists a completely positive map $\Psi : \mathfrak{A} \to \mathcal{B}(\mathfrak{h})$ which extends Φ.*

Proof. Consider the directed net FD of all finite-dimensional subspaces $\mathfrak{f}$ of $\mathfrak{h}$. Denote $P_{\mathfrak{f}}$ the projection onto $\mathfrak{f}$ defined on $\mathfrak{h}$ and call $\Phi_{\mathfrak{f}}(a) = P_{\mathfrak{f}}\Phi(a)|_{\mathfrak{f}}$, $a \in \mathcal{S}$, the compression (or reduction) of Φ to $\mathfrak{f}$. From the previous discussion, we know that there exists a completely positive map $\Psi_{\mathfrak{f}} : \mathfrak{A} \to \mathcal{B}(\mathfrak{f})$ which extends $\Phi_{\mathfrak{f}}$, since $\mathcal{B}(\mathfrak{f})$ is isomorphic to an algebra of finite-dimensional matrices. Defining $\Psi_{\mathfrak{f}}$ to be

0 on the orthogonal complement of $\mathfrak{f}$, we extend the range of this map to $\mathcal{B}(\mathfrak{h})$. Moreover $\|\Phi_{\mathfrak{f}}\| \leq \|\Phi(1)\|$, for all $\mathfrak{f} \in FD$, so that this net is compact in the w^*-topology by an application of the Banach-Alaglou Theorem. As a result, there exists a limit point, a completely positive map Ψ, such that $\|\Psi\| \leq \|\Phi\|$. We prove that Ψ extends Φ. Indeed, let $a \in \mathcal{S}$, $u, v \in \mathfrak{h}$ and denote $\mathfrak{f}$ the vector space generated by u and v. Then, for any other finite-dimensional subspace $\mathfrak{f}_1$ of $\mathfrak{h}$ which contains $\mathfrak{f}$ it holds $\langle v, \Phi(a)u \rangle = \langle v, \Psi_{\mathfrak{f}_1}(a)u \rangle$. Thus, since $\mathfrak{f}_1$ is cofinal, we obtain $\langle v, \Phi(a)u \rangle = \langle v, \Psi(a)u \rangle$.

3 Completely bounded maps

The sum of completely positive maps is again completely positive as well as the composition of two of such maps. Furthermore, any *-homomorphism of algebras is completely positive. Thus, given any representation $(\pi, \mathfrak{k})$ of the C^*-algebra $\mathfrak{A}$, π is completely positive. To summarize, the set $\mathbf{CP}(\mathfrak{A}, \mathfrak{B})$ of completely positive maps from $\mathfrak{A}$ to $\mathfrak{B}$ defines a cone.

In a C^*-algebra $\mathfrak{A}$, the cone $\mathfrak{A}^+$ of positive elements defines a norm-closed convex cone. If $h \in \mathfrak{A}$ is a self-adjoint element, the functional calculus shows easily that h can be written as the difference of two positive elements. Indeed, it suffices to use the decomposition of any real number x in its positive $x^+ = \sup\{x, 0\}$ and negative parts $x^- = \sup\{-x, 0\}$. Furthermore, using the Cartesian decomposition of an arbitrary element $a \in \mathfrak{A}$, namely, $a = h + ik$, where h and k are self-adjoints, one obtains

$$a = (h^+ - h^-) + i(k^+ - k^-),$$

where $h^\pm$, $k^\pm$ are positive elements of $\mathfrak{A}$. Thus $\mathfrak{A}$ is the complex linear span of $\mathfrak{A}^+$.

We want to extend this property to $\mathbf{CP}(\mathfrak{A}, \mathfrak{B})$, for two C^*-algebras. That is, we want to study the complex linear span of the above cone.

Definition 3.1. *With the notations previous to Proposition 2.4, let $\Phi : \mathfrak{A} \rightarrow \mathfrak{B}$ be a linear map. We say that Φ is **completely bounded** if $\|\Phi\|_{cb} := \sup_n \|\Phi_n\| < \infty$. The normed space of completely bounded maps from $\mathfrak{A}$ to $\mathfrak{B}$ is denoted $\mathbf{CB}(\mathfrak{A}, \mathfrak{B})$.*

It is easily seen that $\mathbf{CB}(\mathfrak{A}, \mathfrak{B})$ is indeed a Banach space and any $\Phi \in \mathbf{CB}(\mathfrak{A}, \mathfrak{B})$ can be decomposed into a linear combination of completely positive maps. Indeed, this theory in its current development, has obtained deeper results which the interested reader can follow in the book of Paulsen [28]. We limit ourselves to give below a partial account of those important properties.

Proposition 3.2. *Let S be an operator system in a C^*-algebra with unit and $\Phi : S \rightarrow \mathfrak{B}$ a completely positive map, where $\mathfrak{B}$ is another C^*-algebra. Then Φ is completely bounded and $\|\Phi\|_{cb} = \|\Phi(1)\|$.*

Proof. It is clear that $\|\Phi(1)\| \leq \|\Phi\| \leq \|\Phi\|_{cb}$. So that we only need to prove that $\|\Phi\|_{cb} \leq \|\Phi(1)\|$. Denote $\mathbf{1}_n$ the unit of $\mathcal{M}_n(\mathfrak{A})$. Let $A = (a_{i,j})$ be in $\mathcal{M}_n(\mathcal{S})$ and $\|A\| \leq 1$. The matrix,

$$\begin{pmatrix} \mathbf{1}_n & A \\ A^* & \mathbf{1}_n \end{pmatrix},$$

is positive, hence so is

$$\Phi_{2n} \begin{pmatrix} \mathbf{1}_n & A \\ A^* & \mathbf{1}_n \end{pmatrix} = \begin{pmatrix} \Phi_n(\mathbf{1}_n) & \Phi_n(A) \\ \Phi_n(A)^* & \Phi_n(\mathbf{1}_n) \end{pmatrix}.$$

Therefore, $\|\Phi_n(A)\| \leq \|\Phi_n(\mathbf{1}_n)\| = \|\Phi(\mathbf{1})\|$.

Like in Theorem 2.3 we obtain that bounded maps are completely bounded if its range is an abelian C^*-algebra.

Theorem 3.3. *Let S be an operator system and $\Phi : S \to \mathfrak{B}$ a bounded linear map, where $\mathfrak{B}$ is a commutative C^*-algebra. Then $\|\Phi\|_{cb} = \|\Phi\|$.*

Proof. Since $\mathfrak{B}$ is commutative, we identify elements b of $\mathfrak{B}$ with continuous functions $b(x)$ defined on a compact Hausdorff space X. Every element $B = (b_{i,j})$ of $\mathcal{M}_n(\mathfrak{B})$ is identified with continuous matrix-valued functions; multiplication is just pointwise muliplication and the involution is the $*$ operation on matrices. $\mathcal{M}_n(\mathfrak{B})$ is a C^*-algebra with the norm $\|B\| = \sup\{\|B(x)\| : x \in X\}$.

Let $x \in X$, and define $\Phi^x : S \to \mathbb{C}$ by $\Phi^x(a) = \Phi(a)(x)$. Thus,

$$\|\Phi_n\| = \sup\{\|\Phi_n^x\| : x \in X\} = \sup\{\|\Phi^x\| : x \in X\} = \|\Phi\|.$$

4 Dilations of CP and CB maps

Throughout this section we assume that the C^*–algebras $\mathfrak{A}$ and $\mathfrak{B}$ have a unit denoted in both cases by the same symbol $\mathbf{1}$.

Theorem 4.1 (Stinespring). *Let $\mathfrak{B}$ be a sub C^*–algebra of the algebra of all bounded operators on a given complex separable Hilbert space $\mathfrak{h}$. Assume $\mathfrak{A}$ to be a C^*–algebra with unit. A linear map $\Phi : \mathfrak{A} \to \mathfrak{B}$ is completely positive if and only if it has the form*

$$\Phi(x) = V^*\pi(x)V, \tag{27}$$

where $(\pi, \mathfrak{k})$ is a representation of $\mathfrak{A}$ on some Hilbert space $\mathfrak{k}$, and V is a bounded operator from $\mathfrak{h} \to \mathfrak{k}$.

Proof. Assume that $\mathfrak{A}$ and $\mathfrak{B}$ are C^*-algebras, with $\mathfrak{B} \subset \mathcal{B}(\mathfrak{h})$, where $\mathfrak{h}$ is a complex separable Hilbert space. Let be given a completely positive map $\Phi : \mathfrak{A} \to \mathfrak{B}$. Take two arbitrary elements $x = \sum_i a_i \otimes u_i$, $y = \sum_j b_j \otimes v_j$ in the algebraic tensor product $\mathfrak{A} \otimes \mathfrak{h}$, where both sums contain a finite number of terms, and define

$$\langle\!\langle x, y \rangle\!\rangle = \sum_{i,j} \langle u_i, \Phi(a_i^* b_j) v_j \rangle.$$

Since Φ is completely positive, $\langle\!\langle x, x \rangle\!\rangle \geq 0$. Denote

$$\mathcal{N} = \{x \in \mathfrak{A} \otimes \mathfrak{h}; \langle\!\langle x, x \rangle\!\rangle = 0\},$$

and introduce on the quotient space $(\mathfrak{A} \otimes \mathfrak{h})/\mathcal{N}$ the scalar product

$$\langle\!\langle x + \mathcal{N}, y + \mathcal{N} \rangle\!\rangle = \langle x, y \rangle.$$

By completion, we obtain a Hilbert space denoted $\mathfrak{k}$.

Our purpose now is to define a *-homomorphism $\pi : \mathfrak{A} \to \mathcal{B}(\mathfrak{k})$. This is done in two steps. Firstly, define $\pi_0(a)$ for $a \in \mathfrak{A}$ on elements of the form x before:

$$\pi_0(a) \left(\sum_i a_i \otimes u_i \right) = \sum_i (a a_i) \otimes u_i.$$

For x and y as before, $\pi_0(a)$ is a linear application in $\mathfrak{A} \otimes \mathcal{H}$ which satisfies

$$\langle\!\langle x, \pi_0(a)y \rangle\!\rangle = \langle\!\langle \pi_0(a^*)x, y \rangle\!\rangle \tag{28}$$
$$\|\pi_0(a)x\|^2 = \langle\!\langle x, \pi_0(a^*a)x \rangle\!\rangle \le \|a^*a\| \langle\!\langle x, \pi_0(1)x \rangle\!\rangle$$
$$\le \|a\|^2 \|x\|^2. \tag{29}$$

From the above relations, π_0 extends into a *-homomorphism $\pi : \mathfrak{A} \to \mathcal{B}(\mathfrak{k})$ and $(\pi, \mathfrak{k})$ is a representation of $\mathfrak{A}$.

Moreover, we can define a linear operator $V : \mathfrak{h} \to \mathfrak{k}$ by

$$Vu = 1 \otimes u + \mathcal{N}.$$

This is a bounded operator since

$$\|Vu\|^2 = \langle u, \Phi(1)u \rangle \le \|\Phi(1)\| \|u\|^2.$$

Finally, Φ may be written in the form

$$\Phi(a) = V^* \pi(a) V,$$

for all $a \in \mathfrak{A}$.

On the other hand, if Φ is given through (27), an elementary computation shows that Φ is completely positive.

Thus we have obtained the celebrated characterization of completely positive maps due to Stinespring [35] (see also [26], [30]). The representation (27) is not unique. We call the couple (π, V) a *Stinespring representation* of Φ. Moreover, the above representation is said to be *minimal* if $\{\pi(x)Vu : x \in \mathfrak{A}, u \in \mathfrak{h}\}$ is dense in $\mathfrak{k}$. For a given completely positive map, the minimal representation is unique up to a unitary equivalence.

Proposition 4.2. *Let $\mathfrak{A}$ be a C^*-algebra and $\Phi : \mathfrak{A} \to \mathcal{B}(\mathfrak{h})$ a completely positive map. Suppose two minimal Stinespring dilations $(\pi_i, V_i, \mathfrak{k}_i)$, $i = 1, 2$, be given for Φ. Then there exists a unitary operator $U : \mathfrak{k}_1 \to \mathfrak{k}_2$ which satisfies $U V_1 = V_2$ and $U \pi_1 U^* = \pi_2$.*

Proof. Vectors like $\sum_{j=1}^{n} \pi_i(a_j) V_i u_j$ form a dense subset $\mathfrak{V}_i$ of $\mathfrak{k}_i$, $(i = 1, 2)$. Thus, the theorem follows from mapping these two dense subsets via an operator U. Define

$$U\left(\sum_{j=1}^{n} \pi_1(a_j) V_1 u_j\right) = \sum_{j=1}^{n} \pi_2(a_j) V_2 u_j,$$

for any integer $n \geq 1$, $a_1, \ldots, a_n \in \mathfrak{A}$, $u_1, \ldots, u_n \in \mathfrak{h}$. The density of $\mathfrak{V}_1$ and $\mathfrak{V}_2$ implies that U is onto. It remains to prove that it is an isometry, which follows from the computation below:

$$\left\|\sum_{j=1}^{n} \pi_1(a_j) V_1 u_j\right\|^2 = \sum_{i,j} \langle u_i V_1^* \pi_1(a_i^* a_j) V_1 u_j\rangle$$

$$= \sum_{i,j} \langle u_i, \Phi(a_i^* a_j) u_j\rangle$$

$$= \left\|\sum_{j=1}^{n} \pi_2(a_j) V_2 u_j\right\|^2.$$

If the completely positive map Φ is σ–weakly continuous and preserves the identity, then its minimal representation (π, V) is such that π is σ–weakly continuous, and V is an isometry: $V^* V = 1$. We denote by $\mathbf{CP}(\mathfrak{A}, \mathfrak{B})$ the set of all σ–weakly continuous completely positive maps $\Phi : \mathfrak{A} \to \mathfrak{B}$ which preserve the identity. Furthermore, in this case $\mathfrak{h}$ may be identified with the subspace $V\mathfrak{h}$ of $\mathfrak{k}$, V^* becoming the projection $P_\mathfrak{h}$ onto this subspace and the representation of Φ can be written

$$\Phi(a) = P_\mathfrak{h} \pi(a)|_\mathfrak{h},$$

for all $a \in \mathfrak{A}$.

For a von Neumann algebra $\mathfrak{A}$, and $\mathfrak{B} = \mathcal{B}(\mathfrak{k})$, Kraus (see [23]) obtained the following characterization of normal completely positive maps.

Theorem 4.3 (Kraus). *Let be given two complex separable Hilbert spaces $\mathfrak{h}$, $\mathfrak{k}$, and a von Neumann algebra $\mathfrak{A}$ of operators of $\mathfrak{h}$. Then a linear map $\Phi : \mathfrak{A} \to \mathcal{B}(\mathfrak{k})$ is normal and completely positive if and only if there exists a sequence $(V_j)_{j \in \mathbb{N}}$ of linear bounded operators from $\mathfrak{k}$ to $\mathfrak{h}$ such that the series $\sum_{j=1}^{\infty} V_j^* a V_j$ strongly converges for any $a \in \mathfrak{A}$ and*

$$\Phi(a) = \sum_{j=1}^{\infty} V_j^* a V_j. \tag{30}$$

Proof. It suffices to show that there exists a representation of a normal π in (27) leading to (30). Firstly, it can be shown that there exists a sequence of vectors $(u_n)_{n \in \mathbb{N}}$ in $\mathfrak{h}$ such that $\sum_n \|u_n\|^2 = 1$ and $\langle \Omega, \pi(a)\Omega\rangle = \sum_n \langle u_n, a u_n\rangle$, where Ω is a cyclic vector for $\pi(\mathfrak{A})$.

Moreover,

$$\|xu_n\|^2 = \langle u_n, (x^*x)u_n \rangle \le \langle \Omega, \pi(x^*x)\Omega \rangle = \|\pi(x)\Omega\|^2$$

Let then, $V_n \pi(x)\Omega = xu_n$, for all $x \in \mathfrak{A}$. Thus we have,

$$\langle \pi(x)\Omega, \pi(a)\pi(x)\Omega \rangle = \sum_j \langle \pi(x)\Omega, V_j^* a V_j \pi(x)\Omega \rangle.$$

Remark 4.4. The above representation can be improved by introducing an additional arbitrary complex and separable Hilbert space $\tilde{\mathfrak{h}}$ with an orthonormal basis $(f_n)_{n \in \mathbb{N}}$. Indeed, defining $V : \mathfrak{k} \to \mathfrak{h} \otimes \tilde{\mathfrak{h}}$ by

$$Vu = \sum_j V_j u \otimes f_j, \quad (u \in \mathfrak{k}),$$

then

$$\Phi(a) = V^*(a \otimes 1)V, \tag{31}$$

where 1 is the identity operator of $\tilde{\mathfrak{h}}$, $a \in \mathfrak{A}$.

Remark 4.5. Following the same procedure used to prove the Kraus representation of a completely positive map Φ, one can obtain a dilation based on random operators. Indeed, take Φ like in Theorem 4.3. Denote E an orthonormal basis in $\mathfrak{h}$ (which is countable, since $\mathfrak{h}$ has been assumed separable). On the space E define the σ-algebra of all subsets and define a probability μ such that $\langle \Omega, \pi(a)\Omega \rangle = \int_E \langle e, ae \rangle \mu(de) = \sum_{e \in E} \langle e, ae \rangle \mu(\{e\}) = \mathbb{E}(\langle \cdot, a \cdot \rangle)$, where Ω is a cyclic vector for $\pi(\mathfrak{A})$.

Now define, like in the proof of 30, $V(e)\pi(x)\Omega = xe$, $x \in \mathfrak{A}$, $e \in E$, which yields,

$$\langle \pi(x)\Omega, \Phi(a)\pi(x)\Omega \rangle = \sum_{e \in E} \mu(\{e\})\langle \pi(x)\Omega, V^*(e)aV(e)\pi(x)\Omega \rangle$$

$$= \mathbb{E}\left(\langle \pi(x)\Omega, V^*(\cdot)aV(\cdot)\pi(x)\Omega \rangle\right)$$

$$= \langle \pi(x)\Omega, \mathbb{E}(V^*aV)\pi(x)\Omega \rangle,$$

where $\mathbb{E}(V^*aV)$ is interpreted as an operator-valued integral, so that

$$\Phi(a) = \int_E V^*(e)aV(e)\mu(de) = \mathbb{E}(V^*aV), \tag{32}$$

for all $a \in \mathfrak{A}$.

Once established the representation for completely positive maps, the next result giving the representation of completely bounded maps is quite natural.

Theorem 4.6. *Let $\mathfrak{A}$ be a C^*-algebra with unit, and let $\Phi : \mathfrak{A} \to B(\mathfrak{h})$ be a completely bounded map. Then there exists a representation $(\pi, \mathfrak{k})$ of $\mathfrak{A}$ and bounded operators $V_i : \mathfrak{h} \to \mathfrak{k}$, $i = 1, 2$, with $\|\Phi\|_{cb} = \|V_1\| \, \|V_2\|$ such that*

$$\Phi(a) = V_1^* \pi(a) V_2, \tag{33}$$

for all $a \in \mathfrak{A}$. If $\|\Phi\|_{cb} = 1$, then V_1 and V_2 may be taken to be isometries.

Proof. We may assume $\|\Phi\|_{cb} = 1$ which implies that Φ is completely contractive. We first consider a general procedure to obtain CP maps from CB maps. Introduce the operator system

$$\mathcal{S} = \left\{ \begin{pmatrix} \lambda 1 & a \\ b^* & \mu 1 \end{pmatrix} : \lambda, \mu \in \mathbb{C}, \ a, b \in \mathfrak{A} \right\}.$$

and define $\boldsymbol{\Phi} : \mathcal{S} \to B(\mathfrak{h} \oplus \mathfrak{h})$ by

$$\boldsymbol{\Phi} \begin{pmatrix} \lambda 1 & a \\ b^* & \mu 1 \end{pmatrix} = \begin{pmatrix} \lambda 1 & \Phi(a) \\ \Phi(b)^* & \mu 1 \end{pmatrix}.$$

Since Φ is completely contractive, a direct computation using (26) shows that $\boldsymbol{\Phi}$ is completely positive and unital. Then, by Arveson Extension Theorem 2.7, the CP map $\boldsymbol{\Phi}$ can be extended to the whole algebra $\mathcal{M}_2 (\mathfrak{A})$. Let $(\pi, \mathbf{V}, \mathfrak{k})$ be a minimal Stinespring representation for $\boldsymbol{\Phi}$. Since $\boldsymbol{\Phi}$ is unital, $\mathbf{V}$ may be taken to be an isometry and π unital. $\mathcal{M}_2 (\mathfrak{A})$ contains a copy of the algebra of 2×2 complex matrices, and we may decompose $\mathfrak{k} = \mathfrak{k} \oplus \mathfrak{k}$ to have $\pi : \mathcal{M}_2 (\mathfrak{A}) \to B(\mathfrak{k} \oplus \mathfrak{k})$ of the form

$$\pi \left(\begin{pmatrix} a & b \\ c & d \end{pmatrix} \right) = \begin{pmatrix} \pi(a) & \pi(b) \\ \pi(c) & \pi(d) \end{pmatrix},$$

where $\pi : \mathfrak{A} \to B(\mathfrak{k})$ is a unital, $*$-homomorphism.

As a result, $\mathbf{V} : \mathfrak{h} \oplus \mathfrak{h} \to \mathfrak{k} \oplus \mathfrak{k}$ is an isometry and

$$\begin{pmatrix} a & \Phi(b) \\ \Phi(c)^* & d \end{pmatrix} = \mathbf{V}^* \begin{pmatrix} \pi(a) & \pi(b) \\ \pi(c) & \pi(d) \end{pmatrix} \mathbf{V}.$$

The isometric property of $\mathbf{V}$ implies that there exists a linear map $V_1 : \mathfrak{h} \to \mathfrak{k}$, which is also an isometry and such that

$$\mathbf{V} \begin{pmatrix} u \\ 0 \end{pmatrix} = \begin{pmatrix} V_1 u \\ 0 \end{pmatrix}.$$

One proves similarly the existence of V_2 such that

$$\mathbf{V} \begin{pmatrix} 0 \\ v \end{pmatrix} = \begin{pmatrix} 0 \\ V_2 v \end{pmatrix}.$$

Thus, we finally obtain

$$\begin{pmatrix} a & \Phi(b) \\ \Phi(c)^* & d \end{pmatrix} = \mathbf{V}^* \begin{pmatrix} \pi(a) & \pi(b) \\ \pi(c) & \pi(d) \end{pmatrix} \mathbf{V} = \begin{pmatrix} V_1^* \pi(a) V_1 & V_1^* \pi(b) V_2 \\ V_2^* \pi(c) V_1 & d \end{pmatrix}.$$

Completely positive maps satisfy a stronger version of Schwartz-type inequalities than the one proved before for 2-positive maps.

Theorem 4.7 (Schwartz-type inequalities). *Let $\mathfrak{A}$ and $\mathfrak{B}$ be two C^*-algebras with unit, $\mathfrak{B} \subseteq B(\mathfrak{h})$ where $\mathfrak{h}$ is a separable complex Hilbert space, and let $\Phi : \mathfrak{A} \to \mathfrak{B}$ be a linear completely positive map such that $\Phi(1) = 1$. Then, for all $a_1, \ldots, a_n \in \mathfrak{A}$, $u_1, \ldots, u_n \in \mathfrak{h}$*

$$\sum_{i,j} \langle u_i, [\Phi(a_i^* a_j) - \Phi(a_i)^* \Phi(a_j)] u_j \rangle \geq 0. \tag{34}$$

In particular, for all $a \in \mathfrak{A}$:

$$\Phi(a^* a) \geq \Phi(a)^* \Phi(a). \tag{35}$$

Moreover, given any positive linear map $\varphi : \mathfrak{A} \to \mathfrak{B}$ such that $\varphi(1) = 1$ and given any normal element $A \in \mathfrak{A}$ it holds

$$\varphi(A^* A) \geq \varphi(A)^* \varphi(A). \tag{36}$$

Proof. Consider a Stinespring representation (π, V) for the map Φ. Since $\Phi(1) = 1$, V is an isometry, so that $V^* V = 1$. Take any collection $a_1, \ldots, a_n \in \mathfrak{A}$, $u_1, \ldots, u_n \in \mathfrak{h}$:

$$\sum_{i,j} \langle u_i, \Phi(a_i^* a_j) u_j \rangle = \langle \sum_i \pi(_i a) V u_i, \sum_j \pi(a_j) V u_j \rangle$$

$$= \left\| \sum_i \pi(a_i) V u_i \right\|^2$$

$$\geq \left\| V^* \sum_i \pi(a_i) V u_i \right\|^2$$

$$= \langle \sum_i \Phi(a_i) u_i, \sum_j \Phi(a_j) u_j \rangle$$

$$= \sum_{i,j} \langle u_i, \Phi(a_i)^* \Phi(a_j) u_j \rangle.$$

The second part is an obvious consequence of the first.

For the third part, to prove the inequality in A with φ positive only, it is worth noticing that we can reduce $\mathfrak{A}$ to be the abelian algebra generated by A. Over that algebra, positive linear maps are completely positive and the second part of the theorem applies.

5 Quantum Dynamical Semigroups and Markov Flows

An homogeneous classical Markov semigroup is characterized by a family $(P_t)_{t \geq 0}$ of Markovian transition kernels defined on a measurable space $(E, \mathcal{E})$ which satisfies Chapman-Kolmogorov equations (or the semigroup property for the composition of kernels). Given a σ-finite measure μ on $(E, \mathcal{E})$, $\mathfrak{A} = L^\infty(E, \mathcal{E}, \mu)$ represents the von Neumman algebra of multiplication operators acting on the Hilbert space $L^2(E, \mathcal{E}, \mu)$. In this case, the predual algebra is $\mathfrak{A}_* = L^1(E, \mathcal{E}, \mu)$.

Moreover, $(P_t)_{t \geq 0}$ is a semigroup of completely positive maps acting on the von Neumann algebra $\mathfrak{A}$. Additionally, this semigroup satisfies the following properties:

- It preserves the unit: $P_t \mathbf{1} = \mathbf{1}$, for all $t \geq 0$.
- $P_0 = I$, the identity mapping.
- Each P_t is σ-weak continuous, that is, for any increasing net f_α of positive elements with upper envelope f in $\mathfrak{A}$,

$$\int_E P_t f(x) g(x) \mu(dx) = \lim_\alpha \int_E P_t f_\alpha(x) g(x) \mu(dx),$$

for all $g \in L^1(E, \mathcal{E}, \mu)$. Indeed, by the Monotone Convergence Theorem first, $P_t f_\alpha(x) \uparrow P_t f(x)$, for all $x \in E$; finally, to conclude, it is enough to apply the Dominated Convergence Theorem to $P_t f_\alpha(x) g(x)$.

All the above properties are crucial to face the extension of Markovian concepts to a non-commutative framework. Moreover, it is well-known that the addition of suitable topological hypotheses on the space $(E, \mathcal{E})$, allows to construct a Markov process associated to a given semigroup. One can take, for instance, E to be a locally compact space with countable basis and $\mathcal{E}$ its Borel σ–field. This leads to a Markovian system

$$(\Omega, \mathcal{F}, (\mathcal{F}_t)_{t \geq 0}, (\mathbb{P}_x)_{x \in E}, (X_t)_{t \geq 0}, E, \mathcal{E})$$

. The semigroup and the process are then related by the equation

$$P_t f(x) = \mathbb{E}_x(f(X_t)),$$

for all $f \in \mathfrak{A}$, $t \geq 0$. Moreover, we choose an arbitrary initial probability ν on $(E, \mathcal{E})$, and denote $\mathbb{P}_\nu = \int \mathbb{P}_x \nu(dx)$.

Now consider the von Neumann algebra $\mathfrak{B} = L^\infty(\Omega, \mathcal{F}, \mathbb{P}_\nu)$. The Markov flow is defined as a *-homomorphism $j_t : \mathfrak{A} \to \mathfrak{B}$ given by

$$j_t(f) = f(X_t),$$

for all $f \in \mathfrak{A}$, $t \geq 0$.

Inspired by these ideas we now turn into the non-commutative framework. We start by defining a *Quantum Dynamical Semigroup*.

Introduced by physicists during the seventies, Quantum Dynamical Semigroups (QDS) are aimed at providing a suitable mathematical framework for studying the evolution of open systems. Typically, an open quantum system involves a dissipative effect modeled through the mutual interaction of different subsystems. One commonly distinguishes between at least the "free system" and the "reservoir".

In general a QDS can be defined over an arbitrary von Neumann algebra, as follows:

Definition 5.1. *A* **Quantum Dynamical Semigroup** *(QDS) (respectively a* **Quantum Markov Semigroup**, *QMS) of a von Neumann algebra μ is a weakly*– continuous one–parameter semigroup $(\mathcal{T}_t)_{t\geq 0}$ of completely positive linear normal maps of μ into itself such that $\mathcal{T}_t(1) \leq 1$ (respectively, $\mathcal{T}_t(1) = 1$. In addition, it is assumed that $\mathcal{T}_0$ coincides with the identity map.*

The class of semigroups defined over the von Neumann algebra $\mu = \mathcal{B}(\mathfrak{h})$ of all bounded operators over a given complex separable Hilbert space h, is better known. In particular, several results on the form of the infinitesimal generator of these QDS are available (see eg. [24], [13], [21]). We denote $\mathcal{L}$ the **infinitesimal generator** of the semigroup $\mathcal{T}$, whose domain is given by the set $D(\mathcal{L})$ of all $X \in \mathcal{B}(\mathfrak{h})$ for which the w^*–limit of $t^{-1}((X) - X)$ exists when $t \to 0$, and we define $\mathcal{L}(X)$ such a limit.

To have a view on the form of the generator, we consider a particular case of QDS.

Definition 5.2. *A quantum dynamical semigroup $\mathcal{T}$ is called* **uniformly continuous** *if*

$$\lim_{t\to 0} \|\mathcal{T}_t - \mathcal{T}_0\| = 0.$$

From the general theory of semigroups it follows that a QDS is uniformly continuous if and only if its generator $\mathcal{L}$ is a bounded operator. Within this framework the canonical form of a generator has been obtained first by Gorini, Kossakowski and Sudarshan in the finite dimensional case, extended later by Lindblad to a general Hilbert space in [24], a celebrated result which we recall below in the version of Parthasarathy ([30], Theorem 30.16).

We start by a modification of complete positivity.

Definition 5.3. *Let $\mathfrak{A}$ denote a C^*-subalgebra of $\mathcal{B}(\mathfrak{h})$ which contains a unit. A bounded linear map $\mathcal{L}(\cdot)$ on $\mathfrak{A}$ is* conditionally completely positive *if for any collection $a_1, \ldots, a_n \in \mathfrak{A}$ and $u_1, \ldots, u_n \in \mathfrak{h}$ such that $\sum_i a_i u_i = 0$, it holds that*

$$\sum_{i,j} \langle u_i, \mathcal{L}(a_i^* a_j) u_j \rangle \geq 0.$$

Theorem 5.4 (Christensen and Evans). *A bounded linear map $\mathcal{L}(\cdot)$ on the C^*- algebra given before such that $\mathcal{L}(a^*) = \mathcal{L}(a)^*$, for any $a \in \mathfrak{A}$ is conditionally completely positive if and only if there exists a completely positive map Φ into its weak closure $\overline{\mathfrak{A}}$ and an element $G \in \overline{\mathfrak{A}}$ such that*

$$\mathcal{L}(a) = G^*a + \Phi(a) + aG, \tag{37}$$

for all $a \in \mathfrak{A}$. Moreover the operator G satisfies the inequality $G + G^* \leq \mathcal{L}(1)$.

Proof. We restrict the proof to the case $\mathfrak{A} = \mathcal{B}(\mathfrak{h})$ for simplicity. The interested reader is referred to the original paper [8] where this result is proved for a general C^*-algebra.

We first take $\mathcal{L}(\cdot)$ given by (37) and prove conditional complete positivity.

Take $a_1, \ldots, a_n \in \mathcal{B}(\mathfrak{h})$, $u_1, \ldots, u_n \in \mathfrak{h}$ such that $\sum_i a_i u_i = 0$. Then

$$\sum_{i,j} \langle u_i, \mathcal{L}(a_i^* a_j) u_j \rangle = \sum_{i,j} \langle a_i G u_i, a_j u_j \rangle$$

$$+ \sum_{i,j} \langle u_i, \Phi(a_i^* a_j) u_j \rangle$$

$$+ \sum_{i,j} \langle a_i u_i, a_j G u_j \rangle$$

$$= \langle \sum_i a_i G u_i, \sum_j a_j u_j \rangle$$

$$+ \sum_{i,j} \langle u_i, \Phi(a_i^* a_j) u_j \rangle$$

$$+ \langle \sum_i a_i u_i, \sum_j a_j G u_j \rangle$$

$$= \sum_{i,j} \langle u_i, \Phi(a_i^* a_j) u_j \rangle$$

$$\geq 0.$$

To prove the converse, fix a unit vector $e \in \mathfrak{h}$ and define

$$G^* u = \mathcal{L}(|u\rangle\langle e|) e - \frac{1}{2} \langle e, \mathcal{L}(|e\rangle\langle e|) e \rangle u,$$

for all $u \in \mathfrak{h}$. Given $a_1, \ldots, a_n \in \mathcal{B}(\mathfrak{h})$, $u_1, \ldots, u_n \in \mathfrak{h}$, let

$$u_{n+1} = e, \tag{38}$$

$$v = -\sum_{j=1}^{n} a_j u_j, \tag{39}$$

$$a_{n+1} = |v\rangle\langle e|. \tag{40}$$

Then $\sum_{j=1}^{n+1} a_j u_j = 0$. Since $\mathcal{L}(\cdot)$ is conditionally completely positive,

$$0 \leq \sum_{i,j=1}^{n} \langle u_i, \mathcal{L}(a_i^* a_j) u_j \rangle$$

$$+ \sum_{i=1}^{n} \langle u_i, \mathcal{L}(|a_i^* v\rangle\langle e|) e \rangle$$

$$+ \sum_{j=1}^{n} \langle e, \mathcal{L}(|e\rangle\langle a_j^* v|) u_j \rangle$$

$$+ \langle e, \mathcal{L}(|e\rangle\langle e|) e \rangle \|v\|^2.$$

Using the definition of G^*, the sum of the last three terms becomes

$$\sum_{i=1}^{n}\langle u_i, G^*a_i^*v\rangle + \sum_{j=1}^{n}\langle G^*a_j^*v, u_j\rangle = -\sum_{i,j}\langle u_i, G^*a_i^*a_ju_j\rangle - \sum_{i,j=1}^{n}\langle u_i, a_i^*a_jGu_j\rangle.$$

If we define $\Phi(a) = \mathcal{L}(a) - G^*a - aG$, the inequality we obtained here before can be written

$$\sum_{i,j=1}^{n}\langle u_i, \Phi(a_i^*a_j)u_j\rangle \geq 0,$$

which means that Φ is completely positive and the Theorem is proved.

Assume $\mathcal{T}$ to be a norm continuous quantum Markov semigroup on $\mathcal{B}(\mathfrak{h})$. By the Schwartz inequalities, for any $a_1, \ldots, a_n \in \mathcal{B}(\mathfrak{h})$, $u_1, \ldots, u_n \in \mathfrak{h}$, and any $t \geq 0$:

$$\sum_{i,j=1}^{n}\langle u_i, (\mathcal{T}_t(a_i^*a_j) - \mathcal{T}_t(a_i)^*\mathcal{T}_t(a_j))u_j\rangle \geq 0$$

The norm continuity of $\mathcal{T}$ implies that $\mathcal{L}(\cdot)$ is defined as a bounded operator on the whole algebra $\mathcal{B}(\mathfrak{h})$, so that the above inequality implies

$$\sum_{i,j=1}^{n}\langle u_i, (\mathcal{L}(a_i^*a_j) - \mathcal{L}(a_i^*)a_j - a_i^*\mathcal{L}(a_i))u_j\rangle \geq 0,$$

from which, if $\sum_i a_iu_i = 0$, it follows easily

$$\sum_{i,j=1}^{n}\langle u_i, \mathcal{L}(a_i^*a_j)u_j\rangle \geq 0.$$

So that $\mathcal{L}(\cdot)$ is conditionally completely positive. As a result, the following characterization follows.

Theorem 5.5. *Given a norm continuous quantum dynamical semigroup $\mathcal{T}$ on $\mathcal{B}(\mathfrak{h})$, there exists an operator G and a completely positive map Φ such that its generator is represented as*

$$\mathcal{L}(x) = G^*x + \Phi(x) + xG, \ (x \in \mathcal{B}(\mathfrak{h})). \tag{41}$$

Since $\mathcal{B}(\mathfrak{h})$ is a von Neumann algebra, the representation before can be improved using Kraus Theorem to represent the completely positive map Φ.

Theorem 5.6 (Lindblad). *Let be given a uniformly continuous quantum dynamical semigroup on the algebra $\mathcal{B}(\mathfrak{h})$ of a complex separable Hilbert space $\mathfrak{h}$. Let ρ be any state in h. Then there exists a bounded self-adjoint operator H and a sequence $(L_k)_{k\in\mathbb{N}}$ of elements in $\mathcal{B}(\mathfrak{h})$ which satisfy*

(1) $\mathrm{tr}\,\rho L_k = 0$ for each k;
(2) $\sum_k L_k^*L_k$ is a strongly convergent sum;
(3) If $\sum_k |c_k|^2 < \infty$ and $c_0 + \sum_k c_kL_k = 0$ for scalars c_k, then $c_k = 0$ for all k;

(4) The generator $\mathcal{L}$ of the semigroup admit the representation

$$\mathcal{L}(X) = i[H, X] - \frac{1}{2}\sum_k (L_k^* L_k X - 2L_k^* X L_k + X L_k^* L_k),$$

for all $X \in \mathcal{B}(\mathfrak{h})$.

This result has been extended by Davies (see [13]) to a class of QDS with unbounded generators.

Generators of QDS commonly appear in Physics articles in its *predual* form. That is, given the von Neumann algebra $\mathfrak{A} = \mathcal{B}(\mathfrak{h})$ its *predual space* consists of $\mathfrak{A}_* = \mathcal{I}^1(h)$ the Banach space of trace-class operators. A quantum dynamical semigroup $\mathcal{T}$ induces a *predual* semigroup $\mathcal{T}_*$ on $\mathfrak{A}_*$ given through the relation

$$\text{tr} \left(\mathcal{T}_{*t}(Y)X\right) = \text{tr} \left(Y\mathcal{T}_t(X)\right),$$

for any $Y \in \mathfrak{A}_*$, $X \in \mathfrak{A}$.

The generator of the predual semigroup is denoted $\mathcal{L}_*$. What is usually called a *master equation* in Open Quantum Systems, is referred to the relation between the predual semigroup and its generator, written in the form

$$\frac{d}{dt}\rho_t = \mathcal{L}_*(\rho_t),$$

where $\rho_t = \mathcal{T}_{*t}(\rho)$, for any $t \geq 0$, ρ being a *state*, that is, an element $\rho \in \mathfrak{A}_*$ with unitary trace.

6 Dilations of quantum Markov semigroups

It is worth noticing that in general a QDS is not a *-homomorphism of algebras. Such a property concerns quantum flows and the concept of *dilation* which we precise below. We first introduce the algebraic notion of *conditional expectation*.

Definition 6.1. *Let $\mathfrak{B}$ be a von Neumann algebra endowed with a state $\mathbb{E}$. A **conditional expectation** on $\mathfrak{B}$ is a linear completely positive map $\mathbb{E}_0 : \mathfrak{B} \to \mathfrak{B}$ which satisfies*

1. *$\mathbb{E}_0(1) = 1$,*
2. *$\mathbb{E} \circ \mathbb{E}_0 = \mathbb{E}$,*
3. *$\mathbb{E}_0\left(a\mathbb{E}_0(b)\right) = \mathbb{E}_0(a)\mathbb{E}_0(b)$, for all $a, b \in \mathfrak{B}$.*

Definition 6.2. *A **dilation** of a given QDS is a system $(\mathfrak{B}, \mathbb{E}, (\mathfrak{B}_t, \mathbb{E}_t, j_t)_{t\geq 0})$ where*

1. *$\mathfrak{B}$ is a von Neumann algebra with a given state $\mathbb{E}$;*
2. *$(\mathfrak{B}_t)_{t\geq 0}$ is an increasing family of von Neumann sub-algebras of $\mathfrak{B}$;*
3. *For any $t \geq 0$, $\mathbb{E}_t$ is a conditional expectation from $\mathfrak{B}$ onto $\mathfrak{B}_t$, such that for all $s, t \geq 0$, $\mathbb{E}_s\mathbb{E}_t = \mathbb{E}_{s\wedge t}$;*

4. All the maps $j_t : \mathfrak{A} \to \mathfrak{B}_t$ are *–homomorphisms which preserve the identity and satisfy the Markov property:

$$\mathbb{E}_s \circ j_t = j_s \circ T_{t-s}.$$

$J = (j_t)_{t \geq 0}$ is known as a **Quantum Markov Flow** associated to the given QDS.

We call the structure $\mathfrak{B} = (\mathfrak{B}, \mathbb{E}, (\mathfrak{B}_t, \mathbb{E}_t)_{t \geq 0})$ a **quantum stochastic basis**.

The canonical form of a quantum Markov flow is given by $j_t(X) = V(t)^* X V(t)$, $(t \geq 0)$, where $V(t) : \mathfrak{A} \to \mathfrak{B}_t$ is a **cocycle** with respect to a given family of **time-shift** operators $(\theta_t)_{t \geq 0}$. To be more precise

Definition 6.3. *Given a quantum stochastic basis* $\mathfrak{B}$, *a family* $(\theta_t)_{t \geq 0}$ *of *–homomorphisms of* $\mathfrak{B}$ *is called a* **covariant shift** *if*

- $\theta_0(Y) = Y$,
- $\theta_t(\theta_s(Y)) = \theta_{t+s}(Y)$,
- $\theta_t^*(\theta_t(Y)) = Y$,
- $\theta_t(\mathbb{E}_0(\theta_s(Y))) = \mathbb{E}_t(\theta_{t+s}(Y))$,

for any $Y \in \mathfrak{B}$.

A family $(V(t))_{t \geq 0}$ of elements in $\mathfrak{B}$ is a **left cocycle** (resp. **right cocycle**) with respect to a given covariant shift whenever

$$V(t + s) = V(s)\theta_s(V(t)), \quad (\text{resp. } V(t+s) = \theta_s(V(t))V(s)), \quad (s, t \geq 0). \quad (42)$$

6.1 A view on classical dilations of QMS

Throughout the following examples we construct *classical probabilistic dilations* of QMS. We explain further how to proceed with *quantum probabilistic dilations*.

Example 6.4. According to the above definition, the quantum dynamical semigroup associated to the Schrödinger equation is

$$T_t(X) = U^*(t)X U(t),$$

where $U(t) = e^{-itH}$ and H is a self-adjoint operator in $\mathcal{B}(\mathfrak{h})$.

Here T_t is itself a *–homomorphism; U is a fortiori a cocycle with respect to the trivial shift $\theta_t = \theta_0 = I$. The differential equation satisfied by U is simply $dU(t) = -iHU(t)dt$ and the generator of T is given by

$$\mathcal{L}(X) = i[H, X], \quad (X \in \mathcal{B}(\mathfrak{h})).$$

Example 6.5. Let be given a filtered probability space $(\Omega, \mathcal{F}, (\mathcal{F}_t)_{t \geq 0}, \mathbb{P})$ where a classical Brownian Motion $W = (W_t)_{t \geq 0}$ is defined and a complex separable Hilbert space $\mathfrak{h}$. For instance, let Ω to be the space $C(\mathbb{R}_+, \mathbb{R})$ of continuous real functions endowed with the Wiener measure $\mathbb{P}$, $W_t(\omega) = \omega(t)$, for any $\omega \in \Omega$, $t \geq 0$. consider the stochastic differential equation:

$$d\psi_t = (-iH + K)\psi_t dt + L\psi_t dW_t; \quad \psi_0 = \psi, \qquad (43)$$

where $H = H^*$, $K = K^*$, L are elements of $\mathcal{B}(\mathfrak{h})$ and ψ is a fixed unitary vector of $\mathfrak{h}$.

This equation has a unique solution $(\psi_t)_{t\geq 0}$, with $\psi_t \in L^2_\mathfrak{h}(\Omega, \mathcal{F}, \mathbb{P})$, $(t \geq 0)$. To have $\mathbb{E}(\|\psi_t\|^2) = 1$ for all $t \geq 0$, K and L need to satisfy additional conditions. Indeed,

$$
\begin{aligned}
d\langle \psi_t, \psi_t \rangle &= \langle d\psi_t, \psi_t \rangle + \langle \psi_t, d\psi_t \rangle + \langle d\psi_t, d\psi_t \rangle \\
&= \langle \psi_t, (2K + L^*L)\psi_t \rangle dt + \langle \psi_t, (L^* + L)\psi_t \rangle dW_t.
\end{aligned}
$$

Thus, $d\mathbb{E}(\langle \psi_t, \psi_t \rangle) = 0$ if and only if

$$K = -\frac{1}{2}L^*L. \qquad (44)$$

Furthermore, call $Z(t) = (-iH - \frac{1}{2}L^*L)t + LW_t$. This is an operator-valued semi-martingale and the stochastic Schrödinger equation may be written

$$d\psi_t = dZ(t)\psi_t, \quad \psi_0 = \psi.$$

Notice that $L^2_\mathfrak{h}(\Omega, \mathcal{F}, \mathbb{P})$ is isomorphic to $\mathfrak{h} \otimes L^2_\mathbb{C}(\Omega, \mathcal{F}, \mathbb{P})$. And we can define an operator-valued (classical) stochastic process $V : \Omega \times \mathbb{R}_+ \to \mathcal{B}(\mathfrak{h})$, such that $V(\cdot, t) : \mathfrak{h} \mapsto \mathfrak{h} \otimes L^2_\mathbb{C}(\Omega, \mathcal{F}_t, \mathbb{P})$, for any $t \geq 0$, which associates to each unitary $\psi \in h$ the unique solution ψ_t of (43). This operator-valued process may be interpreted as a family $(V(t))_{t\geq 0}$ of applications belonging to the algebra $L^\infty_{\mathcal{B}(\mathfrak{h})}(\Omega, \mathcal{F}, \mathbb{P})$, that is $V : \mathbb{R}_+ \to L^\infty_{\mathcal{B}(\mathfrak{h})}(\Omega, \mathcal{F}, \mathbb{P})$, given by $t \mapsto V(\cdot, t)$. The process $V(t)$ is given as the solution to the (right) operator-valued linear stochastic differential equation:

$$dV(t) = (-iH - \frac{1}{2}L^*L)V(t)dt + LV(t)dW_t = dZ(t)V(t), \quad V(0) = I. \quad (45)$$

We now consider the classical shift operator $\theta_t : \Omega \to \Omega$, given by $\theta_t(\omega)(s) = \omega(t + s)$ for any $s, t \geq 0$, $\omega \in \Omega$. This induces a shift, denoted θ_s as well, on the elements $Y \in \mathfrak{B} = L^\infty_{\mathcal{B}(\mathfrak{h})}(\Omega, \mathcal{F}, \mathbb{P})$ given by

$$\theta_s(Y) = Y \circ \theta_s.$$

This shift is covariant with respect to the family $(\mathbb{E}(\cdot/\mathcal{F}_t))_{t\geq 0}$, that is:

- $\theta_0(Y) = Y$, for any $Y \in \mathfrak{B}$;
- $\theta_t(\theta_s(Y)) = \theta_{t+s}(Y)$;
- $\theta_t^*(\theta_t(Y)) = Y$;
- $\theta_t(\mathbb{E}(\theta_s(Y))) = \mathbb{E}(\theta_{t+s}(Y)/\mathcal{F}_t)$.

$V(t)$ is then a right cocycle with respect to this family of applications since, like in the classical case, the following relation is satisfied:

$$V(\omega, t + s) = V(\theta_s(\omega), t) V(\omega, s).$$

We now define the corresponding flow as

$$j_t(X) = V(t)^* X V(t), \quad (X \in \mathcal{B}(\mathfrak{h})).$$

And we obtain

$$
\begin{aligned}
dj_t(X) &= dV(t)^* X V(t) + V(t)^* X dV(t) + dV(t)^* X dV(t) \\
&= V(t)^* dZ(t)^* X V(t) + V(t)^* X dZ(t) V(t) + V(t)^* dZ(t)^* X dZ(t) V(t) \\
&= j_t(\mathcal{L}(X)) dt + j_t(\alpha(X)) dW_t,
\end{aligned}
\tag{46}
$$

where,

$$\mathcal{L}(X) = i[H, X] - \frac{1}{2}(L^* L X - 2L^* X L + X L^* L); \tag{47}$$

$$\alpha(X) = XL + L^* X. \tag{48}$$

Now, the associated quantum dynamical semigroup is given by

$$\mathcal{T}_t(X) = \mathbb{E}(j_t(X)),$$

which has a generator obtained from the structure equation (46), and given by (47).

Thus, stochastic Schrödinger equations are naturally included within the framework of quantum dynamical semigroups. They represent a *classical* dilation of the QDS whose generator is (47). The adjective "classical" is used here for the probabilistic model used to dilate the semigroup. A given QDS may have many different dilations, based upon classical or quantum noises.

This example may be generalized as follows: take a collection $(W^j)_{j \geq 1}$ of independent Brownian motions, and operators $L_j \in \mathcal{B}(\mathfrak{h})$ such that $\sum_j L_j^* L_j$ converges in $\mathcal{B}(\mathfrak{h})$. Consider the stochastic Schrödinger equation:

$$d\psi_t = (-iH - \frac{1}{2} \sum_j L_j^* L_j) \psi_t dt + \sum_j L_j dW_t^j \psi_t, \quad \psi_0 = \psi. \tag{49}$$

The corresponding associated QDS has a generator of the form

$$\mathcal{L}(X) = i[H, X] - \frac{1}{2} \sum_j (L_j^* L_j X - 2L_j^* X L_j + X L_j^* L_j). \tag{50}$$

We profit of this example to explain a useful notation for stochastic differentials due to Belavkin. Introduce first the noises as follows:

$$\Lambda_0^0(t) = t, \tag{51}$$

$$\Lambda_j^0(t) = W_t^j = \Lambda_0^j(t), \tag{52}$$

for all $j \geq 1, t \geq 0$. These are real scalar noises, thus the expression $\Lambda_m^{\ell *}$ has a trivial meaning. However, due to the further extension to operator noises, we adopt the convention $\Lambda_m^{\ell *} = \Lambda_\ell^m$. Moreover we use the customary convention of probabilists for writing characteristic functions: $1_{\{\ldots\}}$ means that we write 1 if condition $\{\ldots\}$ is satisfied and 0 otherwise.

With these notations Itô's formula is written in a compact differential form as

$$d\Lambda_m^k d\Lambda_j^\ell = d\Lambda_j^\ell d\Lambda_m^k = 1_{\{j=k>0, \ell=m=0\}} d\Lambda_0^0. \tag{53}$$

Moreover, if we denote $L_j^0 = L_j = L_0^j$ for all $j \geq 1$ and $L_0^0 = -iH - \frac{1}{2} \sum_j L_j^* L_j$, $L_j^k = 0$ for $j, k \geq 1$ the equation of the cocycle becomes

$$dV(t) = dZ(t)V(t),$$

where

$$dZ(t) = \sum_{j,k=0}^{\infty} L_j^k d\Lambda_k^j(t). \tag{54}$$

Furthermore, to obtain the structure equation for the quantum flow $j_t(X) = V(t)^* X V(t)$ we use the above equation and (53) yields

$$
\begin{aligned}
dj_t(X) &= d(V(t)^* X V(t)) \\
&= V(t)^* dZ(t)^* X V(t) \\
&\quad + V(t)^* X dZ(t)V(t) \\
&\quad + V(t)^* dZ(t)^* X dZ(t)V(t) \\
&= j_t(\sum_{\ell,m} L_m^{\ell *} d\Lambda_m^\ell(t)X) \\
&\quad + j_t(\sum_{\ell,m} X L_m^\ell d\Lambda_m^\ell(t)) \\
&\quad + j_t(\sum_{\ell,m,j,k} L_k^{j *} d\Lambda_k^j(t) X L_m^\ell d\Lambda_m^\ell(t)).
\end{aligned}
$$

As a result,

$$dj_t(X) = \sum_{\ell,m=0}^{\infty} j_t(\theta_\ell^m(X)) d\Lambda_m^\ell(t), \tag{55}$$

for all $X \in \mathcal{B}(\mathfrak{h})$, where $(\theta_\ell^m)_{\ell,m}$ is called the family of *structure maps* of the flow, given by

$$\theta_k^\ell(X) = L_m^{\ell *} X + X L_\ell^m + \sum_{k=1}^{\infty} L_k^{\ell *} X L_k^m, \tag{56}$$

for any $\ell, m \in \mathbb{N}$.

Since the QDS is given by $\mathcal{T}_t(X) = \mathbb{E}(j_t(X))$, it is worth noticing that we recover the generator of the semigroup through the structure map $\theta_0^0(X)$ which is associated to the noise $d\Lambda_0^0(t) = dt$ because the other noises are projected by $\mathbb{E}$ on 0:

$$\mathcal{L}(X) = \theta_0^0(X) = L_0^0{}^* X + X L_0^0 + \sum_{k=1}^{\infty} L_k^0{}^* X L_k^0$$

$$= i[H, X] - \frac{1}{2} \sum_j (L_j^* L_j X - 2 L_j^* X L_j + X L_j^* L_j).$$

Example 6.6. Consider again a classical probability space $(\Omega, \mathcal{F}, \mathbb{P})$ where we suppose defined two independent Poisson processes N^a, N^d with respective intensities λ_a, λ_d. In addition, we take a unitary operator U defined on a given complex separable Hilbert space h. Consider the stochastic differential equation

$$dV(t) = \{(U - I)dN_t^a + (U^* - I)dN_t^d\}V(t), \; V(0) = I, \qquad (57)$$

This equation provides a unitary cocycle V. Moreover, using Belavkin notation we put

$$L_1^1 = U - I, \; L_2^2 = (U^* - I); \; L_k^j = 0, \text{ otherwise,}$$

$$\Lambda_1^1 = N^a, \; \Lambda_2^2 = N^d.$$

The Itô multiplication rule becomes

$$d\Lambda_m^k d\Lambda_j^\ell = d\Lambda_j^\ell d\Lambda_m^k = 1_{\{j=k=\ell=m>0\}} d\Lambda_\ell^\ell \qquad (58)$$

With this choice of notations, the computation of the structure maps gives

$$\theta_1^1(X) = U^* X U - X; \theta_2^2(X) = U X U^* - X, \; (X \in \mathcal{B}(h))$$

that is the equation for the flow is

$$dj_t(X) = j_t(U^* X U - X)dN_t^a + j_t(U X U^* - X)dN_t^d,$$

Since $\mathbb{E}(N_t^a) = \lambda_a t$, $\mathbb{E}(N_t^d) = \lambda_d t$, the generator of the QDS has the form

$$\mathcal{L}(X) = \lambda_a(U^* X U - X) + \lambda_d(U X U^* - X), \; (X \in \mathcal{B}(h)). \qquad (59)$$

In particular, take $h = l^2(\mathbb{Z})$ with its canonical orthonormal basis (e_n). We define the right shift as $Se_n = e_{n+1}$, and the number operator $Ne_n = ne_n$, $(n \in \mathbb{Z})$. Take $U = S$ and $X = f(N)$, where $f(N)e_n = f(n)e_n$ for all $n \in \mathbb{Z}$, f being any bounded function defined on the integers. Then (59) gives

$$\mathcal{L}(f(N)) = \lambda_a(f(N+1) - f(N)) + \lambda_d(f(N-1) - f(N)),$$

which coincides with the generator of a birth and death process.

Example 6.7. Take E to be a compact space and call $\mathcal{E}$ its Borel σ-field and a probability space $(\Omega, \mathcal{F}, \mathbb{P})$. Given a finite measure μ on $(E, \mathcal{E})$ we consider a marked Poisson process with intensity μ which is characterized by a double sequence of random variables $(\xi_n, T_n)_{n \geq 1}$ taking values in $E \times \mathbb{R}_+$ such that

$$N(A, [0, t]) = \sum_{n \geq 1} \delta_{(\xi_n, T_n)}(A \times]0, t]),$$

has a Poisson distribution with intensity parameter $\mu(A)$, for all $A \in \mathcal{E}$, $t \geq 0$.

Consider a family $(U(x))_{x \in E}$ of unitary operators on a given complex separable Hilbert space $\mathfrak{h}$ and a selfadjoint operator H. Moreover, we assume that the map $x \mapsto U(x)$ from E into $\mathcal{B}(\mathfrak{h})$ is measurable (strong or weak measurability are equivalent in this case). We consider a Schrödinger equation where the dynamics is perturbed by random kicks as follows

$$d\psi_t = -iH\psi_t dt + \sum_{n \geq 1} (U(\xi_n) - I)\psi_{T_n} 1_{\{T_n \leq t < T_{n+1}\}}. \tag{60}$$

The above equation may be written for a cocycle as

$$dV(t) = \left[-iH dt + \int_E (U(x) - I)N(dx, dt) \right] V(t). \tag{61}$$

To justify the above expressions, we verify the integrability of the map $x \mapsto U(x) - I$ with respect to the Poisson process. Indeed, for any $t \geq 0$,

$$\mathbb{E}\left(\int_E \|U(x) - I\| N(dx,]0, t]) \right) \leq 2\mu(E) \, t < \infty.$$

Proceeding as we did in the previous examples, we examine the differential equation satisfied by the flow $j_t(X) = V(t)^* X V(t)$, $(t \geq 0, X \in \mathcal{B}(\mathfrak{h}))$. Here we have $dZ(t) = -iH dt + \int_E (U(x) - I)N(dx, dt)$, therefore

$$dj_t(X) = j_t(dZ(t)^* X + X dZ(t) + dZ(t)^* X dZ(t)).$$

We use again the Itô's rule, which becomes here $N(\{x\}, dt)N(\{y\}, dt) = \delta_{x,y} N(\{y\}, dt)$, it follows

$$dj_t(X) = j_t(i[H, X])dt + \int_E j_t(U(x)^* X U(x) - X)N(dx, dt).$$

Finally, taking expectations

$$\mathcal{T}_t(X) = \mathbb{E}\left(\int_0^t j_s(i[H, X] + \int_E (U(x)^* X U(x) - X)\mu(dx))ds \right),$$

which means that the generator $\mathcal{L}$ of the QDS has the form

$$\mathcal{L}(X) = i[H, X] + \int_E (U(x)^* X U(x) - X)\mu(dx), \quad (X \in \mathcal{B}(\mathfrak{h})). \qquad (62)$$

It is worth noticing that the master equation associated to (62) provides an example of a quantum Boltzmann equation. Indeed, the master equation is simply

$$\frac{\partial \rho_t}{\partial t} = \mathcal{L}_*(\rho_t), \quad \rho_0 = \rho \in \mathcal{I}^1(h).$$

Now, introduce a *collision map*

$$\kappa_*(\rho) = \int_E (U(x)\rho U(x)^* - \rho)\mu(dx).$$

This allows to write the predual generator as $\mathcal{L}_*(\rho) = -i[H, \rho] + \kappa_*(\rho)$, and the master equation becomes:

$$\frac{\partial \rho_t}{\partial t} + i[H, \rho_t] = \kappa_*(\rho_t), \quad \rho_0 = \rho. \qquad (63)$$

6.2 Towards quantum dilations of QMS

The above examples provide a partial view on quantum dynamical semigroups, in all these cases the generators are bounded which is not satisfactory from the point of view of physical applications. Moreover, the dilations have been built in a classical manner through Wiener or Poisson processes. More general dilations, that is, those obtained with *quantum noises* are more suitable for the description of open quantum systems. Quantum noises appear naturally within the framework of Fock spaces. Numerous authors (see for instance [26]) have stressed the main advantage of a (boson) Fock space: that structure supports both, the Canonical Commutation Relations (CCR) and a theory of stochastic integration with respect to quantum noises providing a non commutative version of Itô's algebra for differentials.

Furthermore, the study of linear quantum stochastic differential equations with unbounded coefficients have been done by several authors. Namely, Fagnola in [11] established a useful criterion on the existence and uniqueness of solutions to equations of the form

$$dV(t) = V(t) \sum_{\ell,m} L_\ell^m d\Lambda_m^\ell(t),$$

where the processes Λ_m^ℓ are quantum noises. We will use this result in the sequel. Notice that $V(t)$ appears here on the left of the right-hand side of the equation. This is done to simplify the handling of domains which appear as a major problem related to unbounded coefficients. As a result, one uses to write flows like $j_t(X) = V(t)XV(t)^*$ instead of $V(t)^*XV(t)$. So that, these notes have a natural continuation in Franco Fagnola lectures on the Theory of Quantum Stochastic Differential Equations and in our joint work on the qualitative analysis of QMS, both texts appearing within the current series of lecture notes.

References

1. L. Accardi, Y. G. Lu and I. Volovich, *Quantum theory and its stochastic limit*, Springer, Berlin, 2002; MR1925437 (2003h:81116)
2. R. Alicki and M. Fannes, *Quantum dynamical systems*, Oxford Univ. Press, Oxford, 2001.
3. W.B. Arveson, Subalgebras of C^*-algebras, *Acta Math.*, **123** (1969), 141–224.
4. V.P. Belavkin, A quantum non adapted Itô formula and non stationary evolution in Fock scale. *Quantum Probability and Related Topics*, World Scientific, vol.VI, (1992), 137-180.
5. V.P. Belavkin, Nondemolition measurement and nonlinear filtering of quantum stochastic processes. Lect. Notes in Control and Information Sci., Springer-Verlag, (1988), 245-266.
6. V.P. Belavkin, Quantum stochastic calculus and quantum nonlinear filtering. J. of Multivariate Analysis, vol. 42 (2), (1992), 171-201.
7. P. Biane, *Calcul Stochastique non-commutatif,*École d'Été de Probabilités de St. Flour XXIII, P. Bernard (ed.), Springer Lect.Notes in Maths, **1608**, 1-96,1995.
8. E. Christensen and D.E. Evans, *Cohomology of operator algebras and quantum dynamical semigroups*, J.Lon.Math.Soc. **20** (1979), 358–368.
9. E.B. Davies, *Quantum theory of open systems*, Academic Press, London, 1976.
10. E.B. Davies, Quantum dynamical semigroups and the neutron diffusion equation. *Rep. Math. Phys.* **11** (1977), 169–188.
11. F. Fagnola. On quantum stochastic differential equations with unbounded coefficients. *Prob.Th.and Rel.Fields*, 56:501–516, 1990.
12. F. Fagnola, Unitarity of solutions to quantum stochastic differential equations and conservativity of the associated semigroups. *Quantum Probability and Related Topics* **VII** (1992), 139–148.
13. F. Fagnola, R. Rebolledo, C. Saavedra, Quantum flows associated to a class of laser master equations. *J. Math. Phys.* (1) **35**, 1–12, (1994).
14. F. Fagnola and R. Rebolledo, The approach to equilibrium of a class of quantum dynamical semigroups. *Inf. Dim. Anal. Q. Prob. and Rel. Topics*, 1(4):1–12, 1998.
15. F. Fagnola, R. Rebolledo, A probabilistic view on stochastic differential equations derived from Quantum Optics. *Aportaciones Matemáticas*, Soc.Matem.Mexicana, **14**, 193-214 (1998).
16. F. Fagnola, Quantum markov semigroups and quantum flows. *Proyecciones, Journal of Math.*, 18(3):1–144, 1999.
17. A. Frigerio, Quantum dynamical semigroups and approach to equilibrium. *Lett. Math. Phys.* **2** 79–87, (1977).
18. A. Kossakowski V. Gorini and E.C.G. Sudarshan, Completely positive dynamical semigroups of n-level systems. *J. Math. Phys.*, 17:821–825, 1976.
19. H. Haken, *Handbuch der Physik*, volume XXV/2c. Springer–Verlag, Berlin, 1969.
20. E. Hille, *Functional Analysis and Semigroups*. AMS (1957).
21. A.S. Holevo, *Probabilistic and Statistical aspects of Quantum Theory*, North–Holland, 1982.
22. R.L. Hudson and K.R. Parthasarathy, Quantum Itô's formula and stochastic evolutions. *Comm.Math.Phys.*, 93:301–323, 1984.
23. K. Kraus, General states changes in Quantum Theory, *Ann. Phys.*, **64** (1970), 311–335.
24. G. Lindblad, On the generators of quantum dynamical semigroups. *Commun. Math. Phys.*, 48:119–130, 1976.

25. W.H. Louisell, *Quantum Statistical Properties of Radiation*.John Wiley, N.Y., 1973.

26. P.-A. Meyer, *Quantum Probability for Probabilists*. LNM 1538. Springer Verlag, Berlin Heidelberg, New York 1993.

27. M. Orszag, *Quantum Optics*. Springer-Verlag, 2000.

28. V. Paulsen, *Completely Bounded Maps and Operator Algebras*. Cambridge University Press, Cambridge Studies in Adv. Math. **78**, 2002.

29. K.R. Parthasarathy and K.B. Sinha, Markov chains as Evans-Hudson diffusions in Fock space. *Séminaire de Probabilités*, XXIV(1426):362–369, 1988–1989. LNM–Springer.

30. K.R. Parthasarathy, *An Introduction to Quantum Stochastic Calculus*, volume 85 of *Monographs in Mathematics*. Birkhaüser–Verlag, Basel-Boston-Berlin, 1992.

31. A. Pazy, *Semigroup of Linear Operators and Applications to Partial Differential Equations*. Springer–Verlag, 1975.

32. I. Percival, *Quantum State Diffusion*, Cambridge University Press, 176p.(1999).

33. R. Rebolledo, (1996) Sur les semigroupes dynamiques quantiques, Ann.Math. Blaise Pascal, vol. 3, n.1, 125–142.

34. M. Sargent, M.O. Scully, and W.E. Lamb, *Laser Physics*. Addison-Wesley, 1974.

35. W.F. Stinespring, Positive functions on C^*-algebras, Proc.Am.Math.Soc., **6** (1955), 211–216.

Quantum Stochastic Differential Equations and Dilation of Completely Positive Semigroups

Franco Fagnola

Politecnico di Milano, Dipartimento di Matematica "F. Brioschi",
Piazza Leonardo da Vinci 32, 20133 Milano Italy
e-mail: franco.fagnola@polimi.it

1   Introduction 183

2   Fock space notation and preliminaries 184

3   Existence and uniqueness 188

4   Unitary solutions 191

5   Emergence of H-P equations in physical applications 193

6   Cocycle property 196

7   Regularity 199

8   The left equation: unbounded G_β^α 203

9   Dilation of quantum Markov semigroups 208

10 The left equation with unbounded G_β^α: isometry 213

11 The right equation with unbounded F_β^α 216

References 218

Summary. Quantum stochastic differential equations $dU_t = F_\beta^\alpha U_t d\Lambda_\alpha^\beta(t)$ and $dV_t = V_t G_\beta^\alpha d\Lambda_\alpha^\beta(t)$ driven by the fundamental noises of boson Fock quantum stochastic calculus with F_β^α and G_β^α operators on the initial space are discussed. The existence and uniqueness theorems and the conditions on the F_β^α and G_β^α for the solutions to be a process of isometries or coisometries are studied. The application to the dilation of quantum Markov semigroup is illustrated.

1 Introduction

Quantum stochastic differential equation (QSDE) are stochastic differential equations driven by various types of noises (boson, fermion, free, boolean processes like Brownian motions, Poisson processes, Levy processes) arising in Quantum Probability and its applications.

QSDE usually describe some quantum mechanical evolution and, therefore, are linear. However, as an additional natural requirement, one would like to find solutions given by families of unitary operators (or ∗-homomorphic maps) and not arbitrary families of random variables like solutions of classical stochastic differential equations.

The various types of noise usually do not lead to different methods therefore QSDE can be divided into two classes: QSDE for operators on Hilbert spaces and QSDE for maps on operator algebras (C^* or von Neumann algebras). From an analytic point of view the two classes correspond, roughly speaking, to stochastic equations on Hilbert spaces and stochastic equations on Banach spaces.

In these lecture notes we are concerned only with QSDE of the type of Hudson and Parthasarathy and look for solutions given by families of unitary operators.

The noises are multidimensional versions of the creation $(A_t^+; t \geq 0)$, annihilation $(A_t; t \geq 0)$ and number process $(\Lambda_t; t \geq 0)$ on the boson Fock space over $L^2(\mathbb{R}_+)$. Hudson and Parthasarathy [26] introduced stochastic integrals with respect to the three processes and developed a stochastic calculus which is a non-commutative analogue of the classical Itô calculus.

They also solved the QSDE of the form

$$dU_t = \left(F_1 dA_t^+ + F_2 d\Lambda_t + F_3 dA_t + F_4 dt\right) U_t$$
$$dV_t = V_t \left(G_1 dA_t^+ + G_2 d\Lambda_t + G_3 dA_t + G_4 dt\right)$$

where $F_1, \ldots, F_4, G_1, \ldots, G_4$ are bounded operators on a Hilbert space h called the initial space and the families of operators $(U_t; t \geq 0), (V_t; t \geq 0)$ are the solutions. We shall call the first (resp. second) equation the right (resp. left) equation following a usual terminology (see Meyer [31]). The operators U_t and V_t are related to the evolution of a quantum mechanical system therefore it is a natural requirement in all the applications for them to be unitary or, at least, isometries.

This entails some algebraic conditions on the operators $F_1, \ldots, F_4 \ G_1, \ldots, G_4$ playing the role of a priori estimates on the solution. Under these conditions a QSDE appears as a stochastic generalisation of the Schroedinger equation.

The family of operators $(U_t; t \geq 0), (V_t; t \geq 0)$ provide homomorphic dilations $j_t(a) = U_t^* a U_t, k_t(a) = V_t a V_t^*$ of a completely positive evolution on the von Neumann algebra $\mathcal{B}(\text{h})$ of all bounded operators on the Hilbert space h and can be regarded as quantum stochastic processes in the sense of Accardi, Frigerio and Lewis [2].

In these notes we develop a theory of QSDE of the type of Hudson and Parthasarathy discussing the existence and uniqueness results, conditions for solutions $(U_t; t \geq 0), (V_t; t \geq 0)$ to be families of unitaries, properties of solutions (cocycle property and regularity) and showing the application to the dilation of quantum dynamical semigroups.

We do not discuss applications to concrete and special physical models for lack of space. Some of them have been developed in the references listed at the end of the paper and other can be found in the lecture notes by A. Barchielli on Continual Measurements in Quantum Mechanics. Several applications, however, have not yet been studied; these will certainly stimulate the development of other methods and techniques in this growing subject.

2 Fock space notation and preliminaries

Let h be a complex separable Hilbert space, called the *initial space*, and let k be another complex separable Hilbert space. Let $\mathcal{H} = \text{h} \otimes \mathcal{F}$ be the Hilbert space tensor product of h and

$\mathcal{F} = \Gamma(L^2(\mathbb{R}_+; k))$, the symmetric Fock space over $L^2(\mathbb{R}_+; k)$. The symbol $\otimes$ denotes the tensor product for Hilbert spaces and their vectors; the symbol $\odot$ denotes the algebraic tensor product. We shall omit these two symbols whenever this does not lead to confusion. Moreover we identify bounded operators defined on a factor of a tensor product Hilbert space with their ampliation.

Choose and fix an orthonormal basis $(e_k)_{k \geq 1}$ of k and put

$$\mathcal{M} = \{f \in L^2(\mathbb{R}_+; k) \cap L^\infty_{loc}(\mathbb{R}_+; k) \mid \langle e_k, f(t) \rangle = 0$$
$$\text{identically in } t \text{ for all but a finite number of } k\text{'s}\}$$
$$\mathcal{E} = \text{lin}\{e(f) : f \in \mathcal{M}\}.$$

where $e(f) = ((n!)^{-1/2} f^{\otimes n})_{n \geq 0}$ is the exponential vector associated to the test function f. The notion of adaptedness plays a crucial role in the Hudson and Parthasarathy [26] theory of quantum stochastic calculus. This is expressed through the continuous tensor product factorisation property of Fock space: for each $t > 0$ let

$$\mathcal{F}_t = \Gamma(L^2([0, t[; k)), \quad \mathcal{F}^t = \Gamma(L^2([t, \infty[; k)).$$

Then $\mathcal{F} \cong \mathcal{F}_t \otimes \mathcal{F}^t$ via the continuous linear extension of the map $e(f) \mapsto e(f|_{[0,t[}) \otimes e(f|_{[t,\infty[})$, and $\mathcal{F}_t$ and $\mathcal{F}^t$ embed naturally into $\mathcal{F}$ as subspaces by tensoring with the vacuum vector. Let D be a dense subspace of h. Vectors $ue(f)$ and $ve(g)$ with $u, v \in D$ and $f, g \in \mathcal{M}$, $f \neq g$ are linearly independent and the set $h \odot \mathcal{E}$ is dense in $\mathcal{H}$. Therefore we can determine linear (possibly unbounded) operators on $\mathcal{H}$ by defining their action on $h \odot \mathcal{E}$.

Definition 2.1. *An operator process on D is a family $X = (X_t; t \geq 0)$ of operators on $\mathcal{H}$ satisfying:*

(i) $\bigcap_{t \geq 0} \text{Dom}(X_t) \supset D \odot \mathcal{E}$,
(ii) $t \mapsto X_t ue(f)$ *is strongly measurable,*
(iii) $X_t ue(f|_{[0,t[}) \in h \otimes \mathcal{F}_t$, and $X_t ue(f) = [X_t ue(f|_{[0,t[})] \otimes e(f|_{[t,\infty[})$,

for all $u \in D, f \in \mathcal{M}$ and $t > 0$.
Any process satisfying the further condition

(iv) $\int_0^t \|X_s ue(f)\|^2 \, ds < \infty$ *for all $t > 0$,*

is called stochastically integrable.

Note that condition (ii) is equivalent to weak measurability since all the Hilbert spaces here are separable. Hudson and Parthasarathy [26] defined stochastic integrals $\int_0^t X_s \, d\Lambda^\alpha_\beta(s)$ where Λ^α_β is one of the fundamental noise process defined with respect to the fixed basis of k. The integral has domain $D \odot \mathcal{E}$ and the map $t \mapsto \int_0^t X_s \, d\Lambda^\alpha_\beta(s)$ is strongly continuous on this domain.

The quantum noises in $\mathcal{H}$, $\{\Lambda^\alpha_\beta \mid \alpha, \beta \geq 0\}$ are defined by

$$\Lambda^0_\beta(t) = A^+(1_{[0,t]} \otimes |e_\beta\rangle) \quad = A^\dagger_\beta(t) \text{ if } \beta > 0,$$
$$\Lambda^\alpha_\beta(t) = \Lambda(1_{[0,t]} \otimes |e_\beta\rangle\langle e_\alpha|) \quad \text{if } \alpha, \beta > 0,$$
$$\Lambda^\alpha_0(t) = A(1_{[0,t]} \otimes \langle e_\alpha|) \quad = A^\alpha(t) \text{ if } \alpha > 0,$$
$$\Lambda^0_0(t) = t\mathbf{1}$$

where A^+, A, Λ denote respectively the creation, annihilation and gauge operators in $\mathcal{F}$ defined, for each $u \in$ h and each $e(f) \in \mathcal{F}$ by

$$A^+(1_{[0,t]} \otimes |e_m\rangle)ue(f) = \frac{d}{d\epsilon}ue(f + \epsilon 1_{[0,t]}e_m)\Big|_{\epsilon=0}$$

$$\Lambda(1_{[0,t]} \otimes |e_m\rangle\langle e_\ell|)ue(f) = -i\frac{d}{d\epsilon}ue\left(e^{i\epsilon 1_{[0,t]} \otimes |e_m\rangle\langle e_\ell|}f\right)\Big|_{\epsilon=0}$$

$$A(1_{[0,t]} \otimes \langle e_\ell|)ue(f) = \langle e_\ell 1_{[0,t]}, f\rangle ue(f).$$

We refer to the books of Meyer [31] and Parthasarathy [34] or Attal [7] for the theory of quantum stochastic calculus. This could be developed also using the language of white noise analysis through Wick products (see Obata [33]). We prefer, however, to use a language closer to the original (Hudson and Parthasarathy [26]) which is presently more widespread.

A stochastic integral $I_t^X = I_0^X + \int_0^t X_s \, d\Lambda_\beta^\alpha(s)$, with I_0^X operator on h satisfies the so-called *first fundamental formula* of quantum stochastic calculus

$$\langle ve(g), I_t^X ue(f)\rangle = \langle ve(g), I_0^X ue(f)\rangle + \int_0^t g_\beta(s)f^\alpha(s)\langle ve(g), X_s ue(f)\rangle \, ds \quad (1)$$

for all $u \in$ h, $v \in D, f, g \in \mathcal{M}$ and $t > 0$. Here f^α are the components of the k-valued function f, by convention we set $f^0 = 1$ and $f_\alpha(s) = \overline{f^\alpha(s)}$. The second fundamental formula, the *Itô formula*, gives the product $\langle I_t^Y ve(g), I_t^X ue(f)\rangle$ of I_t^X with another stochastic integral $I_t^Y = I_0^Y + \int_0^t Y_s \, d\Lambda_\nu^\mu(s)$ as

$$\langle I_0^Y ve(g), I_0^X ue(f)\rangle + \int_0^t \Big\{ \langle I_s^Y ve(g), X_s ue(f)\rangle g_\beta(s)f^\alpha(s) \quad (2)$$

$$+\langle Y_s ve(g), I_s^X ue(f)\rangle g_\mu(s)f^\nu(s) + \widehat{\delta}_\beta^\nu \langle Y_s ve(g), X_s ue(f)\rangle g_\mu(s)f^\alpha(s) \Big\} \, ds$$

where $\widehat{\delta}$ is the matrix defined by $\widehat{\delta}_\beta^\nu = 1$ if $\nu = \beta > 0$ and $\widehat{\delta}_\beta^\nu = 0$ otherwise. The Itô formula is written shortly as

$$d\Lambda_\mu^\nu d\Lambda_\beta^\alpha = \widehat{\delta}_\beta^\nu d\Lambda_\mu^\alpha.$$

The following inequality can be proved by a simple application of the Itô formula together with the Gronwall lemma. It plays an important role in the construction of the solution of the simplest QSDE.

Proposition 2.2. *Let X_β^α ($\alpha, \beta \geq 0$) be stochastically integrable processes such that*

$$\sum_{\beta \geq 0} \int_0^t \|X_\beta^\alpha(r)ue(f)\|^2 \, dr < +\infty$$

for all $t > 0$, $u \in D$, $f \in \mathcal{M}$ and all α. For each and $t > s \geq 0$ the series $\sum_{\alpha,\beta} \int_s^t X_\beta^\alpha(r) \, d\Lambda_\alpha^\beta(r)ue(f)$ is norm convergent and

$$\left\|\sum_{\alpha,\beta} \int_s^t X_\beta^\alpha(r) \, d\Lambda_\alpha^\beta(r)ue(f)\right\|^2 \leq c_s^t(f) \int_s^t \sum_{\alpha,\beta} \|X_\beta^\alpha(r)ue(f)\|^2 |f_\alpha(r)|^2 dr$$

where $c_s^t(f)$ is the constant

$$c_s^t(f) = 2d(f)e^{\int_s^t (1+|f(r)|^2)dr}$$

with $d(f)$ number of non identically 0 components f^β and $|f(r)|^2 = \sum_{k \geq 1} |f^k(r)|^2$.

Proof. Let Z_s^t denote the left-hand side stochastic integral. By the Itô formula we have

$$\left\|Z_s^t ue(f)\right\|^2 = 2\Re \sum_{\alpha,\beta \geq 0} \int_s^t \langle Z_s^r ue(f), X_\beta^\alpha(r)ue(f)\rangle f_\alpha(r) f^\beta(r) dr$$

$$+ \sum_{\nu,\alpha \geq 0} \sum_{\ell \geq 1} \int_s^t \langle X_\ell^\nu(r)ue(f), X_\ell^\alpha(r)ue(f)\rangle f_\alpha(r) f^\nu(r) dr$$

$$= 2\Re \sum_{\beta \geq 0} \int_s^t \langle Z_s^r ue(f), \sum_{\alpha \geq 0} f_\alpha(r) X_\beta^\alpha(r)ue(f)\rangle f^\beta(r) dr$$

$$+ \sum_{\ell \geq 1} \int_s^t \left\|\sum_{\alpha \geq 0} f_\alpha(r) X_\ell^\alpha(r)ue(f)\right\|^2 dr$$

For each $\beta \geq 0$ let

$$x_\beta(r) = \sum_{\alpha \geq 0} f_\alpha(r) X_\beta^\alpha(r)ue(f).$$

The Schwarz inequality for the scalar product $\langle \cdot, \cdot \rangle$ in $\mathcal{H}$ and for the double integral given by sum on $\beta \geq 0$ together with the integral on $[s, t]$ shows that the term $2\Re(\cdots)$ is not bigger than

$$2 \left(\int_s^t \sum_{\beta \geq 0} \|Z_s^r ue(f)\|^2 |f^\beta(r)|^2 dr \right)^{1/2} \cdot \left(\int_s^t \sum_{\beta \geq 0} \|x_\beta(r)ue(f)\|^2 dr \right)^{1/2}.$$

Thus, since $\sum_{\beta \geq 0} |f^\beta(r)|^2 = (1 + |f(r)|^2)$, the elementary inequality for positive real numbers $2ab \leq a^2 + b^2$ leads to the inequality

$$\left\|Z_s^t ue(f)\right\|^2 \leq \int_s^t \|Z_s^r ue(f)\|^2 (1 + |f(r)|^2) dr + 2 \int_s^t \sum_{\beta \geq 0} \|x_\beta(r)ue(f)\|^2 dr$$

and then, by Gronwall Lemma,

$$\left\|Z_s^t ue(f)\right\|^2 \leq 2e^{\int_s^t (1+|f(r)|^2) dr} \int_s^t \sum_{\beta \geq 0} \|x_\beta(r)ue(f)\|^2 dr.$$

Now the norm inequality

$$\|x_\beta(r)\|^2 \leq d(f) \sum_{\alpha \geq 0} \|X_\beta^\alpha(r)ue(f)\|^2 |f_\alpha(r)|^2$$

implies then the claimed inequality. $\square$

The inequality in Proposition 2.2 allows us to establish estimates on iterated stochastic integrals and then solving the simplest quantum stochastic differential equations by the usual Picard iteration method. The analogs in other stochastic calculi (fermion, free, boolean, ...) lead to similar results for quantum stochastic differential equations driven by other noises.

3 Existence and uniqueness

We now study the left and right QSDE:

$$dV_t = V_t G_\beta^\alpha \, d\Lambda_\alpha^\beta(t), \quad V_0 = \mathbf{1}, \tag{3}$$

$$dU_t = F_\beta^\alpha U_t \, d\Lambda_\alpha^\beta(t), \quad U_0 = \mathbf{1}, \tag{4}$$

where $G = [G_\beta^\alpha]_{\alpha,\beta \geq 0}$ and $F = [F_\beta^\alpha]_{\alpha,\beta \geq 0}$ are matrices of operators on h, and Einstein's summation convention for repeated indices applies, with greek indices running from 0 and roman indices running from 1.

Here we will look for contractive solutions of these equations, that is V or U such that V_t or U_t is a contractive operator for each t. Indeed, this is by far the most interesting case in the applications.

Let $D \subset$ h be a dense subspace.

Definition 3.1. *An operator process V is a solution of (3) on $D \odot \mathcal{E}$ for the operator matrix G if*

(Li) $D \subset \bigcap_{\alpha,\beta} \mathrm{Dom}(G_\beta^\alpha)$,

(Lii) *the linear manifold $\left(\cup_{\alpha,\beta} G_\beta^\alpha(D)\right) \odot \mathcal{E}$ is contained in the domain of V_t for all $t \geq 0$, the processes $(V_t G_\beta^\alpha; t \geq 0)$ are stochastically integrable and*

$$\sum_{\beta \geq 0} \int_0^t \|V_s G_\beta^\alpha u e(f)\|^2 ds < \infty$$

for all $\alpha \geq 0$, $u \in D$, $f \in \mathcal{M}$,

(Liii)

$$V_t = \mathbf{1} + \int_0^t V_s G_\beta^\alpha \, d\Lambda_\alpha^\beta(s).$$

for all $t \geq 0$.

The integrability condition in *(Lii)* is obviously satisfied when the Hilbert space k is finite dimensional.

For the right equation (4) the situation is in general more complex since there is no reason to expect that, for any solution U, the range of each U_t should lie in an algebraic tensor product of the form $D' \odot \mathcal{F}$. For this reason we only define solutions of (4) when each component F_β^α of F is closable. In this case it can be shown (e.g. Fagnola and Wills [22], Section 1) that the standard ampliation $F_\beta^\alpha \odot 1$ to $\mathcal{H}$ is closable.

Let D be as above.

Definition 3.2. *A process U is a solution of (R) on $D \odot \mathcal{E}$ for the operator matrix F if*

(Ri) $\bigcup_t U_t(D \odot \mathcal{E}) \subset \bigcap_{\alpha,\beta} \mathrm{Dom}(\overline{F_\beta^\alpha \odot 1})$,

(Rii) *each process $\overline{F_\beta^\alpha \odot 1}U$ is stochastically integrable, and*

$$\sum_{\beta \geq 0} \int_0^t \|\overline{F_\beta^\alpha \odot 1}U_s u e(f)\|^2 ds < \infty$$

for all $\alpha \geq 0$, $u \in D$, $f \in \mathcal{M}$,

(Riii)

$$U_t = \mathbf{1} + \int_0^t \overline{F_\beta^\alpha \odot 1} U_s \, d\Lambda_\alpha^\beta(s).$$

In order to deal with stochastic equations driven by a possibly infinite number of noises (i.e. with k infinite dimensional), in this Section, we shall assume that the operators G_β^α and F_β^α satisfy the so-called *Mohari-Sinha summability condition*: for all $\alpha \geq 0$ *there exists a constant* $c(\alpha)$ *such that*

$$\sum_{\beta \geq 0} \|G_\beta^\alpha u\|^2 \leq c(\alpha)^2 \|u\|^2, \qquad \sum_{\beta \geq 0} \|F_\beta^\alpha u\|^2 \leq c(\alpha)^2 \|u\|^2 \qquad (5)$$

for all $u \in D$ where $c(\alpha)$ is a positive real constant.

Since the domain D is dense in h, the above condition implies that the operators G_β^α and F_β^α are bounded and the inequality (5) holds for all $u \in$ h. Moreover, note that these inequalities imply that the column of operators $[G_\beta^0 \, G_\beta^1 \ldots]^T$ defines a bounded operator on h $\to \oplus_{\alpha \geq 0}$h for each β. Of course, when the Hilbert space k is finite dimensional, (5) is *equivalent* to boundedness of all the G_β^α and F_β^α.

It is easy to prove the following

Theorem 3.3. *Suppose that (5) holds. There exist operator processes $(V_t; t \geq 0)$ and $(U_t; t \geq 0)$ solving (3) and (4) on h $\odot \mathcal{E}$.*

Proof. Both the operator processes $(V_t; t \geq 0)$ and $(U_t; t \geq 0)$ will be constructed via the Picard iteration method.

We first consider the left equation (3). Define by recurrence the sequence of stochastically integrable processes on h $\odot \mathcal{E}$

$$V_t^{(0)} = \mathbf{1}, \qquad V_t^{(n+1)} = \int_0^t V_s^{(n)} G_\beta^\alpha d\Lambda_\alpha^\beta(s).$$

Applying Proposition 2.2, it is easy to prove by induction that, $(V^{(n)})_{n \geq 0}$ is a sequence of stochastically integrable processes and, for all $u \in$ h, $t \geq 0$ and $f \in \mathcal{M}$, the following inequality holds

$$\left\|V_t^{(n)} u e(f)\right\|^2 \leq \|u\|^2 \left(c_0^t(f)\right)^n \left(\max_{\{\alpha \geq 0 \,|\, f^\alpha \neq 0\}} c(\alpha)\right)^{2n}$$
$$\cdot \left(\int_0^t (1 + |f(s)|^2) ds\right)^n / n! \qquad (6)$$

where $c_0^t(f)$ is the constant as in Proposition 2.2. Therefore the series

$$\sum_{n \geq 0} V_t^{(n)} u e(f)$$

is convergent in the norm topology on $\mathcal{H}$ for all $u \in$ h and $f \in \mathcal{M}$. By defining $V_t u e(f)$ as its limit we find an operator process V. It is easy to check that it is stochastically integrable on h $\odot \mathcal{E}$.

Moreover, for all $n \geq 0$, we have

$$\sum_{m=0}^n V_t^{(m)} = \mathbf{1} + \sum_{m=1}^n V_t^{(m)} = \mathbf{1} + \int_0^t \sum_{m=0}^{n-1} V_s^{(m)} G_\beta^\alpha d\Lambda_\alpha^\beta(s).$$

Letting n tend to infinity it follows that the process $(V_t; t \geq 0)$ is a solution of (3) on $h \odot \mathcal{E}$.

The proof for (4) is similar. We omit it. $\quad \square$

If we knew that the solution of (3) for $G_\beta^\alpha = (F_\beta^\alpha)^*$ is bounded we could find a solution of (4) simply by taking the adjoint $U_t = (V_t)^*$. Unfortunately, in general, there is no reason for $h \odot \mathcal{E}$ to be contained in the domain of $(V_t)^*$.

The natural uniqueness result is, perhaps surprisingly, slightly different for the right and left equation.

Theorem 3.4. *Suppose that (5) holds. Then:*

1. *the operator processes $(U_t; t \geq 0)$ on $h \odot \mathcal{E}$ solving (4) on $h \odot \mathcal{E}$ is unique,*
2. *the operator processes $(V_t; t \geq 0)$ on $h \odot \mathcal{E}$ solving (3) on $h \odot \mathcal{E}$ is unique among the operator processes satisfying*

$$I(V, t, f) = \sup_{0 \leq s \leq t, \|u\| \leq 1} \|V_s ue(f)\|^2 < +\infty$$

for all $f \in \mathcal{M}$ and $t \geq 0$.

Proof. Let $(U_t^{(1)}; t \geq 0)$ and $(U_t^{(2)}; t \geq 0)$ be two operator processes solving (4) and let $Z_t = U_t^{(1)} - U_t^{(2)}$. Then

$$Z_t = \int_0^t F_\beta^\alpha Z_s d\Lambda_\alpha^\beta(s).$$

By Proposition 2.2, for all $u \in h$, $t > 0$ and $f \in \mathcal{M}$ we have

$$\|Z_t ue(f)\|^2 \leq c_0^t(f) \sum_{\alpha, \beta} \int_0^t \|F_\beta^\alpha Z_s ue(f)\|^2 |f_\alpha(s)|^2 ds$$

$$\leq c_0^t(f) \sum_{\alpha \geq 0} c(\alpha)^2 \int_0^t \|Z_s ue(f)\|^2 |f_\alpha(s)|^2 ds$$

$$\leq c_0^t(f) \left(\max_{\{\alpha \geq 0 \,|\, f^\alpha \neq 0\}} c(\alpha) \right)^2 \int_0^t \|Z_s ue(f)\|^2 (1 + |f(s)|^2) ds.$$

It follows then, from Gronwall's lemma, that $\|Z_t ue(f)\| = 0$.

We prove now the second statement.

The difference Z_t of two solutions $(V_t^{(1)}; t \geq 0)$ and $(V_t^{(2)}; t \geq 0)$ of (3) satisfies now

$$Z_t = \int_0^t Z_s G_\beta^\alpha d\Lambda_\alpha^\beta(s)$$

Thus, for all $u \in h$, $t > 0$ and $f \in \mathcal{M}$, we have

$$\|Z_t ue(f)\|^2 \leq c_0^t(f) \sum_{\alpha, \beta} \int_0^t \|Z_s G_\beta^\alpha ue(f)\|^2 |f_\alpha(s)|^2 ds$$

An n-times iteration of this formula shows that $\|Z_t ue(f)\|^2$ is not bigger than $(c_0^t(f))^n$ times

$$\sum_{\alpha_1, \beta_1, \ldots, \alpha_n, \beta_n} \int_0^t ds_1 \int_0^{s_1} .. \int_0^{s_{n-1}} ds_n \|Z_s G_{\beta_n}^{\alpha_n} .. G_{\beta_1}^{\alpha_1} ue(f)\|^2 |f_{\alpha_n}(s_n)|^2 .. |f_{\alpha_1}(s_1)|^2.$$

By the initial space boundedness condition this is not bigger than a constant $I(Z, t, f) :=$
$2(I(V^{(1)}, t, f) + I(V^{(2)}, t, f))$ times

$$\sum_{\alpha_1, \beta_1, \ldots, \alpha_n, \beta_n} \|G_{\beta_n}^{\alpha_n} \ldots G_{\beta_1}^{\alpha_1} u\|^2 \int_0^t ds_1 \int_0^{s_1} \ldots \int_0^{s_{n-1}} ds_n |f_{\alpha_n}(s_n)|^2 \ldots |f_{\alpha_1}(s_1)|^2.$$

It follows then form (5) that, for all $n \geq 1$, $\|Z_t ue(f)\|^2$ is not bigger than $I(Z, t, f)$ times

$$(c_0^t(f))^n \sum_{\alpha_1, \ldots, \alpha_n} (c(\alpha_1)..c(\alpha_n))^2 \int_0^t ds_1 \int_0^{s_1} .. \int_0^{s_{n-1}} ds_n |f_{\alpha_n}(s_n)|^2 .. |f_{\alpha_1}(s_1)|^2$$

$$\leq (c_0^t(f))^n \sum_{\alpha_1, \ldots, \alpha_n} \left(\max_{\{\alpha \mid f^\alpha \neq 0\}} c(\alpha) \right)^{2n} \left(\int_0^t (1 + |f(s)|^2) ds \right)^n \Big/ n!.$$

The conclusion follows then letting n tend to infinity. $\square$

Remarks. (a) Note that the solution is unique when the initial conditions U_0, V_0 are arbitrary operators on h.

(b) An operator process satisfying condition 2. is called *initial space bounded*. If a process $(X_t; t \geq 0)$ is locally bounded, i.e. $\sup_{0 \leq s \leq t} \|X_s\| < +\infty$ for all $t > 0$, then it is obviously initial space bounded. The solutions of (3) and (4) obtained by Picard iteration are initial space bounded as a consequence of (6).

4 Unitary solutions

We now start studying the properties of solutions.

An operator process $X = (X_t; t \geq 0)$ is *bounded* (resp. *a contraction, an isometry, a coisometry, a unitary*) process if each operator X_t is bounded (resp. a contraction, an isometry, a coisometry, a unitary).

Let $G = [G_\beta^\alpha]_{\alpha, \beta \geq 0}$ be a matrix of operators on h. Put $\text{Dom}[G] = \cap_{\alpha, \beta \geq 0} \text{Dom}(G_\beta^\alpha)$. Then given any self-adjoint operator X on h and subspace $D \subset \text{Dom}(X) \cap \text{Dom}[G]$ such that $G_\beta^\ell(D) \subset \text{Dom}(X)$ for all $\ell \geq 1, \beta \geq 0$,

$$((u^\gamma), (v^\nu)) \mapsto \langle X u^\alpha, G_\beta^\alpha v^\beta \rangle + \langle G_\alpha^\beta u^\alpha, X v^\beta \rangle + \langle G_\alpha^\ell u^\alpha, X G_\beta^\ell v^\beta \rangle \qquad (7)$$

defines a sesquilinear form on $(\oplus_{\alpha \geq 0} D) \times (\oplus_{\alpha \geq 0} D)$ (Einstein's summation convention for repeated indices applies). The domain of this form, in order to avoid summation technicalities, will be the linear manifold of vectors $(u^\gamma) \in \oplus_{\alpha \geq 0} D$ with only finitely many nonzero components. We denote this form by $\theta_G(X)$, and say that $\theta_G(X)$ is defined *as a form on D* if we need to make precise the domain of definition. Note that if $X \in \mathcal{B}(h)$ then $\theta_G(X)$ is well-defined as a form on $\text{Dom}[G]$. Moreover, if k is d-dimensional and the operators G_β^α are bounded, then $\theta_G(X)$ is a bounded form to which there is associated a bounded operator on $(\oplus^{(d+1)} h) \times (\oplus^{(d+1)} h)$. In this case the linear map $\theta_G : \mathcal{B}(h) \to M_{d+1}(\mathcal{B}(h))$ is given by

$$\theta_G(X) = (X \otimes 1_{d+1}) G + G^*(X \otimes 1_{d+1}) + G^* \Delta(X) G,$$

where $\Delta(X) = \text{diag}\{0, X, \ldots, X\} \in M_{d+1}(\mathcal{B}(h))$ and 1_{d+1} is the identity matrix in $\mathbb{C}^{d+1}$.

Properties of solutions U and V like boundedness, contractivity, isometry and unitarity are closely related to properties of $\theta_F(\mathbf{1})$ and $\theta_G(\mathbf{1})$. Indeed, we can prove the following

Theorem 4.1. *Suppose that $F = [F_\beta^\alpha]$ is a matrix of bounded operators on h and (5) holds. Let U be the unique solution of (4). Then the following are equivalent:*

1. *the process U is a contraction (resp. an isometry),*
2. *we have $\theta_F(\mathbf{1}) \leq 0$ (resp. $\theta_F(\mathbf{1}) = 0$).*

Proof. $1 \Rightarrow 2$. Let $\xi = \sum_k u_k e(f_k)$ for some finite sets $\{u_k\}$ of h and $\{f_k\}$ of $\mathcal{M}$. Then

$$\int_0^t \Big\{ \langle U_s \varphi^\alpha(s), F_\beta^\alpha U_s \varphi^\beta(s) \rangle + \langle F_\alpha^\beta U_s \varphi^\alpha(s), U_s \varphi^\beta(s) \rangle \tag{8}$$

$$+ \langle F_\alpha^i U_s \varphi^\alpha(s), F_\beta^i U_s \varphi^\beta(s) \rangle \Big\} ds = \|U_t \xi\|^2 - \|\xi\|^2 \leq 0$$

by (2), where $\varphi^\alpha(s) = \sum_k f_k^\alpha(s) u_k e(f_k)$ is the α-th component of a vector $\varphi(s)$ in $(\oplus_{\alpha \geq 0} \mathsf{h}) \otimes \mathcal{F} = \oplus_{\alpha \geq 0}(\mathsf{h} \otimes \mathcal{F})$. If we choose the f_k to be continuous then we can differentiate the above at 0 to get

$$0 \geq \theta_F(\mathbf{1})(\varphi(0), \varphi(0)).$$

Varying the f_k and u_k then gives the result, and note that if U is an isometry process then the inequality in (8) becomes an equality.

$2 \Rightarrow 1$. For all ξ as above, by the Itô formula, we have

$$\|U_t \xi\|^2 = \|\xi\|^2 + \int_0^t \langle U_s \varphi(s), \theta_F(\mathbf{1}) U_s \varphi(s) \rangle ds$$

(here the scalar product is in $\oplus_{\alpha \geq 0}(\mathsf{h} \otimes \mathcal{F})$ and the operators U_s act as $U_s \varphi(s) = (U_s \varphi^0(s), \dots, U_s \varphi^k(s), \dots)$). The conclusion is immediate. $\square$

By taking the adjoints we find a similar result also holds for the left equation.

Corollary 4.2. *Suppose that $G = [G_\beta^\alpha]$ is a matrix of bounded operators on h and (5) holds. Denote by $G^\dagger$ the adjoint matrix $G^\dagger = [(G_\alpha^\beta)^*]$. Let V be the unique initial space bounded solution of (3). Then the following are equivalent:*

1. *the process V is a contraction,*
2. *we have $\theta_{G^\dagger}(\mathbf{1}) \leq 0$.*

Proof. If the operators V_t are contractions then the adjoint operators $U_t = V_t^*$ are also contractions and satisfy the right QSDE $dU_t = (G_\alpha^\beta)^* U_t d\Lambda_\alpha^\beta$. Therefore, by Theorem 4.1, we have $\theta_{G^\dagger}(\mathbf{1}) \leq 0$.

Conversely, if 2 holds, then the unique contraction operator process satisfying the right QSDE $dU_t = (G_\alpha^\beta)^* U_t d\Lambda_\alpha^\beta$ with $U_0 = \mathbf{1}$ is contractive. The adjoint family $(U_t^*)_{t \geq 0}$ is also a contraction (thus initial space bounded) process on $\mathsf{h} \odot \mathcal{E}$ and satisfies (3). By uniqueness it follows that the operators $V_t = U_t^*$ are also contractions. $\square$

The reader noticed a lack of symmetry between Theorem 4.1 for the right equation and Corollary 4.2 for the left equation. Indeed, condition $\theta_F(\mathbf{1}) \leq 0$ for the right equation should correspond to $\theta_G(\mathbf{1}) \leq 0$ for the left equation. Conditions $\theta_G(\mathbf{1}) \leq 0$ and $\theta_{G^\dagger}(\mathbf{1}) \leq 0$, however, are equivalent. This follows from the identity

$$(\mathbf{1} + \Delta G^*)^* \theta_G(X)(\mathbf{1} + \Delta G^*) = \theta_{G^\dagger}(\mathbf{1}) + \theta_{G^\dagger}(\mathbf{1})^* \theta_{G^\dagger}(\mathbf{1}).$$

We now characterise isometry processes solving (3).

Proposition 4.3. *Suppose that* $G = [G^\alpha_\beta]$ *is a matrix of bounded operators on* h *and (5) holds. Let V be the unique initial space bounded solution of (3). Then the following are equivalent:*

1. *the process V is an isometry,*
2. *we have* $\theta_G(\mathbf{1}) = 0$.

Proof. $1 \Rightarrow 2$. This follows from a differentiation at $t = 0$ argument as $1 \Rightarrow 2$ in Theorem 4.1.

$2 \Rightarrow 1$. Fix $g, f \in \mathcal{M}$. For all $v, u \in$ h let $\psi_t(v, u) = \langle ve(g), ue(f) \rangle - \langle V_t ve(g), V_t ue(f) \rangle$. Condition 2 implies that the maps $\psi_t(\cdot, \cdot)$ satisfy the integral equation (Einstein's summation convention for repeated indices applies)

$$\psi_t(v, u) = \int_0^t \{\psi_s(v, G^\alpha_\beta u) + \psi_s(G^\beta_\alpha v, u) + \psi_s(G^\ell_\alpha v, G^\ell_\beta u)\} g_\beta(s) f^\alpha(s) ds$$

We now show that $\psi_t(v, u) = 0$ for all $v, u \in$ h. Indeed, iterating the above equation, the same arguments of the proof of Theorem 3.4 yield the inequality

$$|\psi_t(v, u)| \le (I(V, t, f) I(V, t, g))^{1/2} \frac{\kappa^n}{n!} \left(\int_0^t (1 + |g(s)|^2 + |f(s)|^2) ds \right)^{2n}$$

for all $n \ge 1$ where κ is a constant depending on t, f, g, G. The conclusion follows letting n tend to infinity. $\square$

It can be shown easily (taking the adjoints) that the process V is a coisometry if and only if $\theta_{G^\dagger}(\mathbf{1}) = 0$ and the process U is a coisometry if and only if $\theta_{F^\dagger}(\mathbf{1}) = 0$. The above results allow to prove immediately the characterisation of unitary solutions of (3) and (4)

Theorem 4.4. *Suppose that* $G = [G^\alpha_\beta]$ *and* $F = [F^\alpha_\beta]$ *are matrices of bounded operators on* h *and (5) holds. Let V be the unique initial space bounded solution of (3) and let U be the unique solution of (4).*

1. *The process V is unitary if and only if* $\theta_G(\mathbf{1}) = 0$ *and* $\theta_{G^\dagger}(\mathbf{1}) = 0$.
2. *The process U is unitary if and only if* $\theta_F(\mathbf{1}) = 0$ *and* $\theta_{F^\dagger}(\mathbf{1}) = 0$.

Remark. If all the G^α_β with $\alpha + \beta > 0$ vanish then the solution of (3) is unitary if and only if $G^0_0 + (G^0_0)^* = 0$ i.e. $G = iH$ with H self-adjoint. The same conclusion holds for the right equation. Thus (3) and (4) are quantum stochastic generalisations of the Schroedinger equation.

5 Emergence of H-P equations in physical applications

Quantum stochastic differential equations are a natural tool in the theory of continual measurements in Quantum Mechanics (see the survey paper by A. Barchielli [8] in these lecture notes and the references therein), the theory of quantum filtering and control (Belavkin [9]) and arise from suitable limits of quantum mechanical evolution equations (see e.g. von Waldenfels [39]).

In this section we outline the deduction of a quantum stochastic differential equation from the *stochastic limit* of the evolution equation of a system coupled with a reservoir. Here we consider a very simple model and refer to the book by L. Accardi, Y.G. Lu and Volovich [5] for a general theory.

The state space of the system is a complex separable Hilbert space h. The free evolution of the system is given by a strongly continuous unitary group $(e^{-itH_S})_{t\in\mathbb{R}}$, where H_S is a self-adjoint operator on h. The state space of the reservoir (boson gas) is the boson Fock space $\mathcal{F}$ over a complex separable Hilbert space k_1 (the one-particle space of the reservoir). The free evolution of one particle in the reservoir is given by a strongly continuous one parameter group $(S_t^0)_{t\in\mathbb{R}}$ of unitary operators on k_1 enjoying the following property: *there exists a dense subspace* k *of* k_1 *such that*

$$\int_{\mathbb{R}} \left| \langle g, S_t^0 f \rangle \right| dt < \infty$$

for all $f, g \in$ k. The free evolution of the reservoir is given by the unitary group obtained by second quantization $(\Gamma(S_t^0))_{t\in\mathbb{R}}$ of the unitaries S_t^0 on k_1. This is a strongly continuous unitary group and its generator H_R, the self-adjoint operator on $\mathcal{F}$ such that $e^{-itH_R} = \Gamma(S_t^0)$ for all $t \in \mathbb{R}$, is the Hamiltonian of the reservoir.

The evolution of the whole system is given by the unitary group generated by the total Hamiltonian

$$H_\lambda = H_S \otimes 1_{\mathcal{F}} + 1_S \otimes H_R + \lambda V$$

where λ is a real positive parameter and V is an interaction operator such that H_λ is self-adjoint for all $\lambda > 0$.

In the simplest models the interaction operator (of dipole type) has the form

$$V_g = i \left(D \otimes A^*(g) - D^* \otimes A(g) \right)$$

where D is a suitable operator on the system space h and $A(g)$, $A^*(g)$ are creation and annihilation operators on $\mathcal{F}$ with $g \in k_1$. Moreover the so-called *generalized rotating wave approximation*

$$e^{itH_S} D e^{-itH_S} = e^{-i\omega_0 t} D$$

where $\omega_0 > 0$ (see [5] Def. 4.10.1 p.125) holds.

The sesquilinear form on k

$$(f|g) := \int_{\mathbb{R}} \langle g, S_t^0 f \rangle \, dt$$

is positive. Indeed, letting $E(d\xi)$ denote the spectral measure associated with the self-adjoint generator of $(S_t^0)_{t\in\mathbb{R}}$, we have

$$
\begin{aligned}
(f|f) &= \int_{\mathbb{R}} \langle f, S_t^0 f \rangle \, dt \\
&= \int_{\mathbb{R}} dt \int_{\mathbb{R}} e^{-it\xi} \langle f, E(d\xi) f \rangle \\
&= \int_{\mathbb{R}} \langle f, E(d\xi) f \rangle \int_{\mathbb{R}} dt \, e^{-it\xi} \\
&= \sqrt{2\pi} \langle f, E(\{0\}) f \rangle \geq 0.
\end{aligned}
$$

for all $f \in$ k. Therefore it defines a pre-scalar product on k. We denote by $\mathcal{K}$ the Hilbert space obtained by quotient and completion; the scalar product will be denoted by $(\cdot|\cdot)$.

Defining

$$U_t^{(\lambda)} = e^{itH_0} e^{-itH_\lambda}$$

a straightforward computation shows that the family of unitaries $(U_t^{(\lambda)})_{t\geq0}$ on $h \otimes \mathcal{F}$ satisfies the differential equation

$$\frac{d}{dt}U_t^{(\lambda)} = -i\lambda V_g(t)U_t^{(\lambda)}, \qquad U_0^{(\lambda)} = 1,$$

where

$$V_g(t) = i\left(D \otimes A^*(S_t g) - D^* \otimes A(S_t g)\right)$$

and $S_t = e^{it\omega_0} S_t^{(0)}$.

Let $W(f)$ ($f \in k_1$) denote the unitary Weyl operators on $\mathcal{F}$ acting on exponential vectors as

$$W(f_1)e(f_2) = e^{-\|f_1\|^2/2 - \langle f_1, f_2 \rangle}e(f_1 + f_2).$$

The basic idea of Accardi, Frigerio and Lu [3] was to study the result of small interactions ($\lambda \to 0$) on a large time scale (time goes to infinity). This was realized by scaling time by λ^2 and space by λ and letting λ tend to 0. As a result of the limiting procedure also the state space of the whole system also changes.

The following result allows us to find the structure of the space of the limit evolution.

Proposition 5.1. *For all* $n, n' \in \mathbb{N}$ *and all* $f_1, \ldots, f_n, f_1', \ldots, f_{n'}' \in \mathcal{K}$, $s_1, t_1, \ldots s_n,$ $t_n, s_1', t_1', \ldots s_{n'}', t_{n'}' \in \mathbb{R}$ *with* $s_k \le t_k, s_k' \le t_k'$ *for all* k *denote*

$$W(f_1, \ldots, f_n) = W\left(\lambda \int_{\lambda^{-2}s_1}^{\lambda^{-2}t_1} S_{r_1} f_1 dr_1\right) \cdots W\left(\lambda \int_{\lambda^{-2}s_n}^{\lambda^{-2}t_n} S_{r_n} f_n dr_n\right).$$

We have then

$$\lim_{\lambda \to 0} \left\langle W(f_1, \ldots, f_n)e(0), W(f_1', \ldots, f_{n'}')e(0)\right\rangle$$
$$= \left\langle W\left(f_1 \otimes 1_{[s_1,t_1]}\right) \cdots W\left(f_n \otimes 1_{[s_n,t_n]}\right)e(0),\right.$$
$$\left. W\left(f_1' \otimes 1_{[s_1',t_1']}\right) \cdots W\left(f_{n'}' \otimes 1_{[s_{n'}',t_{n'}']}\right)e(0)\right\rangle$$

where $W\left(f_1 \otimes 1_{[s_1,t_1]}\right), \ldots, W\left(f_{n'}' \otimes 1_{[s_{n'}',t_{n'}']}\right)$ *are Weyl operators in the boson Fock space over* $L^2(\mathbb{R}_+; \mathcal{K})$.

It follows that the state space of the limit evolution is the tensor product of the initial space h with the boson Fock space $\Gamma(L^2(\mathbb{R}_+; \mathcal{K}))$. The limit of unitaries is given in the following theorem.

Theorem 5.2. *Suppose that the operator D is bounded. For all* $v, u \in$ h, $n, n' \in \mathbb{N}$ *and all* $f_1, \ldots, f_n, f_1', \ldots, f_{n'}' \in \mathcal{K}$, $s_1, t_1, \ldots s_n, t_n, s_1', t_1', \ldots, s_{n'}', t_{n'}' \in \mathbb{R}$ *with* $s_k \le t_k, s_k' \le t_k'$ *we have*

$$\lim_{\lambda \to 0} \left\langle vW(f_1, \ldots, f_n)e(0), U_{\lambda^{-2}t}^{(\lambda)} uW(f_1', \ldots, f_{n'}')e(0)\right\rangle$$
$$= \left\langle vW\left(f_1 \otimes 1_{[s_1,t_1]}\right) \cdots W\left(f_n \otimes 1_{[s_n,t_n]}\right)e(0),\right.$$
$$\left. W\left(f_1' \otimes 1_{[s_1',t_1']}\right) \cdots U_t uW\left(f_{n'}' \otimes 1_{[s_{n'}',t_{n'}']}\right)e(0)\right\rangle$$

where U is the unique unitary process satisfying the quantum stochastic differential equation on h $\otimes \Gamma(L^2(\mathbb{R}_+; \mathcal{K}))$

$$dU_t = \left(DdA_t^*(g) - D^* dA_t(g) - (g|g)_- D^* D dt\right)U_t, \qquad U_0 = \mathbf{1}$$

with

$$(g|g)_- = \int_{-\infty}^{0} \langle g, S_t g \rangle dt.$$

Several of the above assumptions (rotating wave approximation, boundedness of D, ...) can be removed or weakened. Moreover other reservoirs and other types of interaction can be studied. As a result more general quantum stochastic differential equations, of the H-P type we are studying, arise. We refer the interested reader to the book by Accardi, Lu and Volovich [5].

Notice that unitaries $U_t^{(\lambda)}$ form a group. This is no longer the case for unitaries U_t. These, however, satisfy a composition law that will be studied in the next section.

6 Cocycle property

We now discuss the most important and useful property of contraction processes U and V: the cocycle property. This is the key ingredient (see Accardi [1]) for constructing homomorphic dilations of quantum Markov (dynamical, in the physical terminology) semigroups by unitary conjugation with U or V. Indeed, this was the original scope for studying unitary solutions of quantum stochastic differential equation.

We start by recalling the definition of operator cocycles.

For each $t \geq 0$ let σ_t be the right shift on $L^2(\mathbb{R}_+; k)$ by $(\sigma_t f)(x) = f(x - t)$ if $x > t$ and $(\sigma_t f)(x) = 0$ if $x \leq t$. Let $\Gamma(\sigma_t)$ be the operators in $\mathcal{F}$ defined by second quantization of σ_t,

$$\Gamma(\sigma_t)\mathrm{e}(f) = \mathrm{e}(\sigma_t f) \tag{9}$$

for all $f \in L^2(\mathbb{R}_+; k)$. The operators σ_t and $\Gamma(\sigma_t)$ are isometries for every $t \geq 0$. Notice that, for all $s, t \geq 0$ we have

$$\Gamma(\sigma_s)^* \Gamma(\sigma_{t+s}) = \Gamma(\sigma_t), \quad \Gamma(\sigma_s)\Gamma(\sigma_t) = \Gamma(\sigma_{s+t}).$$

For each $s \geq 0$ and each bounded operator X on $\mathcal{H}$ the operator $\Gamma(\sigma_s)X\Gamma(\sigma_s)^*$ maps $\mathsf{h} \otimes \mathcal{F}^s$ into itself. Indeed we have the diagram

$$\Gamma(\sigma_s)^* \ X \ \Gamma(\sigma_s)$$
$$\mathsf{h} \otimes \mathcal{F}^s \longrightarrow \mathcal{H} \to \mathcal{H} \longrightarrow \mathsf{h} \otimes \mathcal{F}^s$$

The canonical extension of $\Gamma(\sigma_s)X\Gamma(\sigma_s)^*$ to $\mathcal{H}$ via ampliation will be denoted by $\Theta_s(X)$. Clearly $(\Theta_s)_{s \geq 0}$ is a semigroup of identity preserving normal *-homomorphisms on $\mathcal{B}(\mathcal{H})$. For all $x \in \mathcal{B}(\mathsf{h})$ and all $s \geq 0$ we have $\Theta_s(x) = x$. Moreover, since the map Θ_s is normal, it can be extended to self-adjoint operators affiliated with $\mathcal{B}(\mathsf{h})$ like position and momentum operators $A_\ell^\dagger(t) + A^\ell(t)$, $i(A_\ell^\dagger(t) - A^\ell(t))$, number operator N_t and so on. Therefore, by linearity, it can be extended to the operators $\Lambda_\alpha^\beta(t)$ and

$$\Theta_s(\Lambda_\alpha^\beta(t)) = \Lambda_\alpha^\beta(t + s) - \Lambda_\alpha^\beta(s)$$

for $t, s \geq 0$.

Definition 6.1. *A bounded operator process* $(X_t; t \geq 0)$ *on* $\mathcal{H}$ *is called a* left cocycle *(resp.* right cocycle) *if for every* $t, s \geq 0$ *we have*

$$X_{t+s} = X_s \Theta_s(X_t), \quad (resp. \ X_{t+s} = \Theta_s(X_t)X_s) \tag{10}$$

Solutions to quantum stochastic differential equations (QSDE) are cocycles. In order to prove this fact we start with the following

Lemma 6.2. *Let $(X_t)_{t\geq 0}$ be a bounded operator process on h such that*

$$X_t = \int_0^t M_r d\Lambda_\alpha^\beta(r)$$

for some $\alpha, \beta \geq 0$ and some bounded, stochastically integrable, operator process $(M_r)_{r\geq 0}$ on h. For all $t, s \geq 0$ we have

$$\Theta_s(X_t) = \int_s^{t+s} \Theta_s(M_{r-s}) \, d\Lambda_\alpha^\beta(r).$$

Proof. By definition of the shift semigroup Θ

$$\langle ve(g), \Theta_s(X_t)ue(f)\rangle = \langle ve(\sigma_s^* g), X_t ue(\sigma_s^* f)\rangle e^{\langle g1_{[0,s]}, f\rangle}$$

for all $t, s \geq 0$ and so, since $X_t = \int_0^t M_r d\Lambda_\alpha^\beta(r)$,

$$\langle ve(g), \Theta_s(X_t)\, ue(f)\rangle$$

$$= \int_0^t (\sigma_s^* g)_\alpha(r)(\sigma_s^* f)^\beta(r) \langle ve(\sigma_s^* g), M_r ue(\sigma_s^* f)\rangle \, dr \cdot e^{\langle g1_{[0,s]}, f\rangle}$$

$$= \int_0^t g_\alpha(r+s) f^\beta(r+s) \langle ve(g), \Theta_s(M_r)ue(f)\rangle \, dr \cdot e^{\langle g1_{[0,s]}, f\rangle}$$

$$= \left\langle ve(g), \int_s^{t+s} \Theta_s(M_{r-s})d\Lambda_\alpha^\beta(r)ue(f)\right\rangle.$$

This proves the claimed identity. $\square$

Proposition 6.3. *Let D be a dense subspace of h and let $G = [G_\beta^\alpha]$ (resp. $F = [F_\beta^\alpha]$) be a matrix of operators on h. Suppose that there exists a unique bounded processes V solving the left equation (3) (resp. the right equation (4)) on $D \odot \mathcal{E}$. Then V (resp. U) is a left (resp. right) cocycle.*

Proof. We check the cocycle property for V; the proof for U is similar. Fix $s > 0$ and let X and Y the be bounded processes defined by

$$X_t = \begin{cases} V_t & \text{if } t \leq s, \\ V_s\Theta_s(V_{t-s}), & \text{if } t > s, \end{cases} \quad \text{and} \quad Y_t = V_t.$$

By Lemma 6.2 the process X and satisfies

$$X_t = V_s + \int_s^{t+s} V_s\Theta_s(V_{r-s})G_\beta^\alpha d\Lambda_\alpha^\beta(r)$$

$$= V_s + \int_s^{t+s} X_r G_\beta^\alpha d\Lambda_\alpha^\beta(r)$$

for $t \geq s$. Therefore both X and Y are bounded solutions of (3). The conclusion follows applying Theorem 3.4. $\square$

The cocycle property, on the other hand, turns out to be a useful tool in the analysis of quantum stochastic differential equations. Indeed, by combining the cocycle property and the time reversal, we can establish a simple relationship between the QSDE

$$dV_t = V_t G_\beta^\alpha \, d\Lambda_\alpha^\beta, \quad \text{and} \quad dX_t = X_t (G_\alpha^\beta)^* \, d\Lambda_\alpha^\beta$$

$$dU_t = U_t F_\beta^\alpha \, d\Lambda_\alpha^\beta, \quad \text{and} \quad dY_t = (F_\alpha^\beta)^* Y_t \, d\Lambda_\alpha^\beta.$$

Since conditions for the existence of a (isometric, unitary, ...) solution for the right equation (4) are naturally stronger, this turns out to be a useful tool allowing to shortcut several domain problems.

We now introduce the precise definition of *time reversal* on $\mathcal{H}$.

Let ρ_t be the unitary time reversal on the interval $[0, t]$ defined on $L^2(\mathbb{R}_+; k)$ by

$$(\rho_t f)(s) = f(t - s) \quad \text{if} \quad s \leq t \quad \text{and} \quad f(s) \quad \text{if} \quad s > t.$$

Let $\Gamma(\rho_t)$ be the operator on $\Gamma(L^2(\mathbb{R}_+; \mathbb{C}^d))$ defined by second quantization

$$\Gamma(\rho_t)\mathrm{e}(f) = \mathrm{e}(\rho_t f).$$

The operators ρ_t are unitary and satisfy

$$\rho_t \rho_t = \mathbf{1}, \quad \Gamma(\rho_t)\Gamma(\rho_t) = \mathbf{1}.$$

Let $\mathcal{R}_t$ be the operator on $\mathcal{B}(\mathcal{H})$ defined by

$$\mathcal{R}_t : \mathcal{B}(\mathcal{H}) \to \mathcal{B}(\mathcal{H}), \quad \mathcal{R}_t(x) = \Gamma(\rho_t)x\Gamma(\rho_t)^*.$$

It can be shown (see e.g. Meyer [31] or Fagnola [18] Sect 5.2) that when $(V_t; t \geq 0)$ is a left (resp. right) cocycle then the operator process $(\widetilde{V}_t; t \geq 0)$ defined by

$$\widetilde{V}_t = \mathcal{R}_t(V_t^*) \tag{11}$$

is a left (resp. right) cocycle. The cocycle $\widetilde{V}$ is called the *dual cocycle* of V.

When the cocycle V is the unique bounded solution of a QSDE the dual cocycle $\widetilde{V}$ also satisfy a QSDE and the relationship between the two is given by the following

Proposition 6.4. *Suppose that the G_β^α, F_β^α are bounded operators on* h *and that (3), (4) have unique bounded solutions V, U. Then the dual cocycles $\widetilde{V}, \widetilde{U}$ are the unique bounded solutions of the QSDE*

$$d\widetilde{V}_t = \widetilde{V}_t (G_\alpha^\beta)^* \, d\Lambda_\alpha^\beta, \quad d\widetilde{U}_t = (F_\alpha^\beta)^* \widetilde{U}_t \, d\Lambda_\alpha^\beta.$$

Proof. (Sketch) Differentiating $\langle v\mathrm{e}(g), \widetilde{V}_t u\mathrm{e}(f) \rangle$ it is not hard to find the QSDE satisfied by $\widetilde{V}$ (see Fagnola [18] Prop. 5.12). $\square$

The dual cocycle allows, roughly speaking, to take the adjoint of a left (resp. right) QSDE and end up with a left (resp. right) QSDE without exchanging the operator coefficients G_β^α, F_β^α and the solution. This turns out to be a useful feature because facts on the two equations often are not symmetric.

The proof of Proposition 6.4 has been simplified by J.M. Lindsay and S.J. Wills [30] by using the so-called *semigroup representation* of solutions to quantum stochastic differential equations. This is based on a family of semigroups associated with cocycles U and V described here in Sect. 8, Proposition 8.2.

7 Regularity

Let U, (resp. V) be contractive solutions of a right (resp. left) H-P equation. In the applications (see, for instance, Barchielli [8]) of quantum stochastic calculus it is a quite natural question to ask whether we can define operator processes like $(\Lambda_\alpha^\beta(t)U_t)_{t\geq 0}$ as operator processes on h. This is the case when the vector $U_t u e(f)$ belongs to $\text{Dom}(\Lambda_\alpha^\beta(t))$ for all $t \geq 0$, $u \in$ h, $f \in \mathcal{M}$.

In this section we establish this type of properties that we call *regularity* following the terminology of classical PDE. Indeed, a function solving a PDE is often called regular if it is differentiable, i.e. it belongs to the domain of some differential operator, or it satisfies some growth condition at infinity, i.e. it belongs to the domain of a multiplication operator by $|x|^k$, or both.

We start by recalling some basic facts on domains of operators $\Lambda_\alpha^\beta(t)$ that have been introduced in Section 2 as operators on the exponential domain $\mathcal{E}$.

Let $\mathcal{F}(K)$ be the boson Fock space over a complex separable Hilbert space K. We denote by $F(K)$ the linear manifold in $\mathcal{F}(K)$ spanned by finite particle vectors $f^{\otimes n}$ with $n \geq 0$, $f \in K$. We first recall the definition of number operator (or differential second quantization) following Parthasarathy [34] Ch.II Sect.20.

Let H be a self-adjoint operator on K and let $(\mathrm{e}^{-isH})_{s\in\mathbb{R}}$ be the one-parameter unitary group on K generated by $-iH$. By second quantization

$$\Gamma(\mathrm{e}^{-isH})\mathrm{e}(f) = \mathrm{e}(\mathrm{e}^{-isH}f)$$

we find a strongly continuous one-parameter unitary group $(\Gamma(\mathrm{e}^{-isH}))_{s\in\mathbb{R}}$ on $\mathcal{F}(K)$. We denote its Stone generator by $\Lambda(H)$. This is self-adjoint by definition. Moreover, it can be shown (see [34] Prop. 20.7) that the linear manifold of exponential vectors is an essential domain for $\Lambda(H)$.

When, with the notation in Sect. 2, $K = L^2(\mathbb{R}_+; \mathrm{k})$ and $H = 1_{[0,t]} \otimes |e_\ell\rangle\langle e_\ell|$ we find that $\Lambda(H)$ is the unique self-adjoint extension of the operators $\Lambda_\ell^\ell(t)$ defined in Sect. 2.

When $H = 1_{[0,t]} \otimes 1_\mathrm{k}$, we denote by N_t the generator of the unitary group $(\Gamma(\mathrm{e}^{-isH}))_{s\in\mathbb{R}}$ and call *number operator process* the family $(N_t)_{t\geq 0}$.

The following proposition is easily checked.

Lemma 7.1. *For all $t \geq 0$ we have*

$$N_t = \sum_{\ell\geq 1} \Lambda_\ell^\ell(t)$$

the series being strongly convergent on $\text{Dom}(N_t)$.

Proof. By the definition of N_t, for all $g, f \in \mathcal{M}$, we have

$$\langle \mathrm{e}(g), N_t\mathrm{e}(f)\rangle = i\frac{d}{ds}\left\langle \mathrm{e}(g), \Gamma\left(\mathrm{e}^{-is1_{[0,t]}\otimes 1_\mathrm{k}}\right)\mathrm{e}(f)\right\rangle\bigg|_{s=0}$$

$$= i\frac{d}{ds}\mathrm{e}^{\langle g, \mathrm{e}^{-is1_{[0,t]}}f\rangle}\bigg|_{s=0}$$

$$= \langle g, 1_{[0,t]}f\rangle\mathrm{e}^{\langle g,f\rangle}.$$

Moreover, by the standard rules of quantum stochastic calculus, we have also

$$\left\langle \mathrm{e}(g), \left(\sum_{\ell\geq 1}\Lambda_\ell^\ell(t)\right)\mathrm{e}(f)\right\rangle = \sum_{\ell\geq 1}\int_0^t g_\ell(s)f^\ell(s)ds\, \mathrm{e}^{\langle g,f\rangle} = \langle g, 1_{[0,t]}f\rangle\mathrm{e}^{\langle g,f\rangle}.$$

It follows that the claimed identity holds on the exponential domain $\mathcal{E}$. Therefore it holds on $\mathrm{Dom}(N_t)$ since this is an essential domain for N_t and the operators $\Lambda_\ell^\ell(t)$ are self-adjoint. □

The domain of annihilation $A^\ell(t)$ and creation $A_\ell^\dagger(t)$ operators can be extended to $\mathrm{Dom}(N_t)$ by the following

Lemma 7.2. *For all $\xi \in \mathrm{Dom}(N_t)$ we have*

$$\sum_{\ell \geq 1} \|A^\ell(t)\xi\|^2 = \|N_t^{1/2}\xi\|^2.$$

Proof. By the above remarks on the domains of $A_\ell(t)$ and N_t it suffices to check the above inequality for all vector ξ of the form

$$\sum_{k=1}^n z_k e(f_k)$$

with $n \in \mathbb{N}^*$, $z_1, \ldots, z_n \in \mathbb{C}$, $f_1, \ldots, f_n \in \mathcal{M}$ (there is an abuse of notation here!). In this case we have

$$\sum_{\ell \geq 1} \|A^\ell(t)\xi\|^2 = \sum_{\ell \geq 1} \sum_{j,k=1}^n \langle z_j f_j^\ell, 1_{[0,t]}\rangle \langle 1_{[0,t]}, z_k f_k^\ell \rangle \langle e(f_j), e(f_k)\rangle$$

$$= \sum_{j,k=1}^n \langle e(f_j), e(f_k)\rangle \sum_{\ell \geq 1} \langle z_j f_j, 1_{[0,t]} e_\ell \rangle \langle 1_{[0,t]} e_\ell, z_k f_k\rangle$$

$$= \sum_{j,k=1}^n \langle e(f_j), e(f_k)\rangle \int_0^t \langle z_j f_j(s), z_k f_k(s)\rangle_k ds.$$

Moreover we have also

$$\|N_t^{1/2}\xi\|^2 = \sum_{j,k=1}^n \overline{z_j} z_k \sum_{\ell \geq 1} \langle e(f_j), \Lambda_\ell^\ell(t) e(f_k)\rangle$$

$$= \sum_{j,k=1}^n \overline{z_j} z_k \sum_{\ell \geq 1} \int_0^t \overline{f_j^\ell}(s) f_k^\ell(s) ds \, \langle e(f_j), e(f_k)\rangle$$

$$= \sum_{j,k=1}^n \langle e(f_j), e(f_k)\rangle \int_0^t \langle z_j f_j(s), z_k f_k(s)\rangle_k ds.$$

This proves the Lemma. □

We now find the quantum stochastic differential equation satisfied by the powers of the number operator process.

Lemma 7.3. *For all $n \geq 1$ we have*

$$dN_t^n = \sum_{\ell \geq 1} ((N_t + 1)^n - N_t^n) \, d\Lambda_\ell^\ell(t).$$

Proof. The above formula is clearly true for $n = 1$. Suppose it has been established for an integer n, then, by the Ito formula 2, for all $f, g \in \mathcal{M}$, we have

$$\langle e(g), N_t^{n+1} e(f) \rangle = \langle N_t e(g), N_t^n e(f) \rangle$$

$$= \int_0^t \Big\{ \langle e(g), N_s^n e(f) \rangle + \langle N_s e(g), ((N_s+1)^n - N_s^n) e(f) \rangle$$

$$+ \langle e(g), ((N_s+1)^n - N_s^n) e(f) \rangle \Big\} g_\ell(s) f^\ell(s) ds$$

$$= \int_0^t \Big\{ \langle e(g), (N_s^n + (N_s+1)((N_s+1)^n - N_s^n)) e(f) \rangle \Big\} g_\ell(s) f^\ell(s) ds.$$

This proves the claimed identity. $\square$

Fix a positive integer n. For all $\varepsilon > 0$ let

$$Z_t^\varepsilon = (N_t + 1)^n e^{-\varepsilon(N_t+1)}. \tag{12}$$

Clearly, the operators Z_t^ε are bounded (it can be easily shown that $\|Z_t^\varepsilon\| \le n^n (\varepsilon e)^{-n}$).

Proposition 7.4. *The adapted process* $(Z_t^\varepsilon)_{t \ge 0}$ *satisfies*

$$dZ_t^\varepsilon = \sum_{\ell \ge 1} Y_t^\varepsilon Z_t^\varepsilon d\Lambda_\ell^\ell(t), \quad Z_0^\varepsilon = e^{-\varepsilon} \mathbf{1}.$$

where $(Y_t^\varepsilon)_{t \ge 0}$ *is the bounded contraction process*

$$Y_t^\varepsilon = \big((N_t + 2)(N_t + 1)^{-1}\big)^n e^{-\varepsilon} - 1$$

satisfying

$$-1 \le Y_t^\varepsilon \le (2^n - 1) \mathbf{1}.$$

Proof. The function

$$\varphi : \mathbb{R} \to \mathbb{R}, \quad \varphi(x) = (x + 1)^n e^{-\varepsilon(x+1)}$$

is analytic. Therefore it suffices to apply Lemma 7.3. $\square$

Proposition 7.5. *Let* U *be a contractive solution of the right equation (4). Suppose that the operators* F_β^α *are bounded. For all* $u \in \mathsf{h}$, $f \in \mathcal{M}$ *and* $t \ge 0$ *we have*

$$\|Z_t^\varepsilon U_t u e(f)\|^2 \le e^{2^{2n} \|f\|^2} \|u\|^2. \tag{13}$$

Proof. Computing the stochastic differential of $Z_t^\varepsilon U_t$ by the Ito formula we have

$$dZ_t^\varepsilon U_t = F_\beta^\alpha Z_t^\varepsilon U_t d\Lambda_\alpha^\beta + Y_t^\varepsilon Z_t^\varepsilon U_t d\Lambda_\ell^\ell + F_\beta^\alpha Y^\varepsilon Z_t^\varepsilon U_t d\Lambda_\ell^\ell d\Lambda_\alpha^\beta$$

$$= F_\beta^\alpha Z_t^\varepsilon U_t d\Lambda_\alpha^\beta + \hat{\delta}_\beta^\alpha Y_t^\varepsilon Z_t^\varepsilon U_t d\Lambda_\alpha^\beta + \hat{\delta}_\alpha^\ell F_\beta^\alpha Y^\varepsilon Z_t^\varepsilon U_t d\Lambda_\ell^\beta$$

$$= \big(F_\beta^\alpha + (\hat{\delta}_\beta^\alpha + \hat{\delta}_\alpha^\ell F_\beta^\ell) Y_t^\varepsilon \big) Z_t^\varepsilon U_t d\Lambda_\alpha^\beta.$$

Therefore, computing again by the Ito formula, we find

$$\langle Z_t^\varepsilon U_t u e(f), Z_t^\varepsilon U_t u e(f) \rangle = e^{-2\varepsilon \|f\|^2} \|u\|^2$$

$$+ \int_0^t \langle Z_s^\varepsilon U_s u e(f), M_\beta^\alpha(s) Z_s^\varepsilon U_s u e(f) \rangle f_\alpha(s) f^\beta(s) ds$$

where M_β^α are the operator processes on h given by

$$M_\beta^\alpha(s) = F_\beta^\alpha + F_\alpha^{\beta^*} + F_\alpha^{\ell*}F_\beta^\ell + \left(\hat\delta_\alpha^\ell + F_\alpha^\ell\right)^* \left(\hat\delta_\beta^\ell + F_\beta^\ell\right)\left((Y_s^\varepsilon)^2 + 2Y_s^\varepsilon\right).$$

The operators F_β^α determine an operator $F = [F_\beta^\alpha]$ on $\otimes_{\alpha \geq 0}h$ (isometrically isomorphic to $h \otimes (\mathbb{C} \oplus k)$). In the same way, for each $s \geq 0$, the $M_\beta^\alpha(s)$ determine an operator $M(s)$ on $h \otimes (\mathbb{C} \oplus k) \otimes \mathcal{F}$. Notice that, by Proposition 7.4 we have

$$-\mathbb{1} \leq \left((Y_s^\varepsilon)^2 + 2Y_s^\varepsilon\right) \leq (4^n - 1)\,\mathbb{1}.$$

Therefore we obtain the inequality

$$M(s) \leq (F + F^* + F^*\Delta(\mathbb{1}_h)F) \otimes \mathbb{1}_\mathcal{F}$$

$$+ (4^n - 1)\left(\Delta(\mathbb{1}_h) + F\right)^*\Delta(\mathbb{1}_h)(\Delta(\mathbb{1}_h) + F) \otimes \mathbb{1}_\mathcal{F}$$

where $\Delta(\mathbb{1}_h)$ denotes, as in Sect. 4, the projection $\mathrm{diag}(0, \mathbb{1}_h, \dots)$ on $h \otimes k$. The contractivity condition for U (Theorem 4.1) yields

$$\theta_F(\mathbb{1}) = (F + F^* + F^*\Delta(\mathbb{1}_h)F) \leq 0.$$

Moreover, it imples that

$$(\Delta(\mathbb{1}_h) + F)^*\Delta(\mathbb{1}_h)(\Delta(\mathbb{1}_h) + F) \leq (\Delta(\mathbb{1}_h) + F)^*(\Delta(\mathbb{1}_h) + F) \leq \Delta(\mathbb{1}_h).$$

Therefore we have

$$M(s) \leq (4^n - 1)\,\Delta(\mathbb{1}_h) \otimes \mathbb{1}_\mathcal{F}.$$

Turning back to the computation with the Ito formula we find the inequality

$$\|Z_t^\varepsilon U_t ue(f)\|^2 \leq e^{-2\varepsilon\|f\|^2}\|u\|^2 + (4^n - 1)\int_0^t \|Z_s^\varepsilon U_s ue(f)\|^2\,|f(s)|^2 ds.$$

The conclusion follows then from the Gronwall's Lemma. $\quad\square$

The following Lemma is an immediate consequence of the spectral theorem

Lemma 7.6. *Let X be a positive self-adjoint operator on a Hilbert space K and let $X^{(\varepsilon)}$ be the bounded operator $(X + \mathbb{1})e^{-\varepsilon(X+\mathbb{1})}$ $(\varepsilon > 0)$. A vector $u \in K$ belongs to the domain of X if and only if*

$$\sup_{\varepsilon > 0}\left\|X^{(\varepsilon)}u\right\|^2 < \infty.$$

Theorem 7.7. *Let U be a contractive solution of the right equation (4). Suppose that the operators F_β^α are bounded. For all $n \geq 1$, $t, s \geq 0$, $u \in h$ and $f \in \mathcal{M}$, the vector $U_t ue(f)$ belongs to the domain of the operators N_s^n, $(\Lambda_\alpha^\beta(s))^n$ and*

$$A^\ell(t)U_t = \int_0^t \left(\hat\delta_\beta^\ell + F_\beta^\ell\right)U_s d\Lambda_0^\beta(s) + \int_0^t F_\beta^\alpha A^\ell(s)U_s d\Lambda_\alpha^\beta(s) \tag{14}$$

$$A_\ell^\dagger(t)U_t = \int_0^t U_s dA_\ell^\dagger(s) + \int_0^t F_\beta^\alpha A_\ell^\dagger(s)U_s d\Lambda_\alpha^\beta(s) \tag{15}$$

Proof. Let $u, v \in h$ and $f, g \in \mathcal{M}$. By the Ito formula we have

$$\left\langle A_\ell^\dagger(t)ve(g), U_t ue(f)\right\rangle = \int_0^t \left\langle A_\ell^\dagger(s)ve(g), F_\beta^\alpha U_s ue(f)\right\rangle g_\alpha(s) f^\beta(s) ds$$

$$+ \int_0^t \left\langle ve(g), U_s ue(f)\right\rangle f^\ell(s) ds$$

$$+ \int_0^t \left\langle ve(g), F_\beta^\ell U_s ue(f)\right\rangle f^\beta(s) ds.$$

Now, by Proposition 7.5 and Lemma 7.6, $U_s ue(f)$ belongs to the domain of N_t. Therefore, by Lemma 7.2, $U_s ue(f)$ belongs to the domain of $A_\ell(s)$. Therefore, since $(F_\beta^\alpha)^*$ commutes with $A_\ell^\dagger(s)$, we find

$$\left\langle A_\ell^\dagger(s)ve(g), F_\beta^\alpha U_s ue(f)\right\rangle = \left\langle A_\ell^\dagger(s)(F_\beta^\alpha)^* ve(g), U_s ue(f)\right\rangle$$

$$= \left\langle (F_\beta^\alpha)^* ve(g), A_\ell(s) U_s ue(f)\right\rangle$$

$$= \left\langle ve(g), F_\beta^\alpha A_\ell(s) U_s ue(f)\right\rangle.$$

Therefore we have

$$\langle ve(g), A_\ell(t) U_t ue(f)\rangle = \int_0^t \left\langle ve(g), F_\beta^\alpha A_\ell(s) U_s ue(f)\right\rangle g_\alpha(s) f^\beta(s) ds$$

$$+ \int_0^t \left\langle ve(g), (\hat\delta_\beta^\ell + F_\beta^\ell) U_s ue(f)\right\rangle f^\beta(s) ds.$$

This proves (14). The proof of (15) is similar; we omit it. □
    Similar formulae can be proved for $\Lambda_\ell^\ell(t)$.

8 The left equation: unbounded G_β^α

This section is aimed at illustrating the theory for the left QSDE (3) with unbounded G_β^α. We shall show that the algebraic conditions $\theta_G(\mathbf{1}) = 0$ and $\theta_{G^\dagger}(\mathbf{1}) = 0$ are no longer sufficient for V to be a unitary cocycle. Indeed, thinking of the case when the only nonzero G_β^α is G_0^0, it is clear that these algebraic conditions mean (see Remark after Theorem 4.4) that iG_0^0 is symmetric. However, for V to be unitray, iG_0^0 must be self-adjoint and new analytical conditions appear.
    A natural necessary condition on the G_β^α for V to be a contraction process solving (3) is easily deduced by the differentiation argument of Proposition 4.1.

Proposition 8.1. *Let $G = [G_\beta^\alpha]$ be a matrix of operators on $\mathsf h$, and suppose that there exists a contraction process V that is a strong solution to* (L) *on some dense subspace $D \subset \mathsf h$ for this G. Then $\theta_G(\mathbf{1}) \leq 0$ as a form on D. If V is an isometry process then $\theta_G(\mathbf{1}) = 0$ on D.*

We refer to Fagnola [16], Mohari and Parthasarathy [32] for the original proofs or Fagnola and Wills [22] for a proof with the same notation we use here. An alternative proof via the characterisation of the generators of completely positive contraction flows is given in Lindsay and Parthasarathy [29] and Lindsay and Wills [30].

Remarks. (a) If G is a matrix of operators on $\mathsf h$ such with $\mathrm{Dom}(G_\beta^\alpha) \supseteq D$ such that $\theta_G(\mathbf{1}) \leq 0$ then $[\delta_m^\ell 1 + G_m^\ell]_{\ell,m=1}^d$ defines a contraction on $\oplus_{\ell=1}^d \mathsf h$, and so in particular each G_m^ℓ has a unique continuous extension to an element of $\mathcal{B}(\mathsf{h})$. If $\theta_G(\mathbf{1}) = 0$ then $[\delta_m^\ell \mathbf{1} + G_m^\ell]_{\ell,m=1}^d$ is an isometry.

(b) The inequality $\theta_G(\mathbf{1}) \leq 0$ yields (put $u^\ell = 0$ for $\ell = 1, \ldots, d$ and $u^0 = u \in D$ in (7))

$$\sum_{\ell=1}^d \|G_0^\ell u\|^2 \leq -2\Re\langle u, G_0^0 u\rangle.$$

Therefore, by the Schwarz inequality, for all $z \in \mathbb{C}^d$ and $u \in D$, we have

$$\left\|\sum_{\ell=1}^d z_\ell G_0^\ell u\right\| \leq \left(\sum_{\ell=1}^d |z_\ell|\right)^{1/2} \left(\sum_{\ell=1}^d \|G_0^\ell u\|^2\right)^{1/2}$$

$$\leq 2|z| \cdot \|u\|^{1/2}\|G_0^0 u\|^{1/2} \leq \varepsilon\|G_0^0 u\| + \varepsilon^{-1}|z|^2\|u\|$$

for all $\varepsilon \in]0,1[$. Thus $\sum_\ell z_\ell G_0^\ell$ is relatively bounded with respect to G_0^0 with relative bound less than 1.

Note that a contraction process V solving (3) is strongly continuous on $\mathcal{H}$ (i.e. the maps $t \to V_t\xi$ are continuous for $\xi \in \mathcal{H}$). Moreover conditions on the G_β^α for the existence of a unique contraction solution V to (3) imply that V is a (strongly continuous) left cocycle (see Proposition 6.3). Therefore we can find a family of strongly continuous semigroups on h associated with V in a natural way.

For all $f, g \in \mathcal{E}$ and all $t \geq 0$ define bounded operators $P_t^{g,f}$ on h by

$$\langle v, P_t^{g,f} u\rangle = e^{-\langle g,f\rangle}\langle ve(g), V_t ue(f)\rangle \tag{16}$$

for all $v, u \in \mathsf{h}$. Indeed, since

$$\left|e^{-\langle g,f\rangle}\langle ve(g), V_t ue(f)\rangle\right| \leq e^{-\Re\langle g, f1_{[0,t]}\rangle + (\|g1_{[0,t]}\|^2 + \|f1_{[0,t]}\|^2)/2}\|v\| \cdot \|u\|$$

$$= e^{\|(g-f)1_{[0,t]}\|^2/2}\|v\| \cdot \|u\|,$$

it follows that there exists bounded operators $P_t^{g,f}$ on h with $\|P_t^{g,f}\| \leq \exp(\|(g - f)1_{[0,t]}\|^2/2)$ such that (16) holds for all $v, u \in \mathsf{h}$. Moreover, if the cocycle V is a solution of (3) on $D \odot \mathcal{E}$, then

$$\langle v, P_t^{g,f} u\rangle = \langle v, u\rangle + \int_0^t \langle v, P_s^{g,f}\{G_\beta^\alpha g_\alpha(s) f^\beta(s)\}u\rangle ds.$$

Therefore, when the operators G_β^α are bounded, and the functions are constant in a right neighbourhood $[0, R]$ of 0, this means that the operators $P_t^{g,f}$ for $0 \leq t \leq R$) are the operators at time t of the uniformly continuous semigroup generated by $G_\beta^\alpha g_\alpha(0) f^\beta(0)$.

This fact can be generalised to strongly continuous contractive left cocycles. Indeed, we have the following

Proposition 8.2. *Let V be a left cocycle, let $f, g \in \mathcal{E}$ constant on an interval $[0, r]$ and let $P_t^{g,f}$ be the bounded operators on h defined by (16). For all $t, s > 0$ such that $t + s \leq r$ we have*

$$P_{t+s}^{g,f} = P_t^{g,f} P_s^{g,f}.$$

Moreover, if the cocycle V is strongly continuous on $\mathcal{H}$ then the map $t \to P_t^{g,f}$ is strongly continuous on h.

We refer to Fagnola [24] for the proof.

Remark. (a) It is not hard to show that V is strongly continuous on $\mathcal{H}$ if and only if the maps $t \to P_t^{g,f}$ for g, f constant in a neighbourhood of 0 are strongly continuous. Moreover, by a well-known property of semigroups (see e.g. Pazy [36]), strong continuity is equivalent to weak continuity.

(b) Semigroups similar to the $P^{g,f}$ were introduced in Fagnola and Sinha [21] to study QSDE for quantum flows and extensively used by Lindsay and Parthasarathy [29], Lindsay and Wills [30] and Accardi and Kozyrev [4] in the study of quantum flows and Evans-Hudson QSDE.

(c) The set $\mathcal{M}$ of test functions for exponential vectors $e(f)$ in $\mathcal{E}$ could be replaced by another set $\mathcal{M}_S$ such that $\mathrm{lin}\{e(f) : f \in \mathcal{M}_S\}$ is dense in $\mathcal{F}$ and $e(f1_{[0,t]}) \in \mathcal{M}_S$ for all $t \geq 0$ whenever $e(f) \in \mathcal{M}_S$. A set with the first property is called *totalizing*. A convenient choice for the totalizing set $\mathcal{M}_S$ is the set of step (i.e. constant on the intervals of a partition of $\mathbb{R}_+$) functions f with values in $\{0, 1\}^d$. M. Skeide [38] has shown that this set is totalizing. Exploiting this fact one could show that the family of semigroups $(P_t^{g,f}; t \geq 0)$ with $g, f \in \mathcal{M}_S$ determine a unique cocycle V.

From the above discussion and the Remark after Proposition 8.1 it is clear that the following is a natural hypothesis on the G_β^α.

Hypothesis HGC

(i) The operator G_0^0 is the infinitesimal generator of a strongly continuous contraction semigroup on h and D is a core for G_0^0 contained in $\mathrm{Dom}(G_\beta^\alpha)$ for all α, β.

(ii) For all $\ell, m \in \{1, \ldots, d\}$ the operator G_m^ℓ is bounded,

(iii) We have $\theta_G(\mathbf{1}) \leq 0$ on D i.e.

$$\langle u^\alpha, G_\beta^\alpha u^\beta \rangle + \langle G_\alpha^\beta u^\alpha, u^\beta \rangle + \langle G_\alpha^\ell u^\alpha, G_\beta^\ell u^\beta \rangle \leq 0 \qquad (17)$$

for all $u^0, \ldots, u^d \in D$.

Remarks. (a) The hypothesis **HGC** and the Remark (b) after Proposition 8.1 imply, by a well-known perturbation result in semigroup theory (see Dunford and Schwartz [14], Th. 19 p. 631) that also $G_0^0 + \sum_\ell z_\ell G_0^\ell$ generates a strongly continuous semigroup on h with D as a core. The same conclusion holds for sums of the above perturbation and scalar multiples of bounded operators as the G_m^ℓ ($\ell, m = 1, \ldots, d$).

(b) It is easy to show that the semigroup $P^{0,0}$ corresponding to the dual cocycle $\tilde{V}$ is the adjoint of $P^{0,0}$. Thus the adjoint operator $(G_0^0)^*$ for the dual cocycle $\tilde{V}$ plays the role of G_0^0 for V and there is a natural dual to the above perturbation result. There is no reason, however, for G_0^0 and $(G_0^0)^*$ to have a common essential domain.

We now show that, under the hypothesis **HGC**, there exists a contraction process solving (3). The idea of the proof, as in the Hille-Yosida theorem, is to take bounded approximations of the unbounded operators G_β^α by means of the resolvents $R(n; G_0^0) = (n\mathbf{1} - G_0^0)^{-1}$. The well-known properties of resolvent operators for all $u \in h$, $v \in \mathrm{Dom}(G_0^0)$ yield

$$\lim_{n \to \infty} nR(n; G_0^0)u = u, \qquad \lim_{n \to \infty} nG_0^0 R(n; G_0^0)v = G_0^0 v$$

in the norm topology on h.

We first prove a preliminary Lemma (see Fagnola [16] Proposition 3.3). Recall that $(e_1, \ldots, e_d)$ denotes the canonical orthonormal basis of $\mathbb{C}^d$.

Lemma 8.3. *Let* $G = [G_\beta^\alpha]$ *be a matrix of operators on* h *satisfying the hypothesis* **HGC**. *For each* $n \geq 1$ *let* I_n, G_n *be the operators on* $\oplus^{(d+1)} h$ *with domain* $\oplus^{(d+1)} D$ *defined by*

$$I_n = nR(n; G_0^0) \oplus 1_{(\oplus^{(d)} h)}, \qquad G(n) = I_n^* G I_n \tag{18}$$

where $1_{(\oplus^{(d)} h)}$ *is the identity operator on* $\oplus^{(d)} h$. *Then the matrix* $G(n)$ *of operators on* h *has an extension which is a bounded operator on* $\oplus^{(d+1)} h$ *and*

(i) $\|G(n)\| \leq 2 \left(n + 3\sqrt{n} + 1 \right)$,
(ii) $\theta_{G(n)}(1) \leq 0$,
(iii) *for all* $\xi \in \oplus^{(d+1)} D$, *we have*

$$\lim_{n \to \infty} G(n)\xi = G\xi. \tag{19}$$

The proof can be found in Ref. [16].

We now state the existence theorem

Theorem 8.4. *Let* G *be a matrix of operators on* h *satisfying the hypothesis* **HGC** *and let* $(G(n); n \geq 1)$ *be a sequence matrices of bounded operators on* h *such that* $\theta_{G(n)}(1) \leq 0$ *and, for all* $\xi \in \oplus^{(d+1)} D$, *(19) holds. For all integer* n *let* $V(n) = (V_t(n); t \geq 0)$ *be the unique contraction process solving (3) on* $h \odot \mathcal{E}$ *for the operator matrix* $G(n)$. *There exist a weakly convergent subsequence* $(n_k)_{k \geq 1}$ *such that the contraction process* $(V_t; t \geq 0)$ *defined by*

$$V_t = w - \lim_{k \to \infty} V_t(n_k) \tag{20}$$

solves the QSDE (3) on $D \odot \mathcal{E}$.

Proof. (Sketch) The sequence of contraction processes $(V_t(n); t \geq 0)_{n \geq 1}$ is equicontinuous in t on $D \odot \mathcal{E}$. Indeed, for all $u \in D$, $f \in \mathcal{E}$, $t > s > 0$, by Proposition 2.2 $\|(V_t(n) - V_s(n))ue(f)\|^2$ is not bigger than

$$\left\| \int_s^t V_r(n) G(n)_\beta^\alpha d\Lambda_\alpha^\beta(r) ue(f) \right\|^2$$

$$\leq c_s^t(f, d) \sum_{\alpha, \beta} \int_s^t \left\| V_r(n) G(n)_\beta^\alpha ue(f) \right\|^2 |f_\beta(r)|^2 dr$$

$$\leq c_s^t(f, d) e^{\|f\|^2} \sum_{\alpha, \beta} \left\| G(n)_\beta^\alpha u \right\|^2 \int_s^t |f_\beta(r)|^2 dr.$$

The sequences $(G(n)_\beta^\alpha u)_{n \geq 1}$ in h are bounded by (19) therefore the functions

$$t \to \langle y, V_t(n) ue(f) \rangle,$$

for $u \in D$, $f \in \mathcal{E}$ are equicontinuous and equibounded.

Hence, by the Ascoli-Arzelà theorem and the standard diagonalisation argument, there exists a subsequence converging on a countable subset of $D \odot \mathcal{E}$ uniformly for t in any bounded interval. The proof can be easily completed because $\mathcal{H}$ is separable (see Fagnola [16] Prop. 3.4 for the details). $\square$

Under the hypothesis **HGC** the solution is also unique (see Mohari and Parthasarathy [32] under more restrictive hypotheses).

Theorem 8.5. *Let G be a matrix of operators on* h *satisfying hypothesis* **HGC**. *There exists a unique solution of the QSDE (3) on* $D \odot \mathcal{E}$.

Proof. A solution V of on $D \odot \mathcal{E}$ the QSDE (3) can be constructed by taking the $G(n)$ as in Lemma 8.3 and applying Theorem 8.4.

We now prove uniqueness. Let $(X_t; t \geq 0)$ be the difference of two contractions solving (3) on $D \odot \mathcal{E}$. For all $u, v \in D$, $g, f \in \mathcal{M}$ we have then

$$\langle ve(g), X_t ue(f) \rangle = \int_0^t \langle ve(g), X_t \{G_\beta^\alpha g_\alpha(s) f^\beta(s)\} ue(f) \rangle ds.$$

Since both the sides of this identities are analytic functions in g and f, for all integers $n, m \geq 0$, $\langle vg^{\otimes m}, X_t uf^{\otimes n} \rangle$ is equal to

$$\int_0^t \left(\left\langle vg^{\otimes m}, X_s G_0^0 uf^{\otimes n} \right\rangle \right. $$
$$+ \left\langle vg^{\otimes m}, X_s G_k^0 uf^{\otimes(n-1)} \right\rangle f^k(s) + \left\langle vg^{\otimes(m-1)}, X_s G_0^\ell uf^{\otimes n} \right\rangle g_\ell(s)$$
$$\left. + \left\langle vg^{\otimes(m-1)}, X_s G_k^\ell uf^{\otimes(n-1)} \right\rangle g_\ell(s) f^k(s) \right) ds \qquad (21)$$

for all $u, v \in D$ and all pairs m, n of integer numbers with the convention $g^{\otimes n} = f^{\otimes n} = 0$ if $n < 0$ and $g^{\otimes 0} = f^{\otimes 0} = e(0)$.

We now prove that the left-hand side vanishes by induction on $p = n + m$. Let $p = 0$. For every $\lambda > 0$ the bilinear form on h

$$(v, u) \to \int_0^\infty \exp(-\lambda t) \langle ve(0), X_t ue(0) \rangle \, dt$$

is bounded because $\|X_s\| \leq 2$ and

$$\left| \int_0^\infty \exp(-\lambda t) \langle ve(0), X_t ue(0) \rangle \, dt \right| \leq 2\lambda^{-1} \|v\| \cdot \|u\|.$$

Hence there exists a bounded operator R_λ on h such that

$$\langle v, R_\lambda u \rangle = \int_0^\infty \exp(-\lambda t) \langle ve(0), X_t ue(0) \rangle \, dt.$$

The identity (21) for $n = m = 0$, $u, v \in D$, yields

$$\lambda \langle v, R_\lambda u \rangle = \lambda \int_0^\infty \exp(-\lambda t) dt \int_0^t \left\langle ve(0), X_s G_0^0 ue(0) \right\rangle ds$$
$$= \lambda \int_0^\infty \left\langle ve(0), X_s G_0^0 ue(0) \right\rangle ds \int_s^\infty \exp(-\lambda t) dt$$
$$= \left\langle v, R_\lambda G_0^0 u \right\rangle$$

We have then $R_\lambda(\lambda \mathbf{1} - G_0^0)u = 0$ for every $u \in D$. Since D is a core for G_0^0, the linear manifold $(\lambda \mathbf{1} - G_0^0)(D)$ is dense in h. Thus R_λ vanishes. Therefore $\langle ve(0), X(t)ue(0) \rangle$ also vanishes for every $t \geq 0$.

This proves our statement for $p = 0$. Suppose that it has been established for a positive p. Then, for all m, n with $m + n = p + 1$, the induction hypothesis allows us to write (21) as

$$\langle vg^{\otimes m}, X_t u f^{\otimes n} \rangle = \int_0^t \langle vg^{\otimes m}, X_s G_0^0 u f^{\otimes n} \rangle \, ds.$$

The same argument of the previous case $n = m = 0$ then shows that $\langle vg^{\otimes m}, X_t u f^{\otimes n} \rangle = 0$ for all $t \geq 0$.

This completes the proof. □

It is worth noticing here that, contrary to the case in which G is bounded and $\theta_G(\mathbf{1}) = 0$ on D, the unique solution of the QSDE (3) needs not to be an isometry process even if $\theta_G(\mathbf{1}) = 0$ (see Fagnola [15] for a simple example related to birth-and-death processes). The additional condition needed will be discussed in the Section 10.

The uniqueness result allows us to prove as Proposition 6.3 the following

Corollary 8.6. *Let G be a matrix of operators on* h *satisfying hypothesis* **HGC**. *The unique solution V of the QSDE (3) on $D \odot \mathcal{E}$ is a left cocycle.*

9 Dilation of quantum Markov semigroups

In this section we (try to) discuss briefly the application of QSDE to a dilation problem. Indeed, this was one of the most important original motivations for studying QSDE.

A *quantum dynamical semigroup* on the von Neumann algebra $\mathcal{B}(h)$ of all bounded operators on a Hilbert space h is a w^*-continuous semigroup $\mathcal{T} = (\mathcal{T}_t; t \geq 0)$ of completely positive, normal maps $\mathcal{T}_t$ on $\mathcal{B}(h)$ such that $\mathcal{T}_t(\mathbf{1}) \leq \mathbf{1}$. A quantum dynamical semigroup is called *Markov* or *identity preserving* if $\mathcal{T}_t(\mathbf{1}) = \mathbf{1}$.

We recall that, since $\mathcal{B}(h)$ is the dual space of the Banach space of all trace class operators on h, the w^*-continuity of $\mathcal{T}$ simply means that the maps $t \to \text{trace}(\rho \mathcal{T}_t(x))$ are continuous for each trace class operator ρ on h and all $x \in \mathcal{B}(h)$. Moreover "normal" means that, for every increasing net $(x_\alpha)_\alpha$ in $\mathcal{B}(h)$ with least upper bound $x \in \mathcal{B}(h)$, the least upper bound of $\mathcal{T}_t(x_\alpha)$ is $\mathcal{T}_t(x)$.

Complete positivity is a stronger property than positivity. It means that, for all integers $n \geq 1$, given a matrix $X = [x_{\ell,m}]$ of operators on $\oplus^n h$ such that X is positive, then the matrix $[\mathcal{T}_t(x_{\ell,m})]$ of operators on $\oplus^n h$ is also positive.

Quantum Markov semigroups are the natural mathematical model in the study of irreversible evolutions of quantum open systems (see Davies [12]). Irreversibility, from the mathematical point of view, means that the maps $\mathcal{T}_t$ describing the evolution of the system with state space h are not automorphisms of $\mathcal{B}(h)$.

A natural question arises: is it possible to realise a quantum Markov semigroup as the "projection" of another reversible evolution of a bigger system? More precisely: does there exist a bigger space $\mathcal{K}$, a projection $E : \mathcal{K} \to h$ and a family $(k_t; t \geq 0)$ of automorphisms of $\mathcal{B}(\mathcal{K})$ such that

$$\mathcal{T}_t(x) = E k_t(x) E^*$$

for all $x \in \mathcal{B}(h)$?

This is (a formulation of) the *dilation* problem (see Bhat and Skeide [10] for a detailed discussion and recent results).

The Hilbert space $\mathcal{K}$ is usually taken as the tensor product of the system space h with a noise space (or heat bath in the physical terminology) given as a Fock space. Accardi, Lu and Volovich [5] gave physical reasons for this choice by the theory of the "stochastic limit".

Quantum Markov semigroups are also a natural generalisation of classical Markov semigroups on the L^∞ space of some measurable space $(E, \mathcal{E})$ with a σ-finite measure. Indeed, the above definition, can be given in an arbitrary von Neumann algebra. However we choose the simplest non-commutative framework by taking $\mathcal{B}(h)$ having in mind the applications to quantum open systems.

The following result, a quantum analogue of the classical Feynman-Kac formula, was first proved by Accardi [1]. It can be proved essentially by the same arguments of the proof of Proposition 8.2 (see e.g. Fagnola [18] Sect 2.3).

Theorem 9.1. *Let V be a strongly continuous (left or right) unitary cocycle on $\mathcal{H}$. The maps $\mathcal{T}_t, \widetilde{\mathcal{T}}_t$ on $\mathcal{B}(h)$ defined by*

$$\langle v, \mathcal{T}_t(x)u \rangle = \langle V_t ve(0), (x \otimes \mathbf{1}_{\mathcal{F}}) V_t ue(0) \rangle,$$

$$\langle v, \widetilde{\mathcal{T}}_t(x)u \rangle = \langle V_t^* ve(0), (x \otimes \mathbf{1}_{\mathcal{F}}) V_t^* ue(0) \rangle$$

($\mathbf{1}_{\mathcal{F}}$ denotes the identity on $\mathcal{F}$) for $u, v \in h$ are quantum Markov semigroups on $\mathcal{B}(h)$.

The unitary cocycle V then solves the dilation problem for $\mathcal{T}$ (resp. $\widetilde{\mathcal{T}}$) with Hilbert space $\mathcal{K} = \mathcal{H} = h \otimes \mathcal{F}$, projection $Eue(f) = \exp(\|f\|^2/2)ue(0)$ and automorphisms $k_t(x) = V_t^*(x \otimes \mathbf{1}_{\mathcal{F}})V_t$ (resp. $\tilde{k}_t(x) = V_t(x \otimes \mathbf{1}_{\mathcal{F}})V_t^*$).

When the quantum Markov semigroup is given through its generator the construction of its dilation via a unitary cocycle solving a QSDE is straightforward. We sketch the idea in the simplest case of a norm-continuous, i.e. such that

$$\lim_{t \to 0} \sup_{x \in \mathcal{B}(h)\, \|x\| \leq 1} \|\mathcal{T}_t(x) - x\| = 0,$$

quantum Markov semigroup. In this case, by Lindblad's [28] theorem, the infinitesimal generator $\mathcal{L}$ can be represented in the form

$$\mathcal{L}(x) = K^* x + \sum_{\ell \geq 1} L_\ell^* x L_\ell + x K \qquad (22)$$

where $L_\ell, K \in \mathcal{B}(h)$ and the series $\sum_{\ell \geq 1} L_\ell^* L_\ell$ is strongly convergent (i.e. $\sum_{\ell \geq 1} \|L_\ell v\|^2$ converges for each $v \in h$) and $\mathcal{L}(\mathbf{1}) = 0$. This representation of the infinitesimal generator follows from complete positivity and normality.

Suppose, for simplicity, that $L_\ell = 0$ for $\ell > d$ and let

$$G_0^0 = K, \quad G_0^\ell = L_\ell, \quad G_\ell^0 = -(L_\ell)^*,$$

and $G_m^\ell = 0$ for $\ell, m \in \{1, \ldots, d\}$. It is not hard to check that $\theta_G(\mathbf{1}) = 0 = \theta_{G^\dagger}(\mathbf{1}) = 0$ and there exists a unique unitary solution V of (3). This is a left cocycle dilating the quantum Markov semigroup $\mathcal{T}$ as we outlined in the above discussion.

It is worth noticing here that the choice of the matrix of operators G is not unique. Indeed, we could take G_0^α as above for $\alpha \in \{0, 1, \ldots, d\}$, take operators G_m^ℓ on h such that the matrix of operators $[\, \delta_m^\ell + G_m^\ell \,]$ on $\oplus^{(d)}$h is unitary and define $G_m^0 = -(G_0^m)^* - (G_0^\ell)^* G_m^\ell$.

Remark. The action of a quantum Markov semigroup on a commutative subalgebra (an L^∞ space) of $\mathcal{B}(h)$ may coincide with the action of a classical Markov semigroup. In this case one can use the tools of classical stochastic analysis to study the behaviour of the classical Markov semigroup which often gives valuable hints on the behaviour of the original quantum Markov semigroup (see Parthasarathy and Sinha [35] for jump processes, Fagnola [17], Fagnola and Monte [19] for diffusion processes).

The relationship between the quantum Markov semigroup $\mathcal{T}$ and the cocycle V is established through the generator. If $\mathcal{T}$ is not norm-continuous its generator is unbounded, however, in many interesting cases it can be represented in a generalised Lindablad's form (see Davies [13]).

This happens for the class of w^*-continuous quantum Markov semigroups on $\mathcal{B}(\mathsf{h})$ whose generator is associated with quadratic forms $\mathcal{L}(x)$ ($x \in \mathcal{B}(\mathsf{h})$)

$$\mathcal{L}(x)[v, u] = \langle Kv, xu \rangle + \sum_{\ell=1}^{\infty} \langle L_\ell v, x L_\ell u \rangle + \langle v, xKu \rangle$$

where the operators K, L_ℓ satisfy the following hypotheses:

Hypothesis HQDS

(i) the operator K is the infinitesimal generator of a strongly continuous contraction semigroup $(P_t)_{t \geq 0}$ on h,
(ii) L_ℓ are operators on h with $\mathrm{Dom}(L_\ell) \supseteq \mathrm{Dom}(K)$,
(iii) $\mathcal{L}(\mathbf{1}) \leq 0$, $\mathbf{1}$ being the identity operator on h.

These semigroups arise in the study of irreversible evolutions of quantum open systems (see Accardi, Lu and Volovich [5], Alicki and Lendi [6], Gisin and Percival [25], Schack, Brun and Percival [37]). Often there are only finitely many non zero L_ℓ.

It is well-known (see e.g. Davies [13] Sect.3, Fagnola [18] Sect. 3.3) that, given a domain $D \subseteq \mathrm{Dom}(K)$, which is a core for K, it is possible to built up a quantum dynamical semigroup, called the *minimal* quantum dynamical semigroup associated with K and the L_ℓ, and denoted $\mathcal{T}^{(\min)}$, satisfying the equations:

$$\langle v, \mathcal{T}_t(x)u \rangle = \langle v, xu \rangle + \int_0^t \mathcal{L}(\mathcal{T}_s(x))[v, u]ds,$$

$$\langle v, \mathcal{T}_t(x)u \rangle = \langle P_t v, x P_t u \rangle \tag{23}$$

$$+ \sum_{\ell \geq 1} \int_0^t \langle L_\ell P_{t-s} v, \mathcal{T}_s(x) L_\ell P_{t-s} u \rangle ds$$

for $u, v \in D$. Indeed, the above equations are equivalent. More precisely a w^*-continuous family $(X_t; t \geq 0)$ of elements of $\mathcal{B}(\mathsf{h})$ such that $\|X_t\| \leq \|x\|$ for a fixed $x \in \mathcal{B}(\mathsf{h})$ satisfies the first equation if and only if it satisfies the second. The idea of the proof is simple: differentiate $s \to \langle P_{t-s} v, X_s P_{t-s} u \rangle$ and integrate on $[0, t]$ (see Fagnola [18] Prop. 3.18).

The minimal quantum dynamical semigroup associated with K and the L_ℓ can be defined on positive operators $x \in \mathcal{B}(\mathsf{h})$ as follows:

$$\mathcal{T}_t^{(\min)}(x) = \sup_{n \geq 1} \mathcal{T}_t^{(n)}(x)$$

where the maps $\mathcal{T}_t^{(n)}$ are defined recursively by

$$\left\langle v, \mathcal{T}_t^{(0)}(x)u \right\rangle = \langle P_t v, x P_t u \rangle$$

$$\left\langle v, \mathcal{T}_t^{(n+1)}(x)u \right\rangle = \langle P_t v, x P_t u \rangle \tag{24}$$

$$+ \sum_{\ell=1}^{\infty} \int_0^t \left\langle L_\ell P_{t-s} v, \mathcal{T}_s^{(n)}(x) L_\ell P_{t-s} u \right\rangle ds$$

for $x \in \mathcal{B}(h)$, $u, v \in D$.

The equations (23), however, do not necessarily determine a unique semigroup. The minimal quantum dynamical semigroup is characterised by the following property.

Proposition 9.2. *Suppose that the hypothesis* **HQDS** *holds. Then, for each positive $x \in \mathcal{B}(h)$ and each w^*-continuous family $(X_t)_{t \geq 0}$ of positive operators on $\mathcal{B}(h)$ satisfying (23), we have $T_t^{(\min)}(x) \leq X_t$ for all $t \geq 0$.*

Proof. Immediate from the inequality $T_t^{(n)}(x) \leq X_t$ for $n, t \geq 0$. □

The above proposition allows us to establish immediately another simple characterisation of the minimal quantum dynamical semigroup that will be applied in the study of the left QSDE.

Proposition 9.3. *Suppose that the hypothesis* **HQDS** *holds and that $\mathcal{E}(\mathbf{1}) = 0$. For all $\eta \in$ $]0, 1[$ the minimal quantum dynamical semigroup $T^{(\eta)}$ associated with the operators K, and ηL_ℓ satisfies $T_t^{(\eta)}(x) \leq T_t^{(\min)}(x)$ for all positive $x \in \mathcal{B}(h)$ and all $t \geq 0$.*

Proof. The minimal quantum dynamical semigroup $T^{(\eta)}$ associated with the operators K, and ηL_ℓ is defined on each positive $x \in \mathcal{B}(h)$ as the least upper bound of the sequence $T_t^{(\eta,n)}(x)$ defined recursively by (24) with ηL_ℓ replacing L_ℓ.

It is easy to show by induction that,

$$T_t^{(\eta,n)}(x) \leq T_t^{(n)}(x)$$

for all $n \geq 1$, $\eta \in]0, 1[$. The conclusion follows letting n tend to ∞. □

Proposition 9.4. *Suppose that the hypothesis* **HQDS** *holds and that $\mathcal{E}(\mathbf{1}) = 0$. For all $\eta \in$ $]0, 1[$ the minimal quantum dynamical semigroup $T^{(\eta)}$ associated with the operators K, and ηL_ℓ is the unique quantum dynamical semigroup satisfying*

$$\langle v, T_t^{(\eta)}(x)u \rangle = \langle P_t v, x P_t u \rangle + \eta^2 \sum_{\ell \geq 1} \int_0^t \langle L_\ell P_{t-s} v, T_s^{(\eta)}(x) L_\ell P_{t-s} u \rangle ds$$

for all positive $x \in \mathcal{B}(h)$ and all $t \geq 0$ and

$$T_t^{(\min)}(x) = \sup_{\eta \in]0,1[} T_t^{(\eta)}(x).$$

Proof. The minimal quantum dynamical semigroup $T^{(\eta)}$ associated with the operators K, and ηL_ℓ is defined recursively by (24) with L_ℓ replaced by ηL_ℓ. Therefore, for all $t \geq 0$ and all positive $x \in \mathcal{B}(h)$, $T_t^{(\eta)}(x)$ is the least upper bound of the increasing sequence $(T_t^{(\eta,n)}(x); n \geq 1)$.

It is easy to show by induction that

$$T_t^{(\eta,n)}(x) \leq T_t^{(n)}(x)$$

for all $n \geq 1$, $\eta \in]0, 1[$ and, moreover, $T_t^{(\eta_1,n)}(x) \leq T_t^{(\eta_2,n)}(x)$. Letting n tend to infinity, it follows that, for all $t \geq 0$ and all positive $x \in \mathcal{B}(h)$, the map $\eta \rightarrow T_t^{(\eta)}(x)$ is also increasing. Since $\sup_{\eta \in]0,1[} T_t^{(\eta)}(x)$ satisfies (23), by Proposition 9.2 we have $T_t^{(\min)}(x) = \sup_{\eta \in]0,1[} T_t^{(\eta)}(x)$.

We now prove uniqueness. Suppose that $(\mathcal{S}_t; t \geq 0)$ is another quantum dynamical semigroup satisfying the same integral equation as $\mathcal{T}$. Then we can prove by induction (again!) on n that $T_t^{(\eta,n)}(x) \leq \mathcal{S}_t(x)$ for all $n \geq 1$, $\eta \in]0, 1[$, $t \geq 0$ and all positive $x \in \mathcal{B}(\mathsf{h})$.

Indeed, it is clear that $T^{(\eta,0)}{}_t(x) \leq \mathcal{S}_t(x)$. Suppose that the desired inequality has been established for an integer n. We have then

$$\langle u, T_t^{(\eta,n+1)}(x)u \rangle = \langle P_t u, x P_t u \rangle$$
$$+ \eta^2 \sum_{\ell \geq 1} \int_0^t \langle L_\ell P_{t-s} u, T_s^{(\eta,n)}(x) L_\ell P_{t-s} u \rangle ds$$
$$\leq \langle P_t u, x P_t u \rangle + \eta^2 \sum_{\ell \geq 1} \int_0^t \langle L_\ell P_{t-s} u, \mathcal{S}_s(x) L_\ell P_{t-s} u \rangle ds$$
$$= \langle u, \mathcal{S}_t(x)u \rangle.$$

It follows that $T_t^{(\eta,n+1)}(x) \leq \mathcal{S}_t(x)$ and, letting n tend to infinity, we obtain $T_t^{(\eta)}(x) \leq \mathcal{S}_t(x)$.

Let $\mathcal{D}_t(x) = \mathcal{S}_t(x) - T_t^{(\eta)}(x)$ and fix a $t > 0$. Clearly, for all $s \in [0, t]$, $u \in \mathsf{h}$ and all positive $x \in \mathcal{B}(\mathsf{h})$ we have

$$\langle u, \mathcal{D}_s(x)u \rangle = \eta^2 \sum_{\ell \geq 1} \int_0^s \langle L_\ell P_{s-r} u, \mathcal{D}_r(x) L_\ell P_{s-r} u \rangle ds$$
$$\leq \left(\sup_{0 \leq r \leq t} \| \mathcal{D}_r(x) \| \right) \eta^2 \sum_{\ell \geq 1} \int_0^s \| L_\ell P_{s-r} u \|^2 ds$$
$$= \left(\sup_{0 \leq r \leq t} \| \mathcal{D}_r(x) \| \right) \eta^2 \left(\| u \|^2 - \| P_s u \|^2 \right)$$
$$\leq \eta^2 \left(\sup_{0 \leq r \leq t} \| \mathcal{D}_r(x) \| \right) \| u \|^2.$$

It follows that

$$\left(\sup_{0 \leq r \leq t} \| \mathcal{D}_r(x) \| \right) \leq \eta^2 \left(\sup_{0 \leq r \leq t} \| \mathcal{D}_r(x) \| \right)$$

and, since $0 < \eta < 1$, $\mathcal{D}_r(x) = 0$ for all $r \leq 1$ and all positive $x \in \mathcal{B}(\mathsf{h})$. Therefore $\mathcal{D}_r$ vanishes on $\mathcal{B}(\mathsf{h})$ since every operator x can be decomposed as the sum of four positive operators. $\square$

Proposition 9.5. *Suppose that the hypothesis* **HQDS** *holds and* $\mathcal{L}(\mathbf{1}) = 0$ *on* D. *Then the following are equivalent:*

(i) *the minimal QDS is Markov (i.e.* $T_t^{(\min)}(\mathbf{1}) = \mathbf{1}$),

(ii) *for all* $\lambda > 0$ *there exists no non-zero* $x \in \mathcal{B}(\mathsf{h})$ *such that* $\mathcal{L}(x) = \lambda x$.

We refer to Davies [13] Th. 3.2 or Fagnola [18] Prop. 3.31 and Th. 3.21 for the proof.

Condition (ii) is a quantum analogue of Feller's non-explosion condition for the minimal semigroup of a continuous time classical Markov chain. Proposition 9.5 shows, in particular, that the minimal classical Markov semigroup is not Markov, then also the naturally associated QDS will not be Markov.

Simple applicable (sufficient) conditions for uniqueness and Markovianity have been obtained by Chebotarev and Fagnola [11]; the following result (Th. 4.4 p.394 in their paper) will be sufficient for our purposes.

Theorem 9.6. *Suppose that the hypothesis* **HQDS** *holds and suppose that there exists a positive self-adjoint operator C in* h *with the following properties:*

(a) $\mathrm{Dom}(K)$ *is contained in* $\mathrm{Dom}(C^{1/2})$ *and is a core for* $C^{1/2}$,

(b) *the linear manifold* $L_\ell(\mathrm{Dom}(K^2))$ *is contained in* $\mathrm{Dom}(C^{1/2})$,

(c) *there exists a positive self-adjoint operator* Φ, *with* $\mathrm{Dom}(K) \subseteq \mathrm{Dom}(\Phi^{1/2})$ *and* $\mathrm{Dom}(C) \subseteq \mathrm{Dom}(\Phi)$, *such that, for all* $u \in \mathrm{Dom}(K)$, *we have*

$$-2\Re\langle u, Ku\rangle = \sum_\ell \|L_\ell u\|^2 = \|\Phi^{1/2}u\|^2,$$

(d) *for all* $u \in \mathrm{Dom}(C^{1/2})$ *we have* $\|\Phi^{1/2}u\| \leq \|C^{1/2}u\|$,

(e) *for all* $u \in \mathrm{Dom}(K^2)$ *the following inequality holds*

$$2\Re\langle C^{1/2}u, C^{1/2}Ku\rangle + \sum_{\ell=1}^\infty \|C^{1/2}L_\ell u\|^2 \leq b\|C^{1/2}u\|^2 \qquad (25)$$

where b is a positive constant depending only on K, L_ℓ, C.

Then the minimal quantum dynamical semigroup is Markov.

As shown in Fagnola [18] Sect. 3.6 the domain of K^2 can be replaced by a linear manifold D which is dense in h, is a core for $C^{1/2}$, is invariant under the operators P_t of the contraction semigroup generated by K, and enjoys the properties:

$$R(\lambda; G)(D) \subseteq \mathrm{Dom}(C^{1/2}), \qquad L_\ell\left(R(\lambda; G)\right) \subseteq \mathrm{Dom}(C^{1/2})$$

where $R(\lambda; G)$ ($\lambda > 0$) are the resolvent operators. Moreover the inequality (25) must be satisfied for all $u \in R(\lambda; G)(D)$.

We refer to the lecture notes [20] for the study of properties of quantum Markov semigroups.

We can now return to the study of the left QSDE.

10 The left equation with unbounded G_β^α: isometry

In this section we suppose that the hypothesis **HGC** holds and we discuss conditions for V to be an isometry.

Besides $\theta_G(\mathbf{1}) = 0$ on D as we found in Proposition 8.1 an additional condition is needed: the minimal quantum dynamical semigroup associated with G_0^0 and the G_0^ℓ must be Markov.

We shall divide the proof in several steps.

Proposition 10.1. *Suppose that the hypothesis* **HGC** *holds and $\theta_G(\mathbf{1}) = 0$ on D. For all $\eta \in]0, 1[$ let $G(\eta)$ be the matrix of operators on $\oplus^{(d+1)}$h defined by*

$$I_\eta = \mathbf{1}_h \oplus \left(\eta \mathbf{1}_{(\oplus^{(d)}h)}\right), \qquad G(\eta) = I_\eta^* G I_\eta. \qquad (26)$$

The left cocycle $V^{(\eta)}$ solving the left QSDE

$$dV_t^{(\eta)} = V_t^{(\eta)} G(\eta)_\beta^\alpha d\Lambda_\alpha^\beta(t), \qquad V_0^{(\eta)} = \mathbf{1} \qquad (27)$$

on $D \odot \mathcal{E}$ dilates the (minimal) quantum dynamical semigroup $T^{(\eta)}$ associated with G_0^0 and the ηG_0^ℓ.

Proof. It is easy to see, by our definition, that the matrix $G(\eta)$ of operators $\oplus^{(d+1)}h$ defined by (26) satisfies the hypothesis **HGC** for all $\eta \in]0,1[$. In particular we have $\theta_{G(\eta)}(\mathbf{1}) \leq 0$ and, by Theorem 8.5, there exits a unique contraction process $V^{(\eta)}$ solving the left QSDE (27) on $D \odot \mathcal{E}$.

The contraction process $V^{(\eta)}$ is a left cocycle by Corollary 8.6. Therefore the identity

$$\langle v, \mathcal{S}_t(x)u \rangle = \langle V_t^{(\eta)} v e(0), (x \otimes \mathbf{1}_{\mathcal{F}}) V_t^{(\eta)} u e(0) \rangle$$

$(v,u \in h, x \in \mathcal{B}(h))$ defines a quantum dynamical semigroup on h by Theorem 9.1. Moreover, by the quantum Itô formula (2), we can see immediately that $\mathcal{S}$ satisfies

$$\langle v, \mathcal{S}_t(x)u \rangle = \langle v, xu \rangle$$
$$+ \int_0^t \left(\langle G_0^0 v, \mathcal{S}_s(x)u \rangle + \eta^2 \sum_{\ell \geq 1} \langle G_0^\ell v, \mathcal{S}_s(x) G_0^\ell u \rangle + \langle v, \mathcal{S}_s(x) G_0^0 u \rangle \right) ds$$

for all positive $x \in \mathcal{B}(h)$ and all $t \geq 0$. This equation can be written in the equivalent form (23) with $L_\ell = G_0^\ell$, $K = G_0^0$ and P the semigroup generated by G_0^0.

Therefore, by Proposition 9.4, $\mathcal{S}$ coincides with the *unique* minimal quantum dynamical semigroup $\mathcal{T}^{(\eta)}$ associated with G_0^0 and the ηG_0^ℓ. This completes the proof. $\square$

Theorem 10.2. *Suppose that the hypothesis* **HGC** *holds and* $\theta_G(\mathbf{1}) = 0$ *on D. Then the unique contraction V solving (3) is a left cocycle dilating the minimal quantum dynamical semigroup $\mathcal{T}$ associated with G_0^0 and the G_0^ℓ.*

Proof. (Sketch) By Proposition 10.1 for all $\eta \in]0,1[$ we have

$$\langle v, \mathcal{T}_t^{(\eta)}(x)u \rangle = \langle V_t^{(\eta)} v e(0), (x \otimes \mathbf{1}_{\mathcal{F}}) V_t^{(\eta)} u e(0) \rangle$$

for all $v, u \in D$. We can show, by the argument of the proof of Theorem 8.4, that there exists a sequence $(\eta_k; k \geq 1)$ converging to 1 such that the contractions $V_t^{(\eta_k)}$ converge weakly to the unique solution V_t of the left QSDE (3) for k going to infinity uniformly for t in bounded intervals. Therefore, for all $u \in h$ and all positive $x \in \mathcal{B}(h)$ we have

$$\langle u, \mathcal{T}_t^{(\min)}(x)u \rangle = \lim_{k \to \infty} \langle u, \mathcal{T}_t^{(\eta_k)}(x)u \rangle$$
$$= \liminf_{k \to \infty} \langle V_t^{(\eta_k)} u e(0), (x \otimes \mathbf{1}_{\mathcal{F}}) V_t^{(\eta_k)} u e(0) \rangle$$
$$\geq \langle V_t u e(0), (x \otimes \mathbf{1}_{\mathcal{F}}) V_t u e(0) \rangle.$$

Moreover, since V dilates a quantum dynamical semigroup associated with G_0^0 and the G_0^ℓ and $\mathcal{T}^{(\min)}$ is the minimal one, it follows from Proposition 9.2 that the converse inequality also holds. This proves the theorem. $\square$

We can now prove the charactersation of isometries solving of the left QSDE (3).

Theorem 10.3. *Suppose that the hypothesis* **HGC** *holds and* $\theta_G(\mathbf{1}) = 0$ *on D and let V be the unique contraction solving (3). The following conditions are equivalent:*

(i) *the process V is an isometry,*
(ii) *the minimal quantum dynamical semigroup associated with G_0^0 and the G_0^ℓ is Markov.*

Proof. Clearly, by Theorem 10.2, (i) implies (ii).

We will prove the converse by showing that

$$\langle V_t vg^{\otimes m}, V_t uf^{\otimes n}\rangle = \langle vg^{\otimes m}, uf^{\otimes n}\rangle \tag{28}$$

for all $m, n \geq 0$, $t \geq 0$, $v, u \in \mathsf{h}$ and $f, g \in \mathcal{M}$.

The above identity holds for $n = m = 0$. Indeed, V dilates the minimal quantum dynamical semigroup associated with G_0^0 and the G_0^ℓ by condition (i) and this semigroup is Markov. Suppose that the identity has been established for all integers n, m such that $n + m \leq p$. Then, for all n, m with $n + m = p + 1$ arguing as in the proof of Theorem 8.5 and using the induction hypothesis, we have

$$\langle V_t vg^{\otimes m}, V_t uf^{\otimes n}\rangle = \langle vg^{\otimes m}, uf^{\otimes n}\rangle + \int_0^t \Big(\langle V_s vg^{\otimes m}, V_s G_0^0 uf^{\otimes n}\rangle$$

$$+ \sum_\ell \langle V_s G_0^\ell vg^{\otimes m}, V_s G_0^\ell uf^{\otimes n}\rangle + \langle V_s G_0^0 vg^{\otimes m}, V_s uf^{\otimes n}\rangle \Big) ds \tag{29}$$

Let $\lambda > 0$ and define an operator $R_\lambda \in \mathcal{B}(\mathsf{h})$ by

$$\langle v, R_\lambda u\rangle = \int_0^\infty \exp(-\lambda t)\, \langle V_t vg^{\otimes m}, V_t uf^{\otimes n}\rangle\, dt.$$

$(v, u \in D)$. Multiplying by $\lambda \exp(-\lambda t)$ both sides of (29), integrating on $[0, +\infty[$ and changing the order of integration in the double integral as in the proof of Theorem 8.5 we find

$$\lambda\langle v, R_\lambda u\rangle = \langle vg^{\otimes m}, uf^{\otimes n}\rangle + \pounds(R_\lambda)[v, u].$$

Letting $c = \lambda^{-1}\langle g^{\otimes m}, f^{\otimes n}\rangle$, since $\pounds(\mathbf{1}) = 0$ we have $\pounds(R_\lambda - c\mathbf{1}) = \lambda(R_\lambda - c\mathbf{1})$. It follows then from Proposition 9.5 (ii) that $\lambda R_\lambda = \langle g^{\otimes m}, f^{\otimes n}\rangle \mathbf{1}$ so that

$$\lambda \int_0^\infty \exp(-\lambda t)\, \langle V_t vg^{\otimes m}, V_t uf^{\otimes n}\rangle\, dt = \langle vg^{\otimes m}, uf^{\otimes n}\rangle \mathbf{1}$$

for all $\lambda > 0$. Now the uniqueness of the Laplace transform leads to (28).

This completes the proof. □

The above theorem characterises isometries V solving the left QSDE (3). In order to study when V is (also) a coisometry (then a unitary), it is not possible to write the QSDE satisfied by V^* and apply the above results because this is a right equation and we do not know whether a solution exists. It seems more reasonable to study the dual cocycle $\widetilde{V}$ which is a candidate solution of another left QSDE and, of course it is an isometry if and only if V is a coisometry.

Unfortunately we do not know the most general conditions allowing to deduce that, $\widetilde{V}$ satisfies the left QSDE

$$d\widetilde{V}_t = \widetilde{V}_t (G_\beta^\alpha)^* d\Lambda_\beta^\alpha(t) \tag{30}$$

on some domain $\widetilde{D} \odot \mathcal{E}$ if and only if V satisfies (3) on $D \odot \mathcal{E}$ when the G_β^α are unbounded.

We bypass this difficulty by first regularising the G_β^α, for example by multiplication with some resolvent operator, writing the left QSDE satisfied by the cocycle and, finally removing the regularisation.

Proposition 10.4. *Suppose that the hypothesis* **HGC** *holds. Let V be the unique contraction cocycle solving (3) on $D \odot \mathcal{E}$ and let $G^\dagger = [(G^\dagger)_\beta^\alpha]$ be the matrix of operators on $\oplus^{(d+1)}\mathsf{h}$ such that*

$$(G^\dagger)^\alpha_\beta = (G^\beta_\alpha)^* \big|_{\widetilde{D}}$$

where $\widetilde{D}$ is a dense subspace of h which is a core for $(G^0_0)^*$. Suppose that there exists a sequence $(R_n; n \geq 1)$ of bounded operators on h such that the operators $G^\alpha_\beta R_n$ are bounded for all $n \geq 1$ and

$$\lim_{n \to \infty} R^*_n v = v$$

for all $u \in \widetilde{D}$ and all $v \in \mathsf{h}$ in the weak topology on h. Then the dual cocycle $\widetilde{V}$ is the unique contraction process satisfying (30).

Proof. The bounded processes $(V_t R_n; t \geq 0)$ satisfy the left QSDE $dV_t R_n = V_t G^\alpha_\beta R_n d\Lambda^\beta_\alpha(t)$ with initial condition R_n. The time reversed processes $(\widetilde{V}_t R^*_n; t \geq 0)$ satisfy the QSDE $d\widetilde{V}_t R^*_n = V_t (G^\alpha_\beta R_n)^* d\Lambda^\beta_\alpha(t)$ with initial condition R^*_n. This can be checked by differentiation as in the proof of Proposition 6.4. Therefore, for all $v \in \mathsf{h}$, $u \in \widetilde{D}$, $f, g \in \mathcal{M}$ we have

$$
\langle \widetilde{V}^*_t v e(g), R^*_n u e(f) \rangle = \langle v e(g), R^*_n u e(f) \rangle
$$
$$
+ \int_0^t \langle \widetilde{V}^*_s v e(g), R^*_n (G^\alpha_\beta)^* u e(f) \rangle \, g_\alpha(s) f^\beta(s) ds
$$

The conclusion follows letting n tend to ∞. $\quad \square$

It is possible to prove that the dual cocycle satisfies the expected QSDE by other regularisations of the G^α_β (see e.g. Fagnola [18] Prop. 5.24). The more convenient one usually depend on the special form of the G^α_β appearing in the QSDE.

11 The right equation with unbounded F^α_β

In this section we outline the main result for proving the existence of solutions to the right QSDE (4). It is clear from Theorem 4.1 that conditions for the existence of a solution must be stronger. Indeed, arguing as in the proof of 2. $\Rightarrow$ 1., if $\theta_F(\mathbf{1}) = 0$ then the solution must be an isometry.

When a cocycle V solves a left QSDE we need $\theta_G(\mathbf{1}) = 0$ and the additional condition on an associated quantum dynamical semigroup that turns out to be satisfied when we can apply Theorem 9.6. This suggests that it should be possible to show the existence of isometries solving the right equation assuming an (a priori) inequality like (25) not only on the single operator G^0_0 (the K in (25)) but on the whole matrix of operators $[G^\alpha_\beta]$.

This has been done by the author and S. Wills [22] who proved the following result.

Theorem 11.1. *Let U be a contraction process and F an operator matrix, and suppose that C is a positive self-adjoint operator on h, and $\delta > 0$ and $b_1, b_2 \geq 0$ are constants such that the following hold:*

(i) *There is a dense subspace $D \subset \mathsf{h}$ such that the adjoint process U^* is a strong solution of $dU^*_t = U^*_t (F^\beta_\alpha)^* d\Lambda^\beta_\alpha(t)$ on $D \odot \mathcal{E}$, and is the unique solution for this $[(F^\beta_\alpha)^*]$ and D.*

(ii) *For each $0 < \varepsilon < \delta$ there is a dense subspace $D_\varepsilon \subset D$ such that $(C_\varepsilon)^{1/2}(D_\varepsilon) \subset D$ and each $(F^\alpha_\beta)^* (C_\varepsilon)^{1/2} \big|_{D_\varepsilon}$ is bounded.*

(iii) $\mathrm{Dom}(C^{1/2}) \subset \mathrm{Dom}[\overline{F^\alpha_\beta}]$ *for all α, β.*

(iv) Dom[F] *is dense in* h, *and for all* $0 < \varepsilon < \delta$ *the form* $\theta_F(C_\varepsilon)$ *on* Dom[F] *satisfies the inequality*

$$\theta_F(C_\varepsilon) \leq b_1 \iota(C_\varepsilon) + b_2 \mathbf{1}$$

where $\iota(C_\varepsilon)$ *is the* $(d+1) \times (d+1)$ *matrix* diag$(C_\varepsilon, \ldots, C_\varepsilon)$ *of operators on* h.

Then U *is a strong solution to the right QSDE (4) on* Dom$(C^{1/2})$ *for the operator matrix* F.

We refer to Fagnola and Wills [22] for the proof.

As a Corollary we can give immediately conditions under which we can prove that U is an isometry or a coisometry process.

Corollary 11.2. *Suppose that the conditions of Theorem 11.1 hold and let* U *be the solution to (4) on* Dom$(C^{1/2})$ *for the given matrix* F. *If either*

(i) Dom$(C^{1/2}) \cap$ Dom[F] *is a core for* $C^{1/2}$ *and* $\theta_F(\mathbf{1}) = 0$, *or*
(ii) $\theta_{\overline{F}}(\mathbf{1}) = 0$,

then U *is an isometry process* .

In order to show that U is a coisometry process note that the adjoint process U^* satisfies a *left* QSDE. Therefore it suffices to apply the results of Section 10 to obtain the following

Corollary 11.3 ([16], [18]). *Suppose that the conditions of Theorem 11.1 hold and let* U *be the solution to (4) for the given matrix* F. *Suppose further that* $(F_0^0)^*$ *is the generator of a strongly continuous contraction semigroup, that the subspace* D *is a core for* $(F_0^0)^*$, *and let* T *be the minimal QDS with generator*

$$\langle u, \mathcal{L}(X)v\rangle = \langle u, X(F_0^0)^* v\rangle + \langle (F_0^0)^* u, Xv\rangle + \sum_{i=1}^{d} \langle (F_i^0)^* u, X(F_i^0)^* v\rangle.$$

The following are equivalent:

(i) U *is a coisometry process.*
(ii) $\theta_{F^*}(\mathbf{1}) = 0$ *on* D *and* T *is conservative.*
(iii) $[\delta_j^i \mathbf{1} + F_j^i]_{i,j=1}^{d}$ *is a coisometry on* $\oplus_{i=1}^{d}$h *and* T *is conservative.*

A weaker notion of solution to a right QSDE, the *mild* solution, has been introduced by Fagnola and Wills [23] taking inspiration from classical SDE.

For U to be a *mild* solution we demand that $U_t(D \odot \mathcal{E})$ is contained in the domain of all the F_β^α with $\alpha + \beta > 0$ and that the smeared operator $\int_0^t U_s\, ds$ maps $D \odot \mathcal{E}$ in the domain of F_0^0. Thus a *mild solution* is a process U such that

$$\bigcup_{t>0} U_t(D \odot \mathcal{E}) \subset \bigcap_{\alpha+\beta>0} \text{Dom}(F_\beta^\alpha \otimes 1),$$

$$\bigcup_{t>0} \int_0^t U_s\, ds(D \odot \mathcal{E}) \subset \text{Dom}(F_0^0 \otimes 1),$$

and

$$U_t = 1 + (F_0^0 \otimes 1)\int_0^t U_s\, ds + \sum_{\alpha+\beta>0}^{d} \int_0^t (F_\beta^\alpha \otimes 1)U_s\, d\Lambda_\alpha^\beta(s).$$

This is an important notion because the operators F_β^α are in a natural way "less un-bounded" than F_0^0 (a sort of square root of the F_0^0) and, therefore, have a bigger domain.

An existence theorem for mild solutions inspired by Theorem 11.1 was proved in Fagnola and Wills [23] (Th. 2.3).

Acknowledgements. This work has been supported in part by the *EU Human Potential Programme* contract HPRN-CT-2002-00279, QP-Applications and the MIUR programme *Quantum Probability and Applications* 2003-2004.

The author would like to thank A. Barchielli for discussions and to S. Attal, A. Joye and C.A. Pillet for the invitation to the "École d'été de Mathématiques 2003" in Grenoble. He would like to express also his gratitude to the referee for pointing out some errors and suggesting improvements.

References

1. L. Accardi, On the quantum Feynman-Kac formula, *Rend. Sem. Mat. Fis. Milano* **XLVIII** 135–179 (1978).
2. L. Accardi, A. Frigerio and J.T. Lewis, Quantum stochastic processes, *Publ. R.I.M.S. Kyoto Univ.* **18** 97–133 (1982).
3. L. Accardi, A. Frigerio, Y.G. Lu, The weak coupling limit as a quantum functional central limit, *Comm. Math. Phys.* 131 537–570 1990.
4. L. Accardi, S. V. Kozyrev, On the structure of Markov flows, in *Irreversibility, probability and complexity*, (Les Treilles/Clausthal, 1999). *Chaos Solitons Fractals* **12** 2639–2655 2001.
5. L. Accardi, Y.G. Lu, I. Volovich, *Quantum Theory and its Stochastic Limit*, (Springer-Verlag, Berlin, 2002).
6. R. Alicki, K. Lendi, *Quantum dynamical semigroups and applications*, Lecture Notes in Physics, 286, (Springer-Verlag, Berlin, 1987).
7. S. Attal, *Quantum Noises*. Lecture Notes of the Summer School on Quantum Open Systems. Grenoble 2003.
8. A. Barchielli, *Continual Measurements in Quantum Mechanics. The approach based on QSDE*. Lecture Notes of the Summer School on Quantum Open Systems. Grenoble 2003.
9. V.P. Belavkin, Measurement, filtering and control in quantum open dynamical systems. *Rep. Math. Phys.* **43** 405–425 1999.
10. B. V. Rajarama Bhat; M. Skeide, Tensor product systems of Hilbert modules and dilations of completely positive semigroups. *Infin. Dimens. Anal. Quantum Probab. Relat. Top.* **3** 519–575 2000.
11. A.M. Chebotarev, F. Fagnola, Sufficient conditions for conservativity of minimal quantum dynamical semigroups. *J. Funct. Anal.* **153** 382–404 1998.
12. E.B. Davies, *Quantum theory of open systems* (Academic Press, London-New York, 1976).
13. E.B. Davies, Quantum dynamical semigroups and the neutron diffusion equation. *Rep. Math. Phys.* **11** 169–188 1977.
14. N. Dunford, J.T. Schwartz, *Linear Operators. I. General Theory.* Pure and Applied Mathematics, Vol. **7** Interscience Publishers, Inc., New York; Interscience Publishers, Ltd., London 1958.
15. F. Fagnola, Pure birth and pure death processes as quantum flows in Fock space. *Sankhyā* **A 53** 288–297 (1991).

16. F. Fagnola, Characterization of isometric and unitary weakly differentiable cocycles in Fock space. *Quantum Probability and Related Topics* **VIII** 143–164, World Scientific 1993.
17. F. Fagnola, Diffusion processes in Fock space. *Quantum Probability and Related Topics* **IX** 189–214, World Scientific 1994.
18. F. Fagnola, Quantum Markov Semigroups and Quantum Markov Flows. *Proyecciones* **18** 1-144 (1999).
19. F. Fagnola and R. Monte, Quantum stochastic differential equations of diffusion type. In preparation.
20. F. Fagnola and R. Rebolledo, Lectures on the Qualitative Analysis of Quantum Markov Semigroups. In L.Accardi and F.F Fagnola eds *Quantum Interacting Particle Systems*, QPPQ: Quantum Probability and White Noise Analysis **14**, 197–239, World Scientific 2002.
21. F. Fagnola and K.B. Sinha, Quantum flows with unbounded structure maps and finite degrees of freedom. *J. London Math. Soc.* **48** 537–551 (1993).
22. F. Fagnola and S.J. Wills, Solving quantum stochastic differential equations with unbounded coefficients. *J. Funct. Anal.* **198** n.2 279–310 (2003).
23. F. Fagnola and S.J. Wills, Mild solutions of quantum stochastic differential equations. *Electron. Comm. Probab.* **5** 158–171 (2000).
24. F. Fagnola, H-P Quantum stochastic differential equations. In: N. Obata, T. Matsui, A. Hora (eds.) *Quantum probability and White Noise Analysis*, QP-PQ, **XVI**, 51–96, World Sci. Publishing, River Edge, NJ, 2002.
25. N. Gisin, I.C. Percival, The quantum-state diffusion model applied to open systems. *J. Phys. A: Math. Gen.* **25** 5677–5691 (1992).
26. R. L. Hudson and K. R. Parthasarathy, Quantum Itô's formula and stochastic evolutions, *Comm. Math. Phys.* **93** 301–323 (1984).
27. J.-L. Journé, Structure des cocycles markoviens sur l'espace de Fock. *Probab. Th. Rel. Fields* **75** 291–316 (1987).
28. G. Lindblad, On the generators of Quantum Dynamical Semigroups, *Commun. Math. Phys.* **48** 119-130 (1976).
29. J.M. Lindsay, and K.R. Parthasarathy, On the generators of quantum stochastic flows. *J. Funct. Anal.* **158** 521–549 (1998).
30. J.M. Lindsay and S.J. Wills, Existence, positivity and contractivity for quantum stochastic flows with infinite dimensional noise. *Probab. Theory Related Fields* **116** 505–543 2000.
31. P.-A. Meyer, *Quantum probability for probabilists*, Lecture Notes in Mathematics 1538, (Springer-Verlag, Berlin, 1993).
32. A. Mohari, K.R. Parthasarathy, On a class of generalizes Evans-Hudson flows related to classical markov processes. *Quantum Probability and Related Topics* **VII** 221-249 (1992).
33. N. Obata, Wick product of white noise operators and quantum stochastic differential equations. *J. Math. Soc. Japan* **51** 613–641 (1999).
34. K.R. Parthasarathy, *An introduction to quantum stochastic calculus*, Monographs in Mathematics 85, (Birkhäuser Verlag, Basel, 1992).
35. K. R. Parthasarathy and K.B. Sinha, Markov chains as Evans-Hudson diffusions in Fock space, *Séminaire de Probabilités* **XXIV** 362–369 (1989). Lecture Notes in Math., 1426, Springer, Berlin, 1990.
36. A. Pazy, *Semigroups of Linear Operators and Applications to Partial Differential Equations*, (Springer-Verlag, Berlin 1975).
37. R. Schack, T.A. Brun and I.C. Percival, Quantum-state diffusion with a moving basis: Computing quantum-optical spectra, *Phys. Rev. A* **55** 2694 1995.

38. M. Skeide, Indicator functions of intervals are totalizing in the symmetric Fock space $\Gamma(L^2(\mathbb{R}_+))$. In L. Accardi, H.-H. Kuo, N. Obata, K. Saito, Si Si, and L. Streit eds *Trends in contemporary infinite dimensional analysis and quantum probability*, vol.3 of *Natural and Mathematical Science Series*. Istituto Italiano di Cultura (ISEAS), Kyoto 2000.

39. W. von Waldenfels, An example of the singular coupling limit. In R. Rebolledo (ed/) *Stochastic analysis and mathematical physics* (Santiago, 1998), 155–166, Trends Math., Birkhuser Boston, Boston, MA, 2000.

Index of Volume II

Adapted domain, 114
Algebra
  Banach, 156
  von Neumann, 157
Algebraic probability space, 154

Banach algebra, 156
Brownian interpretation, 107
Brownian motion, 13
  canonical, 14

Chaotic
  expansion, 106
  representation property, 106
  space, 106
Classical probabilistic dilations, 174
Coherent vector, 96
Completely bounded map, 162
Completely positive map, 158
Conditional expectation, 173
Conditionally CP map, 170
Control, 24

Dilation, 208
Dilations of QDS, 173
Dynkin's formula, 21

Elliptic operator, 27
Ergodic, 8

Feller semigroup
  strong, 7
  weak, 7
First fundamental formula, 186

Fock space
  toy, 84
  multiplicity n, 90

Gaussian process, 13
Generator, 7
Gibbs measure, 45

Hörmander condition, 27

Independent increments, 13
Initial distribution, 5
Integral representation, 85
Itô
  integrable process, 99
  integral, 15, 99
  process, 16

Lyapunov
  function, 21

Markov process, 4
Martingale
  normal, 105
Measure preserving, 8
Mild solution, 217
Mixing, 9
Modification, 13

Normal martingale, 105

Obtuse
  system, 90
Operator process, 185

Operator system, 156

Poisson interpretation, 107
Predictable representation property, 105
Probabilistic interpretation, 87, 107
  p-, 88
Probability space
  algebraic, 154
Process, 2
  distribution, 3
  Gaussian, 13
  Itô integrable, 99
  Ito, 16
  Markov, 4
    strong, 20
  modification, 13
  operator, 185
    adapted, 111
  path, 3
  stationary, 7
Product
  p-, 89
  Poisson, 108
  Wiener, 108

Quantum dynamical semigroup, 170, 208
  minimal, 210
Quantum Markov semigroup, 170, 208
Quantum noises, 111
Quantum probabilistic dilations, 174, 180

Regular quantum semimartingales, 128

Sesqui-symmetric tensor, 91
Spectral function, 48
State
  normal, 155
Stationary increments, 13
Stinespring representation, 164
Stochastic integral, 15
  quantum, 115
Stochastically integrable, 185
Stopping time, 21
Strong Markov process, 20
Structure equation, 107

Tensor
  sesqui-symmetric, 91
Topology
  uniform, 156
Total variation norm, 12
Totalizing set, 205
Toy Fock space, 84
  multiplicity n, 90
Transition probability, 5

Uniform topology, 156
Uniformly continuous QMS, 170

Vacuum, 96
von Neumann algebra, 157

Information about the other two volumes

Contents of Volume I

References XII

Introduction to the Theory of Linear Operators
Alain Joye 1
1   Introduction 1
2   Generalities about Unbounded Operators 2
3   Adjoint, Symmetric and Self-adjoint Operators 5
4   Spectral Theorem 13
    4.1   Functional Calculus 15
    4.2 L^2 Spectral Representation 22
5   Stone's Theorem, Mean Ergodic Theorem and Trotter Formula 29
6   One-Parameter Semigroups 35
References 40

Introduction to Quantum Statistical Mechanics
Alain Joye 41
1   Quantum Mechanics 42
    1.1   Classical Mechanics 42
    1.2   Quantization 46
    1.3   Fermions and Bosons 53
2   Quantum Statistical Mechanics 54
    2.1   Density Matrices 54
3   Boltzmann Gibbs 57
References 67

Elements of Operator Algebras and Modular Theory
Stéphane Attal 69
1   Introduction 70
    1.1   Discussion 70
    1.2   Notations 71
2   C^*-algebras 71
    2.1   First definitions 71
    2.2   Spectral analysis 73
    2.3   Representations and states 79

    2.4   Commutative C^*-algebras 83

    2.5   Appendix 84

3   von Neumann algebras 86

    3.1   Topologies on $\mathcal{B}(\mathcal{H})$ 86

    3.2   Commutant 89

    3.3   Predual, normal states 90

4   Modular theory 92

    4.1   The modular operators 92

    4.2   The modular group 96

    4.3   Self-dual cone and standard form 100

References 105

Quantum Dynamical Systems

Claude-Alain Pillet 107

1   Introduction 107

2   The State Space of a C^*-algebras 110

    2.1   States 110

    2.2   The GNS Representation 119

3   Classical Systems 123

    3.1   Basics of Ergodic Theory 123

    3.2   Classical Koopmanism 127

4   Quantum Systems 130

    4.1 C^*-Dynamical Systems 132

    4.2 W^*-Dynamical Systems 139

    4.3   Invariant States 141

    4.4   Quantum Dynamical Systems 142

    4.5   Standard Forms 147

    4.6   Ergodic Properties of Quantum Dynamical Systems 153

    4.7   Quantum Koopmanism 161

    4.8   Perturbation Theory 165

5   KMS States 168

    5.1   Definition and Basic Properties 168

    5.2   Perturbation Theory of KMS States 178

References 180

The Ideal Quantum Gas

Marco Merkli 183

1   Introduction 184

2   Fock space 185

    2.1   Bosons and Fermions 185

    2.2   Creation and annihilation operators 188

    2.3   Weyl operators 191

    2.4   The C^*-algebras $\mathrm{CAR}_F(\mathfrak{H})$, $\mathrm{CCR}_F(\mathfrak{H})$ 194

    2.5   Leaving Fock space 197

3   The CCR and CAR algebras 198

    3.1   The algebra $\mathrm{CAR}(\mathfrak{D})$ 199

    3.2   The algebra $\mathrm{CCR}(\mathfrak{D})$ 200

    3.3   Schrödinger representation and Stone – von Neumann uniqueness theorem .. 203

    3.4 Q–space representation 207

	3.5	Equilibrium state and thermodynamic limit	209
4		Araki-Woods representation of the infinite free Boson gas	213
	4.1	Generating functionals	214
	4.2	Ground state (condensate)	217
	4.3	Excited states	222
	4.4	Equilibrium states	224
	4.5	Dynamical stability of equilibria	228
		References	233

Topics in Spectral Theory
Vojkan Jakšić 235

1		Introduction	236
2		Preliminaries: measure theory	238
	2.1	Basic notions	238
	2.2	Complex measures	238
	2.3	Riesz representation theorem	240
	2.4	Lebesgue-Radon-Nikodym theorem	240
	2.5	Fourier transform of measures	241
	2.6	Differentiation of measures	242
	2.7	Problems	247
3		Preliminaries: harmonic analysis	248
	3.1	Poisson transforms and Radon-Nikodym derivatives	249
	3.2	Local L^p norms, $0 < p < 1$.	253
	3.3	Weak convergence	253
	3.4	Local L^p-norms, $p > 1$	254
	3.5	Local version of the Wiener theorem	255
	3.6	Poisson representation of harmonic functions	256
	3.7	The Hardy class $H^\infty(\mathbb{C}_+)$.	258
	3.8	The Borel transform of measures	261
	3.9	Problems	263
4		Self-adjoint operators, spectral theory	267
	4.1	Basic notions	267
	4.2	Digression: The notions of analyticity	269
	4.3	Elementary properties of self-adjoint operators	269
	4.4	Direct sums and invariant subspaces	272
	4.5	Cyclic spaces and the decomposition theorem	273
	4.6	The spectral theorem	273
	4.7	Proof of the spectral theorem—the cyclic case	274
	4.8	Proof of the spectral theorem—the general case	277
	4.9	Harmonic analysis and spectral theory	279
	4.10	Spectral measure for A.	280
	4.11	The essential support of the ac spectrum	281
	4.12	The functional calculus	281
	4.13	The Weyl criteria and the RAGE theorem	283
	4.14	Stability	285
	4.15	Scattering theory and stability of ac spectra	286
	4.16	Notions of measurability	287
	4.17	Non-relativistic quantum mechanics	290
	4.18	Problems	291

5   Spectral theory of rank one perturbations 295

    5.1   Aronszajn-Donoghue theorem.. 296

    5.2   The spectral theorem 298

    5.3   Spectral averaging 299

    5.4   Simon-Wolff theorems 300

    5.5   Some remarks on spectral instability 301

    5.6   Boole's equality 302

    5.7   Poltoratskii's theorem.. 304

    5.8   F. & M. Riesz theorem 308

    5.9   Problems and comments 309

References 311

Index of Volume-I 313

Information about the other two volumes

Contents of Volume II 318

Index of Volume II... 321

Contents of Volume III 323

Index of Volume III 327

Index of Volume I

∗-algebra, 72
  morphism, 77
C^*-algebra, 71
  morphism, 77
C_0 semigroup, 35
W^*-algebra, 139
∗-derivation, 132
μ-Liouvillean, 143

Adjoint, 6
Algebra
  ∗, 72
  C^*, 71
  Banach, 72
  von Neumann, 88
Analytic vector, 32, 136, 202
Approximate identity, 84
Aronszajn-Donoghue theorem, 297
Asymptotic abelianness, 229

Baker-Campbell-Hausdorff formula, 193
Banach algebra, 72
Birkhoff ergodic theorem, 125
Bogoliubov transformation, 200
Boltzmann's constant, 57
Boole's equality, 302
Borel transform, 249, 261
Bose gas, 140, 145, 177
Boson, 53, 186

Canonical anti-commutation relations, 190
Canonical commutation relations, 50, 190, 192
Canonical transformation, 43

Cantor set, 310
CAR, CCR algebra
  $CAR_F(\mathfrak{h})$, 195
  $CCR_F(\mathfrak{h})$, 195
  quasi-local, 198
  simplicity, 199, 200
  uniqueness, 199, 200
CAR-algebra, 134, 172
Cayley transform, 8
CCR-algebra, 140, 145, 177
Center, 118
Central support, 118
Chaos, 186
Character, 83
Chemical potential, 58
Commutant, 89
Condensate, 217
Configuration space, 42
Conjugation, 10
Contraction semigroup, 37
Critical density, 227
Cyclic
  subspace, 22, 295
  vector, 22, 195, 273

Deficiency indices, 8
Density matrix, 55, 114, 290
Dynamical system
  C^*, 132
  W^*, 139
  classical, 124
  ergodic, 125, 156
  mixing, 127, 156
  quantum, 142

Ensemble
  canonical, 60
  grand canonical, 63
  microcanonical, 57
Entropy
  Boltzmann, 57
Enveloping von Neumann algebra, 119
Essential support, 281
Evolution group, 29
Exponential law, 203

Factor, 118
Faithful
  representation, 80
Fermi gas, 134, 172
Fermion, 53, 186
Finite particle subspace, 192
Finite quantum system, 133
Fock space, 186
Folium, 119
Free energy, 61
Functional calculus, 16, 25, 281

G.N.S. representation, 82

Hahn decomposition theorem, 240
Hamiltonian, 290
Hamiltonian system, 43
Hardy class, 258
Harmonic oscillator, 50, 205
Heisenberg picture, 51
Heisenberg uncertainty principle, 49, 290
Helffer-Sjöstrand formula, 17
Hille-Yosida theorem, 37

Ideal
  left, 84
  right, 84
  two-sided, 84
Ideal gas, 185
Indistinguishable, 186
Individual ergodic theorem, 125
Infinitesimal generator, 35
Internal energy, 58
Invariant subspace, 22, 272
Invertible, 73
Isometric element, 75

Jensen's formula, 259

Kaplansky density theorem, 111
Kato-Rellich theorem, 285
Kato-Rosenblum theorem, 287
Koopman ergodicity criterion, 129
Koopman lemma, 128
Koopman mixing criterion, 129
Koopman operator, 128

Lebesgue-Radon-Nikodym theorem, 240
Legendre transform, 62
Liouville equation, 43
Liouville's theorem, 43
Liouvillean, 128, 143, 150, 161, 168
Lummer Phillips theorem, 38

Mean ergodic theorem, 32, 128
Measure
  absolutely continuous, 240
  complex, 239
  regular Borel, 238
  signed, 239
  space, 238
  spectral, 274, 280, 295
  support, 238
Measurement, 48
  simultaneous, 49
Measures
  equivalent, 280
  mutually singular, 240
Modular
  conjugation, 96
  operator, 96
Morphism
  $*$-algebra, 77
  C^*-algebra, 77

Nelson's analytic vector theorem, 32
Norm resolvent convergence, 27
Normal element, 75
Normal form, 143

Observable, 42, 46, 123, 290
Operator
  (anti-)symmetrization, 187
  closable, 5, 268
  closed, 2, 268
  core, 31, 268
  creation, annihilation, 50, 188
  dissipative, 37

domain, 2
essentially self-adjoint, 7
extension, 2
field, 192
graph, 3, 268
linear, 2
multiplication, 14, 273
number, 186
positive, 271
relatively bounded, 12, 285
Schrödinger, 47
self-adjoint, 7
symmetric, 5
trace class, 286
Weyl, 193

Partition function, 61, 64
Pauli's principle, 54, 191
Perturbation theory
    rank one, 295
Phase space, 42
Planck law, 226
Poisson bracket, 44
Poisson representation, 256
Poisson transform, 249
Poltoratskii's theorem, 262, 304
Positive
    element, 78
    linear form, 80
Predual, 90
Pressure, 58

Quantum dynamical system, 142
Quasi-analytic extension, 16

RAGE theorem, 284, 290
Reduced Liouvillean, 161
Representation, 80
    Q-space (CCR), 207
    Araki-Woods, 224
    faithful, 80
    Fock, 203
    GNS, 120
    GNS (ground state of Bose gas), 221
    Quasi-equivalent, 206
    regular (of CCR), 201
    Schrödinger, 204
Resolvent, 3
    first identity, 4, 268

norm convergence, 27
set, 3, 268
strong convergence, 194
Resolvent set, 73
Return to equilibrium, 127, 230
Riemann-Lebesgue lemma, 241
Riesz representation theorem, 240

Schrödinger picture, 51
Sector, 186
Self-adjoint element, 75
Simon-Wolff theorems, 300
Spatial automorphism, 133
Spectral averaging, 299
Spectral radius, 74
Spectral theorem, 23, 274, 298
Spectrum, 3, 73, 83, 268
    absolutely continuous, 278
    continuous, 278
    essential, 284
    point, 268
    pure point, 278
    singular, 278
    singular continuous, 278
Spin, 53
Standard form, 148
Standard Liouvillean, 150, 168
Standard unitary, 149
State, 81, 198
    absolutely continuous, 155
    centrally faithful, 118
    coherent, 52
    disjoint, 119
    equilibrium, 124
    extremal, 159
    factor, 231
    faithful, 110, 117
    gauge invariant, 173, 212
    generating functional, 214
    Gibbs, 210
    ground (Bose gas), 220
    invariant, 141
    KMS, 169, 210
    local perturbation, 228
    mixed, 54
    mixing, 232
    normal, 92, 112
    orthogonal, 119
    pure, 46, 56

quasi-equivalent, 119
quasi-free, 147, 173, 212
relatively normal, 119, 198
tracial, 96
Stone's formula, 282
Stone's theorem, 30
Stone-von Neumann uniqueness theorem, 205
Strong resolvent convergence, 194
Support, 117

Temperature, 58, 61
Thermodynamic
first law, 58
limit, 184, 197
second law, 58
Topology
σ-strong, 111
σ-weak, 87, 111
strong, 86
uniform, 86

weak, 86
weak-$\star$, 139
Trotter product formula, 33

Unit, 72
approximate, 84
Unitary element, 75

Vacuum, 186
Von Neumann density theorem, 111
Von Neumann ergodic theorem, 33, 128

Wave operators, 286
complete, 286
Weyl (CCR) relations, 193
Weyl commutation relations, 140
Weyl quantization, 47
Weyl's criterion, 283
Weyl's theorem, 286
Wiener theorem, 241, 255

Contents of Volume III

Topics in Non-Equilibrium Quantum Statistical Mechanics
Walter Aschbacher, Vojkan Jakšić, Yan Pautrat, and Claude-Alain P0   1
1   Introduction   2
2   Conceptual framework   3
3   Mathematical framework   5
   3.1   Basic concepts   5
   3.2   Non-equilibrium steady states (NESS) and entropy production   8
   3.3   Structural properties 10
   3.4 C^*-scattering and NESS 11
4   Open quantum systems 14
   4.1   Definition 14
   4.2 C^*-scattering for open quantum systems 15
   4.3   The first and second law of thermodynamics 17
   4.4   Linear response theory 18
   4.5   Fermi Golden Rule (FGR) thermodynamics 22
5   Free Fermi gas reservoir 26
   5.1   General description 26
   5.2   Examples 30
6   The simple electronic black-box (SEBB) model 34
   6.1   The model 34
   6.2   The fluxes 36
   6.3   The equivalent free Fermi gas 37
   6.4   Assumptions 40
7   Thermodynamics of the SEBB model 43
   7.1   Non-equilibrium steady states 43
   7.2   The Hilbert-Schmidt condition 44
   7.3   The heat and charge fluxes 45
   7.4   Entropy production 46
   7.5   Equilibrium correlation functions 47
   7.6   Onsager relations. Kubo formulas. 49
8   FGR thermodynamics of the SEBB model 50
   8.1   The weak coupling limit 50
   8.2   Historical digression—Einstein's derivation of the Planck law 53

	8.3	FGR fluxes, entropy production and Kubo formulas	54
	8.4	From microscopic to FGR thermodynamics	56
9		Appendix	58
	9.1	Structural theorems	58
	9.2	The Hilbert-Schmidt condition	60
		References	63

Fermi Golden Rule and Open Quantum Systems
Jan Dereziński and Rafał Früboes 67

1		Introduction	68
	1.1	Fermi Golden Rule and Level Shift Operator in an abstract setting	68
	1.2	Applications of the Fermi Golden Rule to open quantum systems	69
2		Fermi Golden Rule in an abstract setting	71
	2.1	Notation	71
	2.2	Level Shift Operator	72
	2.3	LSO for C_0^*-dynamics	73
	2.4	LSO for W^*-dynamics	74
	2.5	LSO in Hilbert spaces	74
	2.6	The choice of the projection $\mathbb{P}$	75
	2.7	Three kinds of the Fermi Golden Rule	75
3		Weak coupling limit	77
	3.1	Stationary and time-dependent weak coupling limit	77
	3.2	Proof of the stationary weak coupling limit	80
	3.3	Spectral averaging	83
	3.4	Second order asymptotics of evolution with the first order term	85
	3.5	Proof of time dependent weak coupling limit	87
	3.6	Proof of the coincidence of M_{st} and M_{dyn} with the LSO	88
4		Completely positive semigroups	88
	4.1	Completely positive maps	89
	4.2	Stinespring representation of a completely positive map	89
	4.3	Completely positive semigroups	90
	4.4	Standard Detailed Balance Condition	91
	4.5	Detailed Balance Condition in the sense of Alicki-Frigerio-Gorini-Kossakowski-Verri	93
5		Small quantum system interacting with reservoir	93
	5.1	W^*-algebras	94
	5.2	Algebraic description	95
	5.3	Semistandard representation	95
	5.4	Standard representation	96
6		Two applications of the Fermi Golden Rule to open quantum systems	97
	6.1	LSO for the reduced dynamics	97
	6.2	LSO for the Liouvillean	99
	6.3	Relationship between the Davies generator and the LSO for the Liouvillean in thermal case.	100
	6.4	Explicit formula for the Davies generator	103
	6.5	Explicit formulas for LSO for the Liouvillean	104
	6.6	Identities using the fibered representation	106
7		Fermi Golden Rule for a composite reservoir	108
	7.1	LSO for a sum of perturbations	108

    7.2 Multiple reservoirs 109
    7.3 LSO for the reduced dynamics in the case of a composite reservoir 110
    7.4 LSO for the Liovillean in the case of a composite reservoir 111
A   Appendix – one-parameter semigroups 112
References 115

Decoherence as Irreversible Dynamical Process in Open Quantum Systems
Philippe Blanchard and Robert Olkiewicz 117
References 158

Notes on the Qualitative Behaviour of Quantum Markov Semigroups
Franco Fagnola and Rolando Rebolledo 161
1   Introduction 161
    1.1 Preliminaries 163
2   Ergodic theorems... 164
3   The minimal quantum dynamical semigroup 167
4   The existence of Stationary States 172
    4.1 A general result.. 172
    4.2 Conditions on the generator 173
    4.3 Examples 178
    4.4 A multimode Dicke laser model 178
    4.5 A quantum model of absorption and stimulated emission................ 181
    4.6 The Jaynes-Cummings model 182
5   Faithful Stationary States and Irreducibility 183
    5.1 The support of an invariant state 184
    5.2 Subharmonic projections. The case $\mathfrak{M} = \mathfrak{L}(\mathfrak{h})$ 185
    5.3 Examples 188
6   The convergence towards the equilibrium.................................. 188
    6.1 Main results.. 189
    6.2 Examples 192
7   Recurrence and Transience of Quantum Markov Semigroups 193
    7.1 Potential... 193
    7.2 Defining recurrence and transience 197
    7.3 The behavior of a d-harmonic oscillator........................... 200
References 202

Continual Measurements in Quantum Mechanics and Quantum Stochastic Calculus
Alberto Barchielli 205
1   Introduction 207
    1.1 Three approaches to continual measurements 207
    1.2 Quantum stochastic calculus and quantum optics 207
    1.3 Some notations: operator spaces 208
2   Unitary evolution and states 209
    2.1 Quantum stochastic calculus 209
    2.2 The unitary system–field evolution 216
    2.3 The system–field state 222
    2.4 The reduced dynamics 224
    2.5 Physical basis of the use of QSC 227
3   Continual measurements... 229

	3.1	Indirect measurements on $S_\mathcal{H}$	229
	3.2	Characteristic functionals	232
	3.3	The reduced description	240
	3.4	Direct detection	246
	3.5	Optical heterodyne detection	251
	3.6	Physical models	256
4		A three–level atom and the shelving effect	257
	4.1	The atom–field dynamics	258
	4.2	The detection process	261
	4.3	Bright and dark periods: the V-configuration	263
	4.4	Bright and dark periods: the Λ-configuration	266
5		A two–level atom and the spectrum of the fluorescence light	268
	5.1	The dynamical model	269
	5.2	The master equation and the equilibrium state	273
	5.3	The detection scheme	276
	5.4	The fluorescence spectrum	282

References 288

Information about the other two volumes 299
Contents of Volume I 300
Index of Volume I 304
Contents of Volume II 308
Index of Volume II 311

Index of Volume III

T fixed points set, 166
Λ configuration, 258
ω-continuous, 123
σ-finite von Neumann algebra, 163
σ-weakly continuous groups, 113

Absorption, 271
Adapted process, 213
   regular, 213
   stochastically integrable, 214
   unitary, 218
Adjoint pair, 216
Affinities, 19, 49, 56
Annihilation, creation and conservation
      processes, 212
Antibunching, 263
Araki's perturbation theory, 11

Bose fields, 212
Broad–band approximation, 228

CAR algebra, 26
   even, 28
CCR, 148, 210
Central limit theorem, 20
Characteristic
   functional, 234
   operator, 234
Classes of bounded elements
   left, 127
   right, 127
Classical quantum states, 126
Cocycle property, 219
Coherent vectors, 209

Completely positive
   map, 89
   semigroup, 90
Conditional expectation, 122, 166
   ψ-compatible, 122
Continual measurements, 229
Correlation function, 20, 47
Counting
   process, 263
   quanta, 229
Current
   charge, 36, 45, 55
   heat, 17, 36, 45, 55
   output, 247

Dark state, 266
Davies generator, 97
Decoherence
   –induced spin algebra, 142
   environmental, 119
   time, 125
Demixture, 224
Density operator, 27
Detailed Balance Condition, 92
   AFGKV, 93
Detuning
   parameter, 260
   shifted, 275
Direct detection, 246
Dynamical system
   C^*, 5
   W^*, 94
   weakly asymptotically Abelian, 11

Effective dipole operator, 279
Emission, 271
Entropy
    production, 9, 18, 19, 24, 39, 44, 46, 55
    relative, 9
Ergodic generator, 114
    globally, 114
Exclusive probability densities, 250
Experimental resolution , 139
Exponential
    domain, 209
    vectors, 209
Exponential law, 29, 37

Fermi algebra, 26
Fermi Golden Rule, 54, 76
    analytic, 76
    dynamical, 76
    spectral, 76
Fermi-Dirac distribution, 27
Field quadratures, 212, 230
Fluctuation algebra, 21
Fluorescence spectrum, 279
Flux
    charge, 36, 45, 55
    heat, 17, 36, 45, 55
Fock
    space, 26, 209
    vacuum, 209
Form-potential, 194
Friedrichs model, 40

Gauge group, 26
Gorini-Kossakowski-Sudershan-Lindblad
    generator, 140

Harmonic operator, 184
Heterodyne detection, 251
    balanced, 253

Indirect measurement, 229
Infinitely divisible law, 238
Input fields, 231
Instrument, 243
Interaction picture, 221
Isometric-sweeping decomposition, 133
Ito table, 214

Jacobs-deLeeuw-Glicksberg splitting, 136

Jordan-Wigner transformation, 31
Junction, 15, 35

Kinetic coefficients, 19, 25
Kubo formula, 20, 25, 50, 56

Laser intensity, 275
Level Shift Operator, 73
Lindblad generator, 90
Linear response, 18, 25
Liouville operator, 226
Liouvillean
    $-L^p$, 8
    $-\omega$, 8, 28
    semi–, 96
    perturbation of, 11
    standard, 8
Localization properties, 236

Mωller morphism, 11, 15
Mandel Q-parameter, 248
Markov map, 89
Master equation, 226
Modular
    conjugation, 28
    group, 39
    operator, 28

Narrow topology, 165
NESS, 8, 43
Nominal carrier frequencies, 227

Onsager reciprocity relations, 20, 25, 50, 56
Open system, 14
Operator
    number, 26
Output characteristic operator, 239
Output fields, 231

Pauli matrices, 30
Pauli's principle, 29
Perturbed convolution semigroup, 151
    of promeasures, 151
    Poisson, 152
Phase diffusion model, 224, 272
Photoelectron counter, 247
Photon scattering, 271
Picture
    Heisenberg, 89
    Schrödinger, 89

Standard, 89
Pointer states, 123
  continuous, 139
Poissonian statistics
  sub–, 249
  super–, 249
Polarization, 227
Potential operator, 194
Power spectrum, 278
Predual space, 163
Promeasure, 149
  Fourier transform of, 150

Quantum Brownian motion, 155
Quantum dynamical semigroup, 91, 122
  on CCR algebras, 153
  minimal, 167
Quantum Markovian semigroup, 22, 52
  irreducible, 185
Quantum stochastic equation, 225
Quasi–monochromatic fields, 227

Rabi frequencies, 260
Reduced
  characteristic operator, 240
  dynamics, 243
  evolution, 122
  Markovian dynamics, 122
Representation
  Araki-Wyss, 28
  GNS, 5, 27
  semistandard, 95
  standard, 94
  universal, 6
Reservoir, 14
Response function, 247
Rotating wave approximation, 227

Scattering matrix, 41
Semifinite weight, 121
Semigroup
  C_0-, 112
  C_0^*-, 113
  one-parameter, 112
  recurrent, 199
  transient, 199
Sesquilinear form, 164
Shelving effect, 256
  electron, 257

Shot noise, 278
Singular coupling limit, 140
Spectral Averaging, 83
Spin system, 31, 138
State, 5
  chaotic, 14
  decomposition, 6
  ergodic, 5, 28
  factor, 8
  factor or primary, 27
  faithful, 163
  invariant, 5, 27, 163
  KMS, 8, 19, 27, 28
  mixing, 5, 28
  modular, 8, 27
  non-equilibrium steady, 8, 43
  normal, 163
  primary, 8
  quasi-free gauge-invariant, 27, 31, 35
  reference, 3, 9
  relatively normal, 5
  time reversal invariant, 15
States
  classical, 224
  disjoint, 6
  mutually singular, 6
  orthogonal, 6
  quantum, 224
  quasi-equivalent, 7, 27, 44
  unitarily equivalent, 7, 27, 44
Subharmonic operator, 184
Superharmonic operator, 184

Test functions, 234
Thermodynamic
  FGR, 24, 56
  first law, 17, 24, 37
  second law, 18, 24
Tightness, 165
Time reversal, 15, 42
TRI, 15
Two-positive operator, 127

Unitary decomposition, 130

V configuration, 257
Van Hove limit, 22, 50
Von Neumann algebra
  enveloping, 5

universal enveloping, 6

Wave operator, 41
Weak Coupling Limit, 22, 50, 77

dynamical, 76
stationary, 76
Weyl operator, 210
Wigner-Weisskopf atom, 40

Lecture Notes in Mathematics

For information about earlier volumes
please contact your bookseller or Springer
LNM Online archive: springerlink.com

Vol. 1681: G. J. Wirsching, The Dynamical System Generated by the 3n+1 Function (1998)

Vol. 1682: H.-D. Alber, Materials with Memory (1998)

Vol. 1683: A. Pomp, The Boundary-Domain Integral Method for Elliptic Systems (1998)

Vol. 1684: C. A. Berenstein, P. F. Ebenfelt, S. G. Gindikin, S. Helgason, A. E. Tumanov, Integral Geometry, Radon Transforms and Complex Analysis. Firenze, 1996. Editors: E. Casadio Tarabusi, M. A. Picardello, G. Zampieri (1998)

Vol. 1685: S. König, A. Zimmermann, Derived Equivalences for Group Rings (1998)

Vol. 1686: J. Azéma, M. Émery, M. Ledoux, M. Yor (Eds.), Séminaire de Probabilités XXXII (1998)

Vol. 1687: F. Bornemann, Homogenization in Time of Singularly Perturbed Mechanical Systems (1998)

Vol. 1688: S. Assing, W. Schmidt, Continuous Strong Markov Processes in Dimension One (1998)

Vol. 1689: W. Fulton, P. Pragacz, Schubert Varieties and Degeneracy Loci (1998)

Vol. 1690: M. T. Barlow, D. Nualart, Lectures on Probability Theory and Statistics. Editor: P. Bernard (1998)

Vol. 1691: R. Bezrukavnikov, M. Finkelberg, V. Schechtman, Factorizable Sheaves and Quantum Groups (1998)

Vol. 1692: T. M. W. Eyre, Quantum Stochastic Calculus and Representations of Lie Superalgebras (1998)

Vol. 1694: A. Braides, Approximation of Free-Discontinuity Problems (1998)

Vol. 1695: D. J. Hartfiel, Markov Set-Chains (1998)

Vol. 1696: E. Bouscaren (Ed.): Model Theory and Algebraic Geometry (1998)

Vol. 1697: B. Cockburn, C. Johnson, C.-W. Shu, E. Tadmor, Advanced Numerical Approximation of Nonlinear Hyperbolic Equations. Cetraro, Italy, 1997. Editor: A. Quarteroni (1998)

Vol. 1698: M. Bhattacharjee, D. Macpherson, R. G. Möller, P. Neumann, Notes on Infinite Permutation Groups (1998)

Vol. 1699: A. Inoue, Tomita-Takesaki Theory in Algebras of Unbounded Operators (1998)

Vol. 1700: W. A. Woyczyński, Burgers-KPZ Turbulence (1998)

Vol. 1701: Ti-Jun Xiao, J. Liang, The Cauchy Problem of Higher Order Abstract Differential Equations (1998)

Vol. 1702: J. Ma, J. Yong, Forward-Backward Stochastic Differential Equations and Their Applications (1999)

Vol. 1703: R. M. Dudley, R. Norvaiša, Differentiability of Six Operators on Nonsmooth Functions and p-Variation (1999)

Vol. 1704: H. Tamanoi, Elliptic Genera and Vertex Operator Super-Algebras (1999)

Vol. 1705: I. Nikolaev, E. Zhuzhoma, Flows in 2-dimensional Manifolds (1999)

Vol. 1706: S. Yu. Pilyugin, Shadowing in Dynamical Systems (1999)

Vol. 1707: R. Pytlak, Numerical Methods for Optimal Control Problems with State Constraints (1999)

Vol. 1708: K. Zuo, Representations of Fundamental Groups of Algebraic Varieties (1999)

Vol. 1709: J. Azéma, M. Émery, M. Ledoux, M. Yor (Eds.), Séminaire de Probabilités XXXIII (1999)

Vol. 1710: M. Koecher, The Minnesota Notes on Jordan Algebras and Their Applications (1999)

Vol. 1711: W. Ricker, Operator Algebras Generated by Commuting Projections: A Vector Measure Approach (1999)

Vol. 1712: N. Schwartz, J. J. Madden, Semi-algebraic Function Rings and Reflectors of Partially Ordered Rings (1999)

Vol. 1713: F. Bethuel, G. Huisken, S. Müller, K. Steffen, Calculus of Variations and Geometric Evolution Problems. Cetraro, 1996. Editors: S. Hildebrandt, M. Struwe (1999)

Vol. 1714: O. Diekmann, R. Durrett, K. P. Hadeler, P. K. Maini, H. L. Smith, Mathematics Inspired by Biology. Martina Franca, 1997. Editors: V. Capasso, O. Diekmann (1999)

Vol. 1715: N. V. Krylov, M. Röckner, J. Zabczyk, Stochastic PDE's and Kolmogorov Equations in Infinite Dimensions. Cetraro, 1998. Editor: G. Da Prato (1999)

Vol. 1716: J. Coates, R. Greenberg, K. A. Ribet, K. Rubin, Arithmetic Theory of Elliptic Curves. Cetraro, 1997. Editor: C. Viola (1999)

Vol. 1717: J. Bertoin, F. Martinelli, Y. Peres, Lectures on Probability Theory and Statistics. Saint-Flour, 1997. Editor: P. Bernard (1999)

Vol. 1718: A. Eberle, Uniqueness and Non-Uniqueness of Semigroups Generated by Singular Diffusion Operators (1999)

Vol. 1719: K. R. Meyer, Periodic Solutions of the N-Body Problem (1999)

Vol. 1720: D. Elworthy, Y. Le Jan, X-M. Li, On the Geometry of Diffusion Operators and Stochastic Flows (1999)

Vol. 1721: A. Iarrobino, V. Kanev, Power Sums, Gorenstein Algebras, and Determinantal Loci (1999)

Vol. 1722: R. McCutcheon, Elemental Methods in Ergodic Ramsey Theory (1999)

Vol. 1723: J. P. Croisille, C. Lebeau, Diffraction by an Immersed Elastic Wedge (1999)

Vol. 1724: V. N. Kolokoltsov, Semiclassical Analysis for Diffusions and Stochastic Processes (2000)

Vol. 1725: D. A. Wolf-Gladrow, Lattice-Gas Cellular Automata and Lattice Boltzmann Models (2000)

Vol. 1726: V. Marić, Regular Variation and Differential Equations (2000)

Vol. 1727: P. Kravanja M. Van Barel, Computing the Zeros of Analytic Functions (2000)

Vol. 1728: K. Gatermann Computer Algebra Methods for Equivariant Dynamical Systems (2000)

Vol. 1729: J. Azéma, M. Émery, M. Ledoux, M. Yor (Eds.)

Séminaire de Probabilités XXXIV (2000)

Vol. 1730: S. Graf, H. Luschgy, Foundations of Quantization for Probability Distributions (2000)

Vol. 1731: T. Hsu, Quilts: Central Extensions, Braid Actions, and Finite Groups (2000)

Vol. 1732: K. Keller, Invariant Factors, Julia Equivalences and the (Abstract) Mandelbrot Set (2000)

Vol. 1733: K. Ritter, Average-Case Analysis of Numerical Problems (2000)

Vol. 1734: M. Espedal, A. Fasano, A. Mikelić, Filtration in Porous Media and Industrial Applications. Cetraro 1998. Editor: A. Fasano. 2000.

Vol. 1735: D. Yafaev, Scattering Theory: Some Old and New Problems (2000)

Vol. 1736: B. O. Turesson, Nonlinear Potential Theory and Weighted Sobolev Spaces (2000)

Vol. 1737: S. Wakabayashi, Classical Microlocal Analysis in the Space of Hyperfunctions (2000)

Vol. 1738: M. Émery, A. Nemirovski, D. Voiculescu, Lectures on Probability Theory and Statistics (2000)

Vol. 1739: R. Burkard, P. Deuflhard, A. Jameson, J.-L. Lions, G. Strang, Computational Mathematics Driven by Industrial Problems. Martina Franca, 1999. Editors: V. Capasso, H. Engl, J. Periaux (2000)

Vol. 1740: B. Kawohl, O. Pironneau, L. Tartar, J.-P. Zolesio, Optimal Shape Design. Tróia, Portugal 1999. Editors: A. Cellina, A. Ornelas (2000)

Vol. 1741: E. Lombardi, Oscillatory Integrals and Phenomena Beyond all Algebraic Orders (2000)

Vol. 1742: A. Unterberger, Quantization and Nonholomorphic Modular Forms (2000)

Vol. 1743: L. Habermann, Riemannian Metrics of Constant Mass and Moduli Spaces of Conformal Structures (2000)

Vol. 1744: M. Kunze, Non-Smooth Dynamical Systems (2000)

Vol. 1745: V. D. Milman, G. Schechtman (Eds.), Geometric Aspects of Functional Analysis. Israel Seminar 1999-2000 (2000)

Vol. 1746: A. Degtyarev, I. Itenberg, V. Kharlamov, Real Enriques Surfaces (2000)

Vol. 1747: L. W. Christensen, Gorenstein Dimensions (2000)

Vol. 1748: M. Ruzicka, Electrorheological Fluids: Modeling and Mathematical Theory (2001)

Vol. 1749: M. Fuchs, G. Seregin, Variational Methods for Problems from Plasticity Theory and for Generalized Newtonian Fluids (2001)

Vol. 1750: B. Conrad, Grothendieck Duality and Base Change (2001)

Vol. 1751: N. J. Cutland, Loeb Measures in Practice: Recent Advances (2001)

Vol. 1752: Y. V. Nesterenko, P. Philippon, Introduction to Algebraic Independence Theory (2001)

Vol. 1753: A. I. Bobenko, U. Eitner, Painlevé Equations in the Differential Geometry of Surfaces (2001)

Vol. 1754: W. Bertram, The Geometry of Jordan and Lie Structures (2001)

Vol. 1755: J. Azéma, M. Émery, M. Ledoux, M. Yor (Eds.), Séminaire de Probabilités XXXV (2001)

Vol. 1756: P. E. Zhidkov, Korteweg de Vries and Nonlinear Schrödinger Equations: Qualitative Theory (2001)

Vol. 1757: R. R. Phelps, Lectures on Choquet's Theorem (2001)

Vol. 1758: N. Monod, Continuous Bounded Cohomology of Locally Compact Groups (2001)

Vol. 1759: Y. Abe, K. Kopfermann, Toroidal Groups (2001)

Vol. 1760: D. Filipović, Consistency Problems for Heath-Jarrow-Morton Interest Rate Models (2001)

Vol. 1761: C. Adelmann, The Decomposition of Primes in Torsion Point Fields (2001)

Vol. 1762: S. Cerrai, Second Order PDE's in Finite and Infinite Dimension (2001)

Vol. 1763: J.-L. Loday, A. Frabetti, F. Chapoton, F. Goichot, Dialgebras and Related Operads (2001)

Vol. 1764: A. Cannas da Silva, Lectures on Symplectic Geometry (2001)

Vol. 1765: T. Kerler, V. V. Lyubashenko, Non-Semisimple Topological Quantum Field Theories for 3-Manifolds with Corners (2001)

Vol. 1766: H. Hennion, L. Hervé, Limit Theorems for Markov Chains and Stochastic Properties of Dynamical Systems by Quasi-Compactness (2001)

Vol. 1767: J. Xiao, Holomorphic Q Classes (2001)

Vol. 1768: M.J. Pflaum, Analytic and Geometric Study of Stratified Spaces (2001)

Vol. 1769: M. Alberich-Carramiñana, Geometry of the Plane Cremona Maps (2002)

Vol. 1770: H. Gluesing-Luerssen, Linear Delay-Differential Systems with Commensurate Delays: An Algebraic Approach (2002)

Vol. 1771: M. Émery, M. Yor (Eds.), Séminaire de Probabilités 1967-1980. A Selection in Martingale Theory (2002)

Vol. 1772: F. Burstall, D. Ferus, K. Leschke, F. Pedit, U. Pinkall, Conformal Geometry of Surfaces in S^4 (2002)

Vol. 1773: Z. Arad, M. Muzychuk, Standard Integral Table Algebras Generated by a Non-real Element of Small Degree (2002)

Vol. 1774: V. Runde, Lectures on Amenability (2002)

Vol. 1775: W. H. Meeks, A. Ros, H. Rosenberg, The Global Theory of Minimal Surfaces in Flat Spaces. Martina Franca 1999. Editor: G. P. Pirola (2002)

Vol. 1776: K. Behrend, C. Gomez, V. Tarasov, G. Tian, Quantum Comohology. Cetraro 1997. Editors: P. de Bartolomeis, B. Dubrovin, C. Reina (2002)

Vol. 1777: E. García-Río, D. N. Kupeli, R. Vázquez-Lorenzo, Osserman Manifolds in Semi-Riemannian Geometry (2002)

Vol. 1778: H. Kiechle, Theory of K-Loops (2002)

Vol. 1779: I. Chueshov, Monotone Random Systems (2002)

Vol. 1780: J. H. Bruinier, Borcherds Products on $O(2,1)$ and Chern Classes of Heegner Divisors (2002)

Vol. 1781: E. Bolthausen, E. Perkins, A. van der Vaart, Lectures on Probability Theory and Statistics. Ecole d' Eté de Probabilités de Saint-Flour XXIX-1999. Editor: P. Bernard (2002)

Vol. 1782: C.-H. Chu, A. T.-M. Lau, Harmonic Functions on Groups and Fourier Algebras (2002)

Vol. 1783: L. Grüne, Asymptotic Behavior of Dynamical and Control Systems under Perturbation and Discretization (2002)

Vol. 1784: L.H. Eliasson, S. B. Kuksin, S. Marmi, J.-C. Yoccoz, Dynamical Systems and Small Divisors. Cetraro, Italy 1998. Editors: S. Marmi, J.-C. Yoccoz (2002)

Vol. 1785: J. Arias de Reyna, Pointwise Convergence of Fourier Series (2002)

Vol. 1786: S. D. Cutkosky, Monomialization of Morphisms from 3-Folds to Surfaces (2002)

Vol. 1787: S. Caenepeel, G. Militaru, S. Zhu, Frobenius and Separable Functors for Generalized Module Categories and Nonlinear Equations (2002)

Vol. 1788: A. Vasil'ev, Moduli of Families of Curves for Conformal and Quasiconformal Mappings (2002)

Vol. 1789: Y. Sommerhäuser, Yetter-Drinfel'd Hopf algebras over groups of prime order (2002)

Vol. 1790: X. Zhan, Matrix Inequalities (2002)

Vol. 1791: M. Knebusch, D. Zhang, Manis Valuations and Prüfer Extensions I: A new Chapter in Commutative Algebra (2002)

Vol. 1792: D. D. Ang, R. Gorenflo, V. K. Le, D. D. Trong, Moment Theory and Some Inverse Problems in Potential Theory and Heat Conduction (2002)

Vol. 1793: J. Cortés Monforte, Geometric, Control and Numerical Aspects of Nonholonomic Systems (2002)

Vol. 1794: N. Pytheas Fogg, Substitution in Dynamics, Arithmetics and Combinatorics. Editors: V. Berthé, S. Ferenczi, C. Mauduit, A. Siegel (2002)

Vol. 1795: H. Li, Filtered-Graded Transfer in Using Noncommutative Gröbner Bases (2002)

Vol. 1796: J.M. Melenk, hp-Finite Element Methods for Singular Perturbations (2002)

Vol. 1797: B. Schmidt, Characters and Cyclotomic Fields in Finite Geometry (2002)

Vol. 1798: W.M. Oliva, Geometric Mechanics (2002)

Vol. 1799: H. Pajot, Analytic Capacity, Rectifiability, Menger Curvature and the Cauchy Integral (2002)

Vol. 1800: O. Gabber, L. Ramero, Almost Ring Theory (2003)

Vol. 1801: J. Azéma, M. Émery, M. Ledoux, M. Yor (Eds.), Séminaire de Probabilités XXXVI (2003)

Vol. 1802: V. Capasso, E. Merzbach, B.G. Ivanoff, M. Dozzi, R. Dalang, T. Mountford, Topics in Spatial Stochastic Processes. Martina Franca, Italy 2001. Editor: E. Merzbach (2003)

Vol. 1803: G. Dolzmann, Variational Methods for Crystalline Microstructure – Analysis and Computation (2003)

Vol. 1804: I. Cherednik, Ya. Markov, R. Howe, G. Lusztig, Iwahori-Hecke Algebras and their Representation Theory. Martina Franca, Italy 1999. Editors: V. Baldoni, D. Barbasch (2003)

Vol. 1805: F. Cao, Geometric Curve Evolution and Image Processing (2003)

Vol. 1806: H. Broer, I. Hoveijn. G. Lunther, G. Vegter, Bifurcations in Hamiltonian Systems. Computing Singularities by Gröbner Bases (2003)

Vol. 1807: V. D. Milman, G. Schechtman (Eds.), Geometric Aspects of Functional Analysis. Israel Seminar 2000-2002 (2003)

Vol. 1808: W. Schindler, Measures with Symmetry Properties (2003)

Vol. 1809: O. Steinbach, Stability Estimates for Hybrid Coupled Domain Decomposition Methods (2003)

Vol. 1810: J. Wengenroth, Derived Functors in Functional Analysis (2003)

Vol. 1811: J. Stevens, Deformations of Singularities (2003)

Vol. 1812: L. Ambrosio, K. Deckelnick, G. Dziuk, M. Mimura, V. A. Solonnikov, H. M. Soner, Mathematical Aspects of Evolving Interfaces. Madeira, Funchal, Portugal 2000. Editors: P. Colli, J. F. Rodrigues (2003)

Vol. 1813: L. Ambrosio, L. A. Caffarelli, Y. Brenier, G. Buttazzo, C. Villani, Optimal Transportation and its Applications. Martina Franca, Italy 2001. Editors: L. A. Caffarelli, S. Salsa (2003)

Vol. 1814: P. Bank, F. Baudoin, H. Föllmer, L.C.G. Rogers, M. Soner, N. Touzi, Paris-Princeton Lectures on Mathematical Finance 2002 (2003)

Vol. 1815: A. M. Vershik (Ed.), Asymptotic Combinatorics with Applications to Mathematical Physics. St. Petersburg, Russia 2001 (2003)

Vol. 1816: S. Albeverio, W. Schachermayer, M. Talagrand, Lectures on Probability Theory and Statistics. Ecole d'Eté de Probabilités de Saint-Flour XXX-2000. Editor: P. Bernard (2003)

Vol. 1817: E. Koelink, W. Van Assche(Eds.), Orthogonal Polynomials and Special Functions. Leuven 2002 (2003)

Vol. 1818: M. Bildhauer, Convex Variational Problems with Linear, nearly Linear and/or Anisotropic Growth Conditions (2003)

Vol. 1819: D. Masser, Yu. V. Nesterenko, H. P. Schlickewei, W. M. Schmidt, M. Waldschmidt, Diophantine Approximation. Cetraro, Italy 2000. Editors: F. Amoroso, U. Zannier (2003)

Vol. 1820: F. Hiai, H. Kosaki, Means of Hilbert Space Operators (2003)

Vol. 1821: S. Teufel, Adiabatic Perturbation Theory in Quantum Dynamics (2003)

Vol. 1822: S.-N. Chow, R. Conti, R. Johnson, J. Mallet-Paret, R. Nussbaum, Dynamical Systems. Cetraro, Italy 2000. Editors: J. W. Macki, P. Zecca (2003)

Vol. 1823: A. M. Anile, W. Allegretto, C. Ringhofer, Mathematical Problems in Semiconductor Physics. Cetraro, Italy 1998. Editor: A. M. Anile (2003)

Vol. 1824: J. A. Navarro González, J. B. Sancho de Salas, $\mathscr{C}^\infty$ – Differentiable Spaces (2003)

Vol. 1825: J. H. Bramble, A. Cohen, W. Dahmen, Multiscale Problems and Methods in Numerical Simulations, Martina Franca, Italy 2001. Editor: C. Canuto (2003)

Vol. 1826: K. Dohmen, Improved Bonferroni Inequalities via Abstract Tubes. Inequalities and Identities of Inclusion-Exclusion Type. VIII, 113 p, 2003.

Vol. 1827: K. M. Pilgrim, Combinations of Complex Dynamical Systems. IX, 118 p, 2003.

Vol. 1828: D. J. Green, Gröbner Bases and the Computation of Group Cohomology. XII, 138 p, 2003.

Vol. 1829: E. Altman, B. Gaujal, A. Hordijk, Discrete-Event Control of Stochastic Networks: Multimodularity and Regularity. XIV, 313 p, 2003.

Vol. 1830: M. I. Gil', Operator Functions and Localization of Spectra. XIV, 256 p, 2003.

Vol. 1831: A. Connes, J. Cuntz, E. Guentner, N. Higson, J. E. Kaminker, Noncommutative Geometry, Martina Franca, Italy 2002. Editors: S. Doplicher, L. Longo (2004)

Vol. 1832: J. Azéma, M. Émery, M. Ledoux, M. Yor (Eds.), Séminaire de Probabilités XXXVII (2003)

Vol. 1833: D.-Q. Jiang, M. Qian, M.-P. Qian, Mathematical Theory of Nonequilibrium Steady States. On the Frontier of Probability and Dynamical Systems. IX, 280 p, 2004.

Vol. 1834: Yo. Yomdin, G. Comte, Tame Geometry with Application in Smooth Analysis. VIII, 186 p, 2004.

Vol. 1835: O.T. Izhboldin, B. Kahn, N.A. Karpenko, A. Vishik, Geometric Methods in the Algebraic Theory of Quadratic Forms. Summer School, Lens, 2000. Editor: J.-P. Tignol (2004)

Vol. 1836: C. Năstăsescu, F. Van Oystaeyen, Methods of Graded Rings. XIII, 304 p, 2004.

Vol. 1837: S. Tavaré, O. Zeitouni, Lectures on Probability Theory and Statistics. Ecole d'Eté de Probabilités de Saint-Flour XXXI-2001. Editor: J. Picard (2004)

Vol. 1838: A.J. Ganesh, N.W. O'Connell, D.J. Wischik, Big Queues. XII, 254 p, 2004.

Vol. 1839: R. Gohm, Noncommutative Stationary Processes. VIII, 170 p, 2004.

Vol. 1840: B. Tsirelson, W. Werner, Lectures on Probability Theory and Statistics. Ecole d'Eté de Probabilités de Saint-Flour XXXII-2002. Editor: J. Picard (2004)

Vol. 1841: W. Reichel, Uniqueness Theorems for Variational Problems by the Method of Transformation Groups (2004)

Vol. 1842: T. Johnsen, A.L. Knutsen, K3 Projective Models in Scrolls (2004)

Vol. 1843: B. Jefferies, Spectral Properties of Noncommuting Operators (2004)

Vol. 1844: K.F. Siburg, The Principle of Least Action in Geometry and Dynamics (2004)

Vol. 1845: Min Ho Lee, Mixed Automorphic Forms, Torus Bundles, and Jacobi Forms (2004)

Vol. 1846: H. Ammari, H. Kang, Reconstruction of Small Inhomogeneities from Boundary Measurements (2004)

Vol. 1847: T.R. Bielecki, T. Björk, M. Jeanblanc, M. Rutkowski, J.A. Scheinkman, W. Xiong, Paris-Princeton Lectures on Mathematical Finance 2003 (2004)

Vol. 1848: M. Abate, J. E. Fornaess, X. Huang, J. P. Rosay, A. Tumanov, Real Methods in Complex and CR Geometry, Martina Franca, Italy 2002. Editors: D. Zaitsev, G. Zampieri (2004)

Vol. 1849: Martin L. Brown, Heegner Modules and Elliptic Curves (2004)

Vol. 1850: V. D. Milman, G. Schechtman (Eds.), Geometric Aspects of Functional Analysis. Israel Seminar 2002-2003 (2004)

Vol. 1851: O. Catoni, Statistical Learning Theory and Stochastic Optimization (2004)

Vol. 1852: A.S. Kechris, B.D. Miller, Topics in Orbit Equivalence (2004)

Vol. 1853: Ch. Favre, M. Jonsson, The Valuative Tree (2004)

Vol. 1854: O. Saeki, Topology of Singular Fibers of Differential Maps (2004)

Vol. 1855: G. Da Prato, P.C. Kunstmann, I. Lasiecka, A. Lunardi, R. Schnaubelt, L. Weis, Functional Analytic Methods for Evolution Equations. Editors: M. Iannelli, R. Nagel, S. Piazzera (2004)

Vol. 1856: K. Back, T.R. Bielecki, C. Hipp, S. Peng, W. Schachermayer, Stochastic Methods in Finance, Bressanone/Brixen, Italy, 2003. Editors: M. Fritelli, W. Runggaldier (2004)

Vol. 1857: M. Émery, M. Ledoux, M. Yor (Eds.), Séminaire de Probabilités XXXVIII (2005)

Vol. 1858: A.S. Cherny, H.-J. Engelbert, Singular Stochastic Differential Equations (2005)

Vol. 1859: E. Letellier, Fourier Transforms of Invariant Functions on Finite Reductive Lie Algebras (2005)

Vol. 1860: A. Borisyuk, G.B. Ermentrout, A. Friedman, D. Terman, Tutorials in Mathematical Biosciences I. Mathematical Neurosciences (2005)

Vol. 1861: G. Benettin, J. Henrard, S. Kuksin, Hamiltonian Dynamics – Theory and Applications, Cetraro, Italy, 1999. Editor: A. Giorgilli (2005)

Vol. 1862: B. Helffer, F. Nier, Hypoelliptic Estimates and Spectral Theory for Fokker-Planck Operators and Witten Laplacians (2005)

Vol. 1863: H. Fürh, Abstract Harmonic Analysis of Continuous Wavelet Transforms (2005)

Vol. 1864: K. Efstathiou, Metamorphoses of Hamiltonian Systems with Symmetries (2005)

Vol. 1865: D. Applebaum, B.V. R. Bhat, J. Kustermans, J. M. Lindsay, Quantum Independent Increment Processes I. From Classical Probability to Quantum Stochastic Calculus. Editors: M. Schürmann, U. Franz (2005)

Vol. 1866: O.E. Barndorff-Nielsen, U. Franz, R. Gohm, B. Kümmerer, S. Thorbjønsen, Quantum Independent Increment Processes II. Structure of Quantum Lévy Processes, Classical Probability, and Physics. Editors: M. Schürmann, U. Franz, (2005)

Vol. 1867: J. Sneyd (Ed.), Tutorials in Mathematical Biosciences II. Mathematical Modeling of Calcium Dynamics and Signal Transduction. (2005)

Vol. 1868: J. Jorgenson, S. Lang, $Pos_n(R)$ and Eisenstein Sereies. (2005)

Vol. 1869: A. Dembo, T. Funaki, Lectures on Probability Theory and Statistics. Ecole d'Eté de Probabilités de Saint-Flour XXXIII-2003. Editor: J. Picard (2005)

Vol. 1870: V.I. Gurariy, W. Lusky, Geometry of Müntz Spaces and Related Questions. (2005)

Vol. 1871: P. Constantin, G. Gallavotti, A.V. Kazhikhov, Y. Meyer, S. Ukai, Mathematical Foundation of Turbulent Viscous Flows, Martina Franca, Italy, 2003. Editors: M. Cannone, T. Miyakawa (2006)

Vol. 1872: A. Friedman (Ed.), Tutorials in Mathematical Biosciences III. Cell Cycle, Proliferation, and Cancer (2006)

Vol. 1873: R. Mansuy, M. Yor, Random Times and Enlargements of Filtrations in a Brownian Setting (2006)

Vol. 1874: M. Yor, M. Émery (Eds.), In Memoriam Paul-André Meyer - Séminaire de Probabilités XXXIX (2006)

Vol. 1875: J. Pitman, Combinatorial Stochastic Processes. Ecole d'Eté de Probabilités de Saint-Flour XXXII-2002. Editor: J. Picard (2006)

Vol. 1876: H. Herrlich, Axiom of Choice (2006)

Vol. 1877: J. Steuding, Value Distributions of L-Functions (2006)

Vol. 1878: R. Cerf, The Wulff Crystal in Ising and Percolation Models, Ecole d'Eté de Probabilités de Saint-Flour XXXIV-2004. Editor: Jean Picard (2006)

Vol. 1879: G. Slade, The Lace Expansion and its Applications, Ecole d'Eté de Probabilités de Saint-Flour XXXIV-2004. Editor: Jean Picard (2006)

Vol. 1880: S. Attal, A. Joye, C.-A. Pillet, Open Quantum Systems I, The Hamiltonian Approach (2006)

Vol. 1881: S. Attal, A. Joye, C.-A. Pillet, Open Quantum Systems II, The Markovian Approach (2006)

Vol. 1882: S. Attal, A. Joye, C.-A. Pillet, Open Quantum Systems III, Recent Developments (2006)

Vol. 1883: W. Van Assche, F. Marcellàn (Eds.), Orthogonal Polynomials and Special Functions, Computation and Application (2006)

Vol. 1884: N. Hayashi, E.I. Kaikina, P.I. Naumkin, I.A. Shishmarev, Asymptotics for Dissipative Nonlinear Equations (2006)

Vol. 1885: A. Telcs, The Art of Random Walks (2006)

Recent Reprints and New Editions

Vol. 1471: M. Courtieu, A.A. Panchishkin, Non-Archimedean L-Functions and Arithmetical Siegel Modular Forms. – Second Edition (2003)

Vol. 1618: G. Pisier, Similarity Problems and Completely Bounded Maps. 1995 – Second, Expanded Edition (2001)

Vol. 1629: J.D. Moore, Lectures on Seiberg-Witten Invariants. 1997 – Second Edition (2001)

Vol. 1638: P. Vanhaecke, Integrable Systems in the realm of Algebraic Geometry. 1996 – Second Edition (2001)

Vol. 1702: J. Ma, J. Yong, Forward-Backward Stochastic Differential Equations and their Applications. 1999. – Corrected 3rd printing (2005)